ANNUAL EDITIONS

Physical Anthropology *07/08*

S0-ADG-205

Sixteenth Edition

EDITOR

Elvio Angeloni

Pasadena City College

Elvio Angeloni received his B.A. from UCLA in 1963, his M.A. in anthropology from UCLA in 1965, and his M.A. in communication arts from Loyola Marymount University in 1976. He has produced several films, including *Little Warrior,* winner of the Cinemedia VI Best Bicentennial Theme, and *Broken Bottles*, shown on PBS. He served as an academic adviser on the instructional television series *Faces of Culture*. He received the Pasadena City College Outstanding Teacher Award for 2006. He is the academic editor of *Annual Editions: Anthropology* and *Annual Editions: Physical Anthropology* and co-editor of *Annual Editions: Archaeology*. His primary area of interest has been indigenous peoples of the American Southwest.

Contemporary Learning Series

2460 Kerper Blvd., Dubuque, IA 52001

Visit us on the Internet
http://www.mhcls.com

Credits

1. **The Evolutionary Perspective**
   Unit photo—Courtesy of the National Library of Medicine
2. **Primates**
   Unit photo—Getty Images
3. **Sex and Society**
   Unit photo—CORBIS/Royalty-Free
4. **The Fossil Evidence**
   Unit photo—Royalty-Free/CORBIS
5. **Late Hominid Evolution**
   Unit photo—Siede Preis/Getty Images
6. **Human Diversity**
   Unit photo—Digital Vision
7. **Living With the Past**
   Unit photo—Getty Images

Copyright

Cataloging in Publication Data
Main entry under title: Annual Editions: Physical Anthropology. 2007/2008.
1. Physical Anthropology—Periodicals. I. Angeloni, Elvio, *comp.* II. Title: Physical Anthropology.
ISBN-13: 978–0–07–339726–9   ISBN-10: 0–07–339726–1   658'.05   ISSN 1074–1844

Sixteenth Edition

Cover image © Harnett/Hanzon/Getty Images and Brand X Pictures/PunchStock
Printed in the United States of America   234567890QPDQPD987   Printed on Recycled Paper

Editors/Advisory Board

Members of the Advisory Board are instrumental in the final selection of articles for each edition of ANNUAL EDITIONS. Their review of articles for content, level, currentness, and appropriateness provides critical direction to the editor and staff. We think that you will find their careful consideration well reflected in this volume.

Preface

In publishing ANNUAL EDITIONS we recognize the enormous role played by the magazines, newspapers, and journals of the public press in providing current, first-rate educational information in a broad spectrum of interest areas. Many of these articles are appropriate for students, researchers, and professionals seeking accurate, current material to help bridge the gap between principles and theories and the real world. These articles, however, become more useful for study when those of lasting value are carefully collected, organized, indexed, and reproduced in a low-cost format, which provides easy and permanent access when the material is needed. That is the role played by ANNUAL EDITIONS.

This sixteenth edition of *Annual Editions: Physical Anthropology* contains a variety of articles relating to human evolution. The writings were selected for their timeliness, relevance to issues not easily treated in the standard physical anthropology textbook, and clarity of presentation.

Whereas textbooks tend to reflect the consensus within the field, *Annual Editions: Physical Anthropology 07/08* provides a forum for the controversial. We do this in order to convey to the student the sense that the study of human development is an evolving entity in which each discovery encourages further research and each added piece of the puzzle raises new questions about the total picture.

Our final criterion for selecting articles is readability. All too often, the excitement of a new discovery or a fresh idea is deadened by the weight of a ponderous presentation. We seek to avoid that by incorporating essays written with enthusiasm and with the desire to communicate some very special ideas to the general public.

Included in this volume are a number of features that are designed to be useful for students, researchers, and professionals in the field of anthropology. While the articles are arranged along the lines of broadly unifying subject areas, the *topic guide* can be used to establish specific reading assignments tailored to the needs of a particular course of study. Other useful features include the *table of contents* abstracts, which summarize each article and present key concepts in bold italics, and a comprehensive *index*.

In addition, each unit is preceded by an *overview* that provides a background for informed reading of the articles, emphasizes critical issues, and presents *key points* to consider in the form of questions. Also included are *World Wide Web* sites that are coordinated to follow the volume's units.

In contrast to the usual textbook, which by its nature cannot be easily revised, this book will be continually updated to reflect the dynamic, changing character of its subject. Those involved in producing *Annual Editions: Physical Anthropology 07/08* wish to make the next one as useful and effective as possible. Your criticism and advice are welcomed. Please complete and return the postage-paid *article rating form* on the last page of the book and let us know your opinions. Any anthology can be improved, and this one will continue to be.

Elvio Angeloni

Editor

Contents

UNIT 1
The Evolutionary Perspective

The concepts in bold italics are developed in the article. For further expansion, please refer to the Topic Guide and the Index.

UNIT 2
Primates

The concepts in bold italics are developed in the article. For further expansion, please refer to the Topic Guide and the Index.

UNIT 3
Sex and Society

UNIT 4
The Fossil Evidence

The concepts in bold italics are developed in the article. For further expansion, please refer to the Topic Guide and the Index.

UNIT 5
Late Hominid Evolution

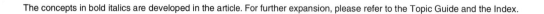

The concepts in bold italics are developed in the article. For further expansion, please refer to the Topic Guide and the Index.

UNIT 6
Human Diversity

UNIT 7
Living With the Past

The concepts in bold italics are developed in the article. For further expansion, please refer to the Topic Guide and the Index.

The concepts in bold italics are developed in the article. For further expansion, please refer to the Topic Guide and the Index.

Topic Guide

This topic guide suggests how the selections in this book relate to the subjects covered in your course. You may want to use the topics listed on these pages to search the Web more easily.

On the following pages a number of Web sites have been gathered specifically for this book. They are arranged to reflect the units of this *Annual Edition*. You can link to these sites by going to the student online support site at *http://www.mhcls.com/online/*.

ALL THE ARTICLES THAT RELATE TO EACH TOPIC ARE LISTED BELOW THE BOLD-FACED TERM.

Aggression
12. Dim Forest, Bright Chimps
19. Apes of Wrath
29. Hard Times Among the Neanderthals

Anatomy
18. What's Love Got to Do With It?
24. Hunting the First Hominid
25. Digital Ancestors Walk Again
27. The Scavenging of "Peking Man"
28. *Erectus* Rising
29. Hard Times Among the Neanderthals
30. Rethinking Neanderthals
31. A Caveful of Clues About Early Humans
32. The Gift *of* Gab
33. We Are All Africans
34. The Littlest Human
37. Does Race Exist? A Proponent's Perspective
38. Does Race Exist? An Antagonist's Perspective

Archeology
23. African Trailblazers
28. *Erectus* Rising
30. Rethinking Neanderthals
31. A Caveful of Clues About Early Humans
34. The Littlest Human

Bipedalism
18. What's Love Got to Do With It?
24. Hunting the First Hominid
25. Digital Ancestors Walk Again
26. Scavenger Hunt

Blood groups
36. Black, White, Other
37. Does Race Exist? A Proponent's Perspective
38. Does Race Exist? An Antagonist's Perspective

Burials
30. Rethinking Neanderthals

Catastrophism
1. The Growth of Evolutionary Science

Chain of being
1. The Growth of Evolutionary Science

Communication
13. Why Are Some Animals So Smart?
16. A Telling Difference
32. The Gift *of* Gab

Culture
11. Got Culture?
13. Why Are Some Animals So Smart?
16. A Telling Difference
30. Rethinking Neanderthals
34. The Littlest Human
41. The Inuit Paradox

Darwin, Charles
1. The Growth of Evolutionary Science
2. Darwin's Influence on Modern Thought
3. Evolution in Action
42. Dr. Darwin

Disease
5. Why Should Students Learn Evolution?
40. The Viral Superhighway
42. Dr. Darwin
43. Curse and Blessing of the Ghetto
44. The Saltshaker's Curse

DNA (deoxyribonucleic acid)
5. Why Should Students Learn Evolution?
9. The 2% Difference
33. We Are All Africans
36. Black, White, Other
43. Curse and Blessing of the Ghetto

Dominance hierarchy
17. What Are Friends For?
18. What's Love Got to Do With It?
19. Apes of Wrath

Ethnicity
36. Black, White, Other
37. Does Race Exist? A Proponent's Perspective
38. Does Race Exist? An Antagonist's Perspective

Evolution
1. The Growth of Evolutionary Science
2. Darwin's Influence on Modern Thought
3. Evolution in Action
4. 15 Answers to Creationist Nonsense
5. Why Should Students Learn Evolution?
6. The Illusion of Design
7. Designer Thinking
8. The Perimeter of Ignorance
9. The 2% Difference
22. The Salamander's Tale
23. African Trailblazers
35. Skin Deep
37. Does Race Exist? A Proponent's Perspective
38. Does Race Exist? An Antagonist's Perspective

Evolutionary perspective
1. The Growth of Evolutionary Science
2. Darwin's Influence on Modern Thought
3. Evolution in Action
4. 15 Answers to Creationist Nonsense
5. Why Should Students Learn Evolution?
6. The Illusion of Design
7. Designer Thinking
8. The Perimeter of Ignorance
21. Had King Henry VIII's Wives Only Known
22. The Salamander's Tale
24. Hunting the First Hominid
35. Skin Deep
37. Does Race Exist? A Proponent's Perspective

Internet References

The following internet sites have been carefully researched and selected to support the articles found in this reader. The easiest way to access these selected sites is to go to our student online support site at *http://www.mhcls.com/online/*.

AE: Physical Anthropology 07/08

The following sites were available at the time of publication. Visit our Web site—we update our student online support site regularly to reflect any changes.

General Sources

American Anthropological Association (AAA)
http://www.aaanet.org/
Maintained by the AAA, this site provides links to AAA's publications (including tables of contents of recent issues, style guides, and others) and to other anthropology sites.

Anthromorphemics
http://www.anth.ucsb.edu/glossary/index2.html
A glossary of anthropological terms is available at this Web site.

Anthropology in the News
http://www.tamu.edu/anthropology/news.html
Texas A&M provides data on news articles that relate to anthropology, including biopsychology and sociocultural anthropology news.

Anthropology on the Web
http://www.as.ua.edu/ant/lib/web1.htm
This Web site provides addresses and tips on acquiring links to regional studies, maps, anthropology tutorials, and other data.

Anthropology 1101 Human Origins Website
http://www.geocities.com/Athens/Acropolis/5579/TA.html
Exploring this site, which is provided by the University of Minnesota, will lead to a wealth of information about our ancient ancestors and other topics of interest to physical anthropologists.

Anthropology Resources on the Internet
http://www.socsciresearch.com/r7.html
Links to Internet resources of anthropological relevance, including Web servers in different fields, are available here. *The Education Index* rated it "one of the best education-related sites on the Web."

Anthropology Resources Page
http://www.usd.edu/anth/
Many topics can be accessed from this University of South Dakota site. South Dakota archaeology, American Indian issues, and paleopathology resources are just a few examples.

Library of Congress
http://www.loc.gov
Examine this extensive Web site to learn about resource tools, library services/resources, exhibitions, and databases in many different subfields of anthropology.

The New York Times
http://www.nytimes.com
Browsing through the archives of the *New York Times* will provide a wide array of articles and information related to the different subfields of anthropology.

Physical Anthropology Resources
http://www.killgrove.org/osteo.html
This Web site provides numerous links to other resources from the Web covering all aspects of physical anthropology.

UNIT 1: The Evolutionary Perspective

Charles Darwin on Human Origins
http://www.literature.org/Works/Charles-Darwin/
This Web site contains the text of Charles Darwin's classic writing, *Origin of Species,* which presents his scientific theory of natural selection.

Enter Evolution: Theory and History
http://www.ucmp.berkeley.edu/history/evolution.html
Find information related to Charles Darwin and other important scientists at this Web site. It addresses preludes to evolution, natural selection, and more. Topics cover systematics, dinosaur discoveries, and vertebrate flight.

Fossil Hominids FAQ
http://www.talkorigins.org/faqs/homs/
Some links to materials related to hominid species and hominid fossils are provided on this site. The purpose of the site is to refute creationist claims that there is no evidence for human evolution.

Harvard Dept. of MCB—Biology Links
http://mcb.harvard.edu/BioLinks.html
This site features sources on evolution and links to anthropology departments and laboratories, taxonomy, paleontology, natural history, journals, books, museums, meetings, and many other related areas.

UNIT 2: Primates

African Primates at Home
http://www.indiana.edu/~primate/primates.html
Don't miss this unusual and compelling site describing African primates on their home turf. "See" and "Hear" features provide samples of vocalizations and beautiful photographs of various types of primates.

Electronic Zoo/NetVet-Primate Page
http://netvet.wustl.edu/primates.htm
This site touches on every kind of primate from A to Z and related information. The long list includes Darwinian theories and the *Descent of Man,* the Ebola virus, fossil hominids, the nonhuman Primate Genetics Lab, the Simian Retrovirus Laboratory, and zoonotic diseases, with many links in between.

Laboratory Primate Newsletter
http://www.brown.edu/Research/Primate/other.html
This series of Web sites on primates includes links to a large number of primate sites.

UNIT 3: Sex and Society

American Anthropologist
http://www.aaanet.org/aa/index.htm
Check out this site, the home page of *American Anthropologist,* for general information about anthropology as well as articles relating to such topics as biological research.

American Scientist

http://www.amsci.org/amsci.html

Investigating this site will help students of physical anthropology to explore issues related to sex and society.

Bonobo Sex and Society

http://songweaver.com/info/bonobos.html

Accessed through Carnegie Mellon University, this site includes a *Scientific American* article discussing a primate's behavior that challenges traditional assumptions about male supremacy in human evolution.

UNIT 4: The Fossil Evidence

The African Emergence and Early Asian Dispersals of the Genus *Homo*

http://www.uiowa.edu/~bioanth/homo.html

Read this classic article to learn about what the Rift Valley in East Africa has to tell us about early hominid species. An excellent bibliography is included.

Anthropology, Archaeology, and American Indian Sites on the Internet

http://dizzy.library.arizona.edu/library/teams/sst/anthro/

This Web page points out a number of Internet sites of interest to different kinds of anthropologists, including physical and biological anthropologists. Visit this page for links to electronic journals and more.

Long Foreground: Human Prehistory

http://www.wsu.edu/gened/learn-modules/top_longfor/lfopen-index.html

This Washington State University site presents a learning module covering three major topics in human evolution: Overview, Hominid Species Timeline, and Human Physical Characteristics. It also provides a helpful glossary of terms and links to other Web sites.

UNIT 5: Late Hominid Evolution

Human Prehistory

http://users.hol.gr/~dilos/prehis.htm

The evolution of the human species, beginning with the *Australopithecus* and continuing with *Homo habilis, Homo erectus,* and *Homo sapiens,* is examined on this site. Also included are data on the people who lived in the Palaeolithic and Neolithic Age and are the immediate ancestors of modern man.

UNIT 6: Human Diversity

Hominid Evolution Survey

http://www.geocities.com/SoHo/Atrium/1381/index.html

This survey of the Hominid family categorizes known hominids by genus and species. Beginning with the oldest known species, data include locations and environments, physical characteristics, technology, social behaviors, charts, and citations.

Human Genome Project Information

http://www.ornl.gov/TechResources/Human_Genome/home.html

Obtain answers about the U.S. Human Genome Project from this site, which details progress, goals, support groups, ethical, legal, and social issues, and genetics information.

OMIM Home Page-Online Mendelian Inheritance in Man

http://www3.ncbi.nlm.nih.gov/omim/

This database from the National Center for Biotechnology Information is a catalog of human genes and genetic disorders. It contains text, pictures, and reference information of great interest to students of physical anthropology.

The Human Diversity Resource Page

http://community-1.webtv.net/SoundBehavior/DIVERSITYFORSOUND/

This page will provide useful resource links for understanding the power of human differences. Differences can challenge assumptions and lead to appreciation. The basic premise of appreciation is understanding. Being open to a level of personal understanding allows for differences to be noticed.

UNIT 7: Living With the Past

Forensic Science Reference Page

http://www.lab.fws.gov

Look over this site from the U.S. Fish and Wildlife Forensics Lab to explore topics related to forensic anthropology.

Zeno's Forensic Page

http://forensic.to/forensic.html

A complete list of resources on forensics is presented on this Web site. It includes general information sources, DNA/serology sources and databases, forensic medicine anthropology sites, and related areas.

We highly recommend that you review our Web site for expanded information and our other product lines. We are continually updating and adding links to our Web site in order to offer you the most usable and useful information that will support and expand the value of your Annual Editions. You can reach us at: *http://www.mhcls.com/annualeditions/.*

UNIT 1

The Evolutionary Perspective

Unit Selections

1. **The Growth of Evolutionary Science**, Douglas J. Futuyma
2. **Darwin's Influence on Modern Thought**, Ernst Mayr
3. **Evolution in Action**, Jonathan Weiner
4. **15 Answers to Creationist Nonsense**, John Rennie
5. **Why Should Students Learn Evolution?**, Brian J. Alters and Sandra Alters
6. **The Illusion of Design**, Richard Dawkins
7. **Designer Thinking**, Mark S. Blumberg
8. **The Perimeter of Ignorance**, Neil deGrasse Tyson

Key Points to Consider

- What is "natural selection "? How does Gregor Mendel's work relate to Charles Darwin's theory?

- In what ways has Charles Darwin influenced modern thought?

- In what ways is human ecological pressure bringing about "evolution in action" in wildlife?

- Why should "creationism " not be taught in the classroom?

- Why should students learn about evolution?

- In nature, how is it that design can occur without a designer, orderliness without purpose?

- What have been some of the greatest intellectual barriers to the acceptance of Darwinism?

- Why have scientists in the past invoked divinity when they reached the boundaries of their understanding?

Student Website

www.mhcls.com/online

Internet References

Further information regarding these websites may be found in this book's preface or online.

Charles Darwin on Human Origins
http://www.literature.org/Works/Charles-Darwin/

Enter Evolution: Theory and History
http://www.ucmp.berkeley.edu/history/evolution.html

Fossil Hominids FAQ
http://www.talkorigins.org/faqs/homs/

Harvard Dept. of MCB—Biology Links
http://mcb.harvard.edu/BioLinks.html

As we reflect upon where science has taken us over the past 300 years, we have been swept along a path of insight into the human condition as well as heightened controversy as to how to handle this potentially dangerous and/or unwanted knowledge of ourselves.

Certainly, Gregor Mendel, in the late nineteenth century, could not have anticipated that his study of pea plants would ultimately lead to the better understanding of over 3,000 genetically caused diseases, such as sickle-cell anemia, Huntington's chorea, and Tay-Sachs. Nor could he have foreseen the present-day controversies over such matters as cloning and genetic engineering.

The significance of Mendel's work, of course, was his discovery that hereditary traits are conveyed by particular units that we now call "genes," a then-revolutionary notion that has been followed by a better understanding of how and why such units change. It is knowledge of the process of "mutation," or alteration of the chemical structure of the gene, that is now providing us with the potential to control the genetic fate of individuals.

This does not mean, however, that we should not continue to look at the role the environment plays in the development of what might be better termed "genetically influenced conditions," such as alcoholism.

The other side of the evolutionary coin, as discussed in "The Growth of Evolutionary Science" and "Darwin's Influence on Modern Thought," is natural selection, a concept provided by Charles Darwin and Alfred Wallace. Natural selection refers to the "weeding out" of unfavorable mutations and the perpetuation of favorable ones.

As our understanding of evolutionary processes continues to be refined (see "Evolution in Action"), it also, unfortunately, continues to be badly understood by the public in general. Thus, we all too often have to confront creationists with the fallacies in their thinking (as in "15 Answers to Creationist Nonsense") as well as remind ourselves why this battle is important (see "Why Should Students Learn Evolution?").

There is something to be gained in this conflict. Consider the claim of "intelligent design theory" that nature is too orderly to have come about by a random process. In "The Illusion of Design," Richard Dawkins argues that natural selection is not a theory of chance, but is instead a process which results in the *appearance* of intentional design. Indeed, says Mark S. Blumberg, what we see in nature is not the absolute perfection that one might expect from a purposeful god, but, rather, the somewhat orderly but less than ideal adaptations on the part of creatures that must make do with what they have. Moreover, adds Neil deGrasse Tyson (in "The Perimeter of Ignorance"), throughout the history of Western scientific development, the only time scholars and scientists invoke divinity as an explanation is precisely when and where they reach the boundaries of their understanding. Neil deGrasse summaries our advance in knowledge best when he notes that:

> As reverent as Newton, Huygens, and other great scientists of earlier centuries may have been, they were also empiricists. They did not retreat from the conclusions their evidence forced them to draw, and when their discoveries conflicted with prevailing articles of faith, they upheld the discoveries.

The Growth of Evolutionary Science

Douglas J. Futuyma

Today, the theory of evolution is an accepted fact for everyone but a fundamentalist minority, whose objections are based not on reasoning but on doctrinaire adherence to religious principles.

—James D. Watson, 1965*

In 1615, Galileo was summoned before the Inquisition in Rome. The guardians of the faith had found that his "proposition that the sun is the center [of the solar system] and does not revolve about the earth is foolish, absurd, false in theology, and heretical, because expressly contrary to Holy Scripture." In the next century, John Wesley declared that "before the sin of Adam there were no agitations within the bowels of the earth, no violent convulsions, no concussions of the earth, no earthquakes, but all was unmoved as the pillars of heaven." Until the seventeenth century, fossils were interpreted as "stones of a peculiar sort, hidden by the Author of Nature for his own pleasure." Later they were seen as remnants of the Biblical deluge. In the middle of the eighteenth century, the great French naturalist Buffon speculated on the possibility of cosmic and organic evolution and was forced by the clergy to recant: "I abandon everything in my book respecting the formation of the earth, and generally all of which may be contrary to the narrative of Moses." For had not St. Augustine written, "Nothing is to be accepted save on the authority of Scripture, since greater is that authority than all the powers of the human mind"?

When Darwin published *The Origin of Species*, it was predictably met by a chorus of theological protest. Darwin's theory, said Bishop Wilberforce, "contradicts the revealed relations of creation to its Creator." "If the Darwinian theory is true," wrote another clergyman, "Genesis is a lie, the whole framework of the book of life falls to pieces, and the revelation of God to man, as we Christians know it, is a delusion and a snare." When *The Descent of Man* appeared, Pope Pius IX was moved to write that Darwinism is "a system which is so repugnant at once to history, to the tradition of all peoples, to exact science, to observed facts, and even to Reason herself, [that it] would seem to need no refutation, did not alienation from God and the leaning toward materialism, due to depravity, eagerly seek a support in all this tissue of fables."[1] Twentieth-century creationism continues this battle of medieval theology against science.

One of the most pervasive concepts in medieval and post-medieval thought was the "great chain of being," or *scala naturae*.[2] Minerals, plants, and animals, according to his concept, formed a gradation, from the lowliest and most material to the most complex and spiritual, ending in man, who links the animal series to the world of intelligence and spirit. This "scale of nature" was the manifestation of God's infinite benevolence. In his goodness, he had conferred existence on all beings of which he could conceive, and so created a complete chain of being, in which there were no gaps. All his creatures must have been created at once, and none could ever cease to exist, for then the perfection of his divine plan would have been violated. Alexander Pope expressed the concept best:

> *Vast chain of being! which from God began,*
> *Natures aethereal, human, angel, man,*
> *Beast, bird, fish, insect, what no eye can see,*
> *No glass can reach; from Infinite to thee,*
> *From thee to nothing.—On superior pow'rs*
> *Were we to press, inferior might on ours;*
> *Or in the full creation leave a void,*
> *Where, one step broken, the great scale's destroy'd;*
> *From Nature's chain whatever link you strike,*
> *Tenth, or ten thousandth, breaks the chain alike.*

Coexisting with this notion that all of which God could conceive existed so as to complete his creation was the idea that all things existed for man. As the philosopher Francis Bacon put it, "Man, if we look to final causes, may be regarded as the centre of the world... for the whole world works together in the service of man... all things seem to be going about man's business and not their own."

"Final causes" was another fundamental concept of medieval and post-medieval thought. Aristotle had distinguished final causes from efficient causes, and the Western world saw no reason to doubt the reality of both. The "efficient cause" of an event is the mechanism responsible for its occurrence: the cause of a ball's movement on a pool table, for example, is the impact of the cue or another ball. The "final cause," however, is the goal, or purpose for its occurrence: the pool ball moves because I wish it to go into the corner pocket. In post-medieval thought there was a final cause—a purpose—for everything; but purpose implies intention, or foreknowledge, by an intellect. Thus the existence of the world, and of all the creatures in it, had a purpose; and that purpose was God's design. This was self-evident, since it was possible to look about the world and see the palpable evidence of God's design everywhere. The heavenly bodies moved in harmonious orbits, evincing the intelligence and harmony of the divine mind; the adaptations of animals and plants to their habitats likewise reflected the devine intelligence, which had fitted all creatures perfectly for their roles in the harmonious economy of nature.

Before the rise of science, then, the causes of events were sought not in natural mechanisms but in the purposes they were meant to serve, and order in nature was evidence of divine intelligence. Since St. Ambrose had declared that "Moses opened his mouth and poured forth what God had said to him," the Bible was seen as the literal word of God, and according to St. Thomas Aquinas, "Nothing was made by God, after the six days of creation, absolutely new." Taking Genesis literally, Archbishop Ussher was able to calculate that the earth was created in 4004 B.C. The earth and the heavens were immutable, changeless. As John Ray put it in 1701 in *The Wisdom of God Manifested in the Works of the Creation*, all living and nonliving things were "created by God at first, and by Him conserved to this Day in the same State and Condition in which they were first made."[3]

The evolutionary challenge to this view began in astronomy. Tycho Brahe found that the heavens were not immutable when a new star appeared in the constellation Cassiopeia in 1572. Copernicus displaced the earth from the center of the universe, and Galileo found that the perfect heavenly bodies weren't so perfect: the sun had spots that changed from time to time, and the moon had craters that strongly implied alterations of its surface. Galileo, and after him Buffon, Kant, and many others, concluded that change was natural to all things.

A flood of mechanistic thinking ensued. Descartes, Kant, and Buffon concluded that the causes of natural phenomena should be sought in natural laws. By 1755, Kant was arguing that the laws of matter in motion discovered by Newton and other physicists were sufficient to explain natural order. Gravitation, for example, could aggregate chaotically dispersed matter into stars and planets. These would join with one another until the only ones left were those that cycled in orbits far enough from each other to resist gravitational collapse. Thus order might arise from natural processes rather than from the direct intervention of a supernatural mind. The "argument from design"—the claim that natural order is evidence of a designer—had been directly challenged. So had the universal belief in final causes. If the arrangement of the planets could arise merely by the laws of Newtonian physics, if the planets could be born, as Buffon suggested, by a collision between a comet and the sun, then they did not exist for any purpose. They merely came into being through impersonal physical forces.

From the mutability of the heavens, it was a short step to the mutability of the earth, for which the evidence was far more direct. Earthquakes and volcanoes showed how unstable terra firma really is. Sedimentary rocks showed that materials eroded from mountains could be compacted over the ages. Fossils of marine shells on mountain-tops proved that the land must once have been under the sea. As early as 1718, the Abbé Moro and the French academician Bernard de Fontenelle had concluded

that the Biblical deluge could not explain the fossilized oyster beds and tropical plants that were found in France. And what of the great, unbroken chain of being if the rocks were full of extinct species?

To explain the facts of geology, some authors—the "catastrophists"—supposed that the earth had gone through a series of great floods and other catastrophes that successively extinguished different groups of animals. Only this, they felt, could account for the discovery that higher and lower geological strata had different fossils. Buffon, however, held that to explain nature we should look to the natural causes we see operating around us: the gradual action of erosion and the slow buildup of land during volcanic eruptions. Buffon thus proposed what came to be the foundation of geology, and indeed of all science, the principle of uniformitarianism, which holds that the same causes that operate now have always operated. By 1795, the Scottish geologist James Hutton had suggested that "in examining things present we have data from which to reason with regard to what has been." His conclusion was that since "rest exists not anywhere," and the forces that change the face of the earth move with ponderous slowness, the mountains and canyons of the world must have come into existence over countless aeons.

If the entire nonliving world was in constant turmoil, could it not be that living things themselves changed? Buffon came close to saying so. He realized that the earth had seen the extinction of countless species, and supposed that those that perished had been the weaker ones. He recognized that domestication and the forces of the environment could modify the variability of many species. And he even mused, in 1766, that species might have developed from common ancestors:

> If it were admitted that the ass is of the family of the horse, and different from the horse only because it has varied from the original form, one could equally well say that the ape is of the family of man, that he is a degenerate man, that man and ape have a common origin; that, in fact, all the families among plants as well as animals have come from a single stock, and that all animals are descended from a single animal, from which have sprung in the course of time, as a result of process or of degeneration, all the other races of animals. For if it were once shown that we are justified in establishing these families; if it were granted among animals and plants there has been (I do not say several species) but even a single one, which has been produced in the course of direct descent from another species... then there would no longer be any limit to the power of nature, and we should not be wrong in supposing that, with sufficient time, she has been able from a single being to derive all the other organized beings.[4]

This, however, was too heretical a thought; and in any case, Buffon thought the weight of evidence was against common descent. No new species had been observed to arise within recorded history, Buffon wrote; the sterility of hybrids between species appeared an impossible barrier to such a conclusion;

and if species had emerged gradually, there should have been innumerable intermediate variations between the horse and ass, or any other species. So Buffon concluded: "But this [idea of a common ancestor] is by no means a proper representation of nature. We are assured by the authority of revelation that all animals have participated equally in the grace of direct Creation and that the first pair of every species issued fully formed from the hands of the Creator."

Buffon's friend and protégé, Jean Baptiste de Monet, the Chevalier de Lamarck, was the first scientist to take the big step. It is not clear what led Lamarck to his uncompromising belief in evolution; perhaps it was his studies of fossil molluscs, which he came to believe were the ancestors of similar species living today. Whatever the explanation, from 1800 on he developed the notion that fossils were not evidence of extinct species but of ones that had gradually been transformed into living species. To be sure, he wrote, "an enormous time and wide variation in successive conditions must doubtless have been required to enable nature to bring the organization of animals to that degree of complexity and development in which we see it at its perfection"; but "time has no limits and can be drawn upon to any extent."

Lamarck believed that various lineages of animals and plants arose by a continual process of spontaneous generation from inanimate matter, and were transformed from very simple to more complex forms by an innate natural tendency toward complexity caused by "powers conferred by the supreme author of all things." Various specialized adaptations of species are consequences of the fact that animals must always change in response to the needs imposed on them by a continually changing environment. When the needs of a species change, so does its behavior. The animal then uses certain organs more frequently than before, and these organs, in turn, become more highly developed by such use, or else "by virtue of the operations of their own inner senses." The classic example of Lamarckism is the giraffe: by straining upward for foliage, it was thought, the animal had acquired a longer neck, which was then inherited by its off-spring.

In the nineteenth century it was widely believed that "acquired" characteristics—alterations brought about by use or disuse, or by the direct influence of the environment—could be inherited. Thus it was perfectly reasonable for Lamarck to base his theory of evolutionary change partly on this idea. Indeed, Darwin also allowed for this possibility, and the inheritance of acquired characteristics was not finally proved impossible until the 1890s.

Lamarck's ideas had a wide influence; but in the end did not convince many scientists of the reality of evolution. In France, Georges Cuvier, the foremost paleontologist and anatomist of his time, was an influential opponent of evolution. He rejected Lamarck's notion of the spontaneous generation of life, found it inconceivable that changes in behavior could produce the exquisite adaptations that almost every species shows, and emphasized that in both the fossil record and among living animals there were numerous "gaps" rather than intermediate forms between species. In England, the philosophy of "natural theology" held sway in science, and the best-known naturalists continued to believe firmly that the features of animals and plants were evidence of God's design. These devout Christians included the foremost geologist of the day, Charles Lyell, whose *Principles of Geology* established uniformitarianism once and for all as a guiding principle. But Lyell was such a thorough uniformitarian that he believed in a steady-state world, a world that was always in balance between forces such as erosion and mountain building, and so was forever the same. There was no room for evolution, with its concept of steady change, in Lyell's world view, though he nonetheless had an enormous impact on evolutionary thought, through his influence on Charles Darwin.

Darwin (1809–1882) himself, unquestionably one of the greatest scientists of all time, came only slowly to an evolutionary position. The son of a successful physician, he showed little interest in the life of the mind in his early years. After unsuccessfully studying medicine at Edinburgh, he was sent to Cambridge to prepare for the ministry, but he had only a half-hearted interest in his studies and spent most of his time hunting, collecting beetles, and becoming an accomplished amateur naturalist. Though he received his B.A. in 1831, his future was quite uncertain until, in December of that year, he was enlisted as a naturalist aboard *H.M.S. Beagle*, with his father's very reluctant agreement. For five years (from December 27, 1831, to October 2, 1836) the *Beagle* carried him about the world, chiefly along the coast of South America, which it was the *Beagle's* mission to survey. For five years Darwin collected geological and biological specimens, made geological observations, absorbed Lyell's *Principles of Geology*, took voluminous notes, and speculated about everything from geology to anthropology. He sent such massive collections of specimens back to England that by the time he returned he had already gained a substantial reputation as a naturalist.

Shortly after his return, Darwin married and settled into an estate at Down where he remained, hardly traveling even to London, for the rest of his life. Despite continual ill health, he pursued an extraordinary range of biological studies: classifying barnacles, breeding pigeons, experimenting with plant growth, and much more. He wrote no fewer than sixteen books and many papers, read voraciously, corresponded extensively with everyone, from pigeon breeders to the most eminent scientists, whose ideas or information might bear on his theories, and kept detailed notes on an amazing variety of subjects. Few people have written authoritatively on so many different topics: his books include not only *The Voyage of the Beagle, The Origin of Species*, and *The Descent of Man,* but also *The Structure and Distribution of Coral Reefs* (containing a novel theory of the formation of coral atolls which is still regarded as correct), *A Monograph on the Sub-class Cirripedia* (the definitive study of barnacle classification), *The Various Contrivances by Which Orchids are Fertilised by Insects, The Variation of Animals and Plants Under Domestication* (an exhaustive summary of information on variation, so crucial to his evolutionary theory), *The Effects of Cross and Self Fertilisation in the Vegetable Kingdom* (an analysis of sexual reproduction and the sterility of hybrids between species), *The Expression of the Emotions in Man and Animals* (on the evolution of human behavior from animal behavior), and *The Formation of Vegetable Mould Through the*

Action of Worms. There is every reason to believe that almost all these books bear, in one way or another, on the principles and ideas that were inherent in Darwin's theory of evolution. The worm book, for example, is devoted to showing how great the impact of a seemingly trivial process like worm burrowing may be on ecology and geology if it persists for a long time. The idea of such cumulative slight effects is, of course, inherent in Darwin's view of evolution: successive slight modifications of a species, if continued long enough, can transform it radically.

When Darwin embarked on his voyage, he was a devout Christian who did not doubt the literal truth of the Bible, and did not believe in evolution any more than did Lyell and the other English scientists he had met or whose books he had read. By the time he returned to England in 1836 he had made numerous observations that would later convince him of evolution. It seems likely, however, that the idea itself did not occur to him until the spring of 1837, when the ornithologist John Gould, who was working on some of Darwin's collections, pointed out to him that each of the Galápagos Islands, off the coast of Ecuador, had a different kind of mockingbird. It was quite unclear whether they were different varieties of the same species, or different species. From this, Darwin quickly realized that species are not the discrete, clear-cut entities everyone seemed to imagine. The possibility of transformation entered his mind, and it applied to more than the mockingbirds: "When comparing... the birds from the separate islands of the Galápagos archipelago, both with one another and with those from the American mainland, I was much struck how entirely vague and arbitrary is the distinction between species and varieties."

In July 1837 he began his first notebook on the "Transmutation of Species." He later said that the Galápagos species and the similarity between South American fossils and living species were at the origin of all his views.

> During the voyage of the *Beagle* I had been deeply impressed by discovering in the Pampean formation great fossil animals covered with armour like that on the existing armadillos; secondly, by the manner in which closely allied animals replace one another in proceeding southward over the continent; and thirdly, by the South American character of most of the productions of the Galápagos archipelago, and more especially by the manner in which they differ slightly on each island of the group; none of these islands appearing to be very ancient in a geological sense. It was evident that such facts as these, as well as many others, could be explained on the supposition that species gradually become modified; and the subject has haunted me.

The first great step in Darwin's thought was the realization that evolution had occurred. The second was his brilliant insight into the possible cause of evolutionary change. Lamarck's theory of "felt needs" had not been convincing. A better one was required. It came on September 18, 1838, when after grappling with the problem for fifteen months, "I happened to read for amusement Malthus on Population, and being well prepared to appreciate the struggle for existence which everywhere goes on from long-continued observation of the habits of animals and plants, it at once struck me that under these circumstances favorable variations would tend to be preserved, and unfavorable ones to be destroyed. The result of this would be the formation of new species. Here, then, I had at last got a theory by which to work."

Malthus, an economist, had developed the pessimistic thesis that the exponential growth of human populations must inevitably lead to famine, unless it were checked by war, disease, or "moral restraint." This emphasis on exponential population growth was apparently the catalyst for Darwin, who then realized that since most natural populations of animals and plants remain fairly stable in numbers, many more individuals are born than survive. Because individuals vary in their characteristics, the struggle to survive must favor some variant individuals over others. These survivors would then pass on their characteristics to future generations. Repetition of this process generation after generation would gradually transform the species.

Darwin clearly knew that he could not afford to publish a rash speculation on so important a subject without developing the best possible case. The world of science was not hospitable to speculation, and besides, Darwin was dealing with a highly volatile issue. Not only was he affirming that evolution had occurred, he was proposing a purely material explanation for it, one that demolished the argument from design in a single thrust. Instead of publishing his theory, he patiently amassed a mountain of evidence, and finally, in 1844, collected his thoughts in an essay on natural selection. But he still didn't publish. Not until 1856, almost twenty years after he became an evolutionist, did he begin what he planned to be a massive work on the subject, tentatively titled *Natural Selection*.

Then, in June 1858, the unthinkable happened. Alfred Russel Wallace (1823–1913), a young naturalist who had traveled in the Amazon Basin and in the Malay Archipelago, had also become interested in evolution. Like Darwin, he was struck by the fact that "the most closely allied species are found in the same locality or in closely adjoining localities and... therefore the natural sequence of the species by affinity is also geographical." In the throes of a malarial fever in Malaya, Wallace conceived of the same idea of natural selection as Darwin had, and sent Darwin a manuscript "On the Tendency of Varieties to Depart Indefinitely from the Original Type." Darwin's friends Charles Lyell and Joseph Hooker, a botanist, rushed in to help Darwin establish the priority of his ideas, and on July 1, 1858, they presented to the Linnean Society of London both Wallace's paper and extracts from Darwin's 1844 essay. Darwin abandoned his big book on natural selection and condensed the argument into a 490-page "abstract" that was published on November 24, 1859, under the title *The Origin of Species by Means of Natural Selection; or, the Preservation of Favored Races in the Struggle for Life*. Because it was an abstract, he had to leave out many of the detailed observations and references to the literature that he had amassed, but these were later provided in his other books, many of which are voluminous expansions on the contents of *The Origin of Species*.

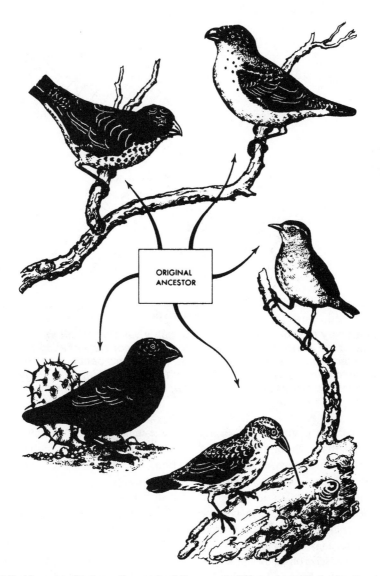

Figure 1. *Some species of Galápagos finches. Several of the most different species are represented here; intermediate species also exist. Clockwise from lower left are a male ground-finch (the plumage of the female resembles that of the tree-finches); the vegetarian tree-finch; the insectivorous tree-finch; the warbler-finch; and the woodpecker-finch, which uses a cactus spine to extricate insects from crevices. The slight differences among these species, and among species in other groups of Galápagos animals such as giant tortoises, were one of the observations that led Darwin to formulate his hypothesis of evolution.* **(From D. Lack, Darwin's Finches [Oxford: Oxford University Press, 1944].)**

The first five chapters of the *Origin* lay out the theory that Darwin had conceived. He shows that both domesticated and wild species are variable, that much of that variation is hereditary, and that breeders, by conscious selection of desirable varieties, can develop breeds of pigeons, dogs, and other forms that are more different from each other than species or even families of wild animals and plants are from each other. The differences between related species then are no more than an exaggerated form of the kinds of variations one can find in a single species; indeed, it is often extremely difficult to tell if natural populations are distinct species or merely well-marked varieties.

Darwin then shows that in nature there is competition, predation, and a struggle for life.

Owing to this struggle, variations, however slight and from whatever cause proceeding, if they be in any de-

gree profitable to the individuals of a species, in their infinitely complex relations to other organic beings and to their physical conditions of life, will tend to the preservation of such individuals, and will generally be inherited by the offspring. The offspring, also, will thus have a better chance of surviving, for, of the many individuals of any species which are periodically born, but a small number can survive. I have called this principle, by which each slight variation, if useful, is preserved, by the term natural selection, in order to mark its relation to man's power of selection.

Darwin goes on to give examples of how even slight variations promote survival, and argues that when populations are exposed to different conditions, different variations will be favored, so that the descendants of a species become diversified

7

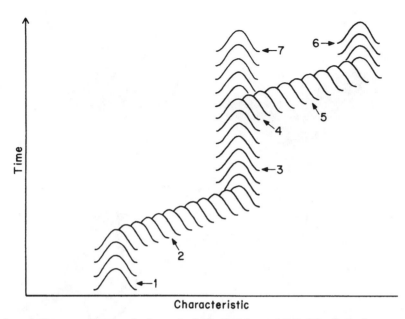

Figure 2. *Processes of evolutionary change. A characteristic that is variable (1) often shows a bell-shaped distribution--individuals vary on either side of the average. Evolutionary change (2) consists of a shift in successive generations, after which the characteristic may reach a new equilibrium (3). When the species splits into two different species (4), one of the species may undergo further evolutionary change (5) and reach a new equilibrium (6). The other may remain unchanged (7) or not. Each population usually remains variable throughout this process, but the average is shifted, ordinarily by natural selection.*

in structure, and each ancestral species can give rise to several new ones. Although "it is probable that each form remains for long periods unaltered," successive evolutionary modifications will ultimately alter the different species so greatly that they will be classified as different genera, families, or orders.

Competition between species will impel them to become more different, for "the more diversified the descendants from any one species become in structure, constitution and habits, by so much will they be better enabled to seize on many and widely diversified places in the polity of nature, and so be enabled to increase in numbers." Thus different adaptations arise, and "the ultimate result is that each creature tends to become more and more improved in relation to its conditions. This improvement inevitably leads to the greater advancement of the organization of the greater number of living beings throughout the world." But lowly organisms continue to persist, for "natural selection, or the survival of the fittest, does not necessarily include progressive development—it only takes advantage of such variations as arise and are beneficial to each creature under its complex relations of life." Probably no organism has reached a peak of perfection, and many lowly forms of life continue to exist, for "in some cases variations or individual differences of a favorable nature may never have arisen for natural selection to act on or accumulate. In no case, probably, has time sufficed for the utmost possible amount of development. In some few cases there has been what we must call retrogression of organization. But the main cause lies in the fact that under very simple conditions of life a high organization would be of no service...."

In the rest of *The Origin of Species*, Darwin considers all the objections that might be raised against his theory; discusses the evolution of a great array of phenomena—hybrid sterility, the

slave-making instinct of ants, the similarity of vertebrate embryos; and presents an enormous body of evidence for evolution. He draws his evidence from comparative anatomy, embryology, behavior, geographic variation, the geographic distribution of species, the study of rudimentary organs, atavistic variations ("throwbacks"), and the geological record to show how all of biology provides testimony that species have descended with modification from common ancestors.

Darwin's triumph was in synthesizing ideas and information in ways that no one had quite imagined before. From Lyell and the geologists he learned uniformitarianism: the cause of past events must be found in natural forces that operate today; and these, in the vastness of time, can accomplish great change. From Malthus and the nineteenth-century economists he learned of competition and the struggle for existence. From his work on barnacles, his travels, and his knowledge of domesticated varieties he learned that species do not have immutable essences but are variable in all their properties and blend into one another gradually. From his familiarity with the works of Whewell, Herschel, and other philosophers of science he developed a powerful method of pursuing science, the "hypothetico-deductive" method, which consists of formulating a hypothesis or speculation, deducing the logical predictions that must follow from the hypothesis, and then testing the hypothesis by seeing whether or not the predictions are verified. This was by no means the prevalent philosophy of science in Darwin's time.[5]

Darwin brought biology out of the Middle Ages. For divine design and unknowable supernatural forces he substituted natural material causes that could be studied by the methods of science. Instead of catastrophes unknown to physical science he invoked forces that could be studied in anyone's laboratory or garden. He replaced a young, static world by one in which there had been constant change for countless aeons. He established that life had a history, and this proved the essential view that differentiated evolutionary thought from all that had gone before.

For the British naturalist John Ray, writing in 1701, organisms had no history—they were the same at that moment, and lived in the same places, doing the same things, as when they were first created. For Darwin, organisms spoke of historical change. If there has indeed been such a history, then fossils in the oldest rocks must differ from those in younger rocks: trilobites, dinosaurs, and mammoths will not be mixed together but will appear in some temporal sequence. If species come from common ancestors, they will have the same characteristics, modified for different functions: the same bones used by bats for flying will be used by horses for running. If species come from ancestors that lived in different environments, they will carry the evidence of their history with them in the form of similar patterns of embryonic development and in vestigial, rudimentary organs that no longer serve any function. If species have a history, their geographical distribution will reflect it: oceanic islands won't have elephants because they wouldn't have been able to get there.

Once the earth and its living inhabitants are seen as the products of historical change, the theological philosophy embodied in the great chain of being ceases to make sense; the plenitude, or fullness, of the world becomes not an eternal manifestation of God's bountiful creativity but an illusion. For most of earth's history, most of the present species have not existed; and many of those that did exist do so no longer. But the scientific challenge to medieval philosophy goes even deeper. If evolution has occurred, and if it has proceeded from the natural causes that Darwin envisioned, then the adaptations of organisms to their environment, the intricate construction of the bird's wing and the orchid's flower, are evidence not of divine design but of the struggle for existence. Moreover, and this may be the deepest implication of all, Darwin brought to biology, as his predecessors had brought to astronomy and geology, the sufficiency of efficient causes. No longer was there any reason to look for final causes or goals. To the questions "What purpose does this species serve? Why did God make tapeworms?" the answer is "To no purpose." Tapeworms were not put here to serve a purpose, nor were planets, nor plants, nor people. They came into existence not by design but by the action of impersonal natural laws.

By providing materialistic, mechanistic explanations, instead of miraculous ones, for the characteristics of plants and animals, Darwin brought biology out of the realm of theology and into the realm of science. For miraculous spiritual forces fall outside the province of science; all of science is the study of material causation.

Of course, *The Origin of Species* didn't convince everyone immediately. Evolution and its material cause, natural selection, evoked strong protests from ecclesiastical circles, and even from scientists.[6] The eminent geologist Adam Sedgwick, for example, wrote in 1860 that species must come into existence by creation,

a power I cannot imitate or comprehend; but in which I can believe, by a legitimate conclusion of sound reason drawn from the laws and harmonies of Nature. For I can see in all around me a design and purpose, and a mutual adaptation of parts which I *can* comprehend, and which prove that there is exterior to, and above, the mere phenomena of Nature a great prescient and designing cause.... The pretended physical philosophy of modern days strips man of all his moral attributes, or holds them of no account in the estimate of his origin and place in the created world. A cold atheistical materialism is the tendency of the so-called material philosophy of the present day.

Among the more scientific objections were those posed by the French paleontologist François Pictet, and they were echoed by many others. Since Darwin supposes that species change gradually over the course of thousands of generations, then, asked Pictet, "Why don't we find these gradations in the fossil record... and why, instead of collecting thousands of identical individuals, do we not find more intermediary forms?... How is it that the most ancient fossil beds are rich in a variety of diverse forms of life, instead of the few early types Darwin's theory leads us to expect? How is it that no species has been seen to evolve during human history, and that the 4000 years which separates us from the mummies of Egypt have been insufficient to modify the crocodile and the ibis?" Pictet protested that, although slight variations might in time alter a species slightly, "all known facts demonstrate... that the prolonged influence of modifying causes has an action which is constantly restrained within sufficiently confined limits."

The anatomist Richard Owen likewise denied "that... variability is progressive and unlimited, so as, in the course of generations, to change the species, the genus, the order, or the class." The paleontologist Louis Agassiz insisted that organisms fall into discrete groups, based on uniquely different created plans, between which no intermediates could exist. He chose the birds as a group that showed the sharpest of boundaries. Only a few years later, in 1868, the fossil *Archaeopteryx*, an exquisite intermediate between birds and reptiles, demolished Agassiz's argument, and he had no more to say on the unique character of the birds.

Within twelve years of *The Origin of Species*, the evidence for evolution had been so thoroughly accepted that philosopher and mathematician Chauncey Wright could point out that among the students of science, "orthodoxy has been won over to the doctrine of evolution." However, Wright continued, "While the general doctrine of evolution has thus been successfully redeemed from theological condemnation, this is not yet true of the subordinate hypothesis of Natural Selection."

Natural selection turned out to be an extraordinarily difficult concept for people to grasp. St. George Mivart, a Catholic scholar and scientist, was not unusual in equating natural selection with chance. "The theory of Natural Selection may (though it need not) be taken in such a way as to lead man to regard the present organic world as formed, so to speak, *accidentally*, beautiful and wonderful as is the confessedly haphazard result."

Many like him simply refused to understand that natural selection is the antithesis of chance and consequently could not see how selection might cause adaptation or any kind of progressive evolutionary change. Even in the 1940s there were those, especially among paleontologists, who felt that the progressive evolution of groups like the horses, as revealed by the fossil record, must have had some unknown cause other than natural selection. Paradoxically, then, Darwin had convinced the scientific world of evolution where his predecessors had failed; but he had not convinced all biologists of his truly original theory, the theory of natural selection.

Natural selection fell into particular disrepute in the early part of the twentieth century because of the rise of genetics—which, as it happened, eventually became the foundation of the modern theory of evolution. Darwin's supposition that variation was unlimited, and so in time could give rise to strikingly different organisms, was not entirely convincing because he had no good idea of where variation came from. In 1865, the Austrian monk Gregor Mendel discovered, from his crosses of pea plants, that discretely different characteristics such as wrinkled versus smooth seeds were inherited from generation to generation without being altered, as if they were caused by particles that passed from parent to offspring. Mendel's work was ignored for thirty-five years, until, in 1900, three biologists discovered his paper and realized that it held the key to the mystery of heredity. One of the three, Hugo de Vries, set about to explore the problem as Mendel had, and in the course of his studies of evening primroses observed strikingly different variations arise, *de novo*. The new forms were so different that de Vries believed they represented new species, which had arisen in a single step by alteration or, as he called it, mutation, of the hereditary material.

In the next few decades, geneticists working with a great variety of organisms observed many other drastic changes arise by mutation: fruit flies (Drosophila), for example, with white instead of red eyes or curled instead of straight wings. These laboratory geneticists, especially Thomas Hunt Morgan, an outstanding geneticist at Columbia University, asserted that evolution must proceed by major mutational steps, and that mutation, not natural selection, was the cause of evolution. In their eyes, Darwin's theory was dead on two counts: evolution was not gradual, and it was not caused by natural selection. Meanwhile, naturalists, taxonomists, and breeders of domesticated plants and animals continued to believe in Darwinism, because they saw that populations and species differed quantitatively and gradually rather than in big jumps, that most variation was continuous (like height in humans) rather than discrete, and that domesticated species could be altered by artificial selection from continuous variation.

The bitter conflict between the Mendelian geneticists and the Darwinians was resolved in the 1930s in a "New Synthesis" that brought the opposing views into a "neo-Darwinian" theory of evolution.[7] Slight variations in height, wing length, and other characteristics proved, under careful genetic analysis, to be inherited as particles, in the same way as the discrete variations studied by the Mendelians. Thus a large animal simply has inherited more particles, or genes, for large size than a smaller member of the species has. The Mendelians were simply studying particularly well marked variations, while the naturalists were studying more subtle ones. Variations could be very slight, or fairly pronounced, or very substantial, but all were inherited in the same manner. All these variations, it was shown, arose by a process of mutation of the genes.

Three mathematical theoreticians, Ronald Fisher and J. B. S. Haldane in England and Sewall Wright in the United States, proved that a newly mutated gene would not automatically form a new species. Nor would it automatically replace the preexisting form of the gene, and so transform the species. Replacement of one gene by a mutant form of the gene, they said, could happen in two ways. The mutation could enable its possessors to survive or reproduce more effectively than the old form; if so, it would increase by natural selection, just as Darwin had said. The new characteristic that evolved in this way would ordinarily be considered an improved adaptation.

Sewall Wright pointed out, however, that not all genetic changes in species need be adaptive. A new mutation might be no better or worse than the preexisting gene—it might simply be "neutral." In small populations such a mutation could replace the previous gene purely by chance—a process he called random genetic drift. The idea, put crudely, is this. Suppose there is a small population of land snails in a cow pasture, and that 5 percent of them are brown and the rest are yellow. Purely by chance, a greater percentage of yellow snails than of brown ones get crushed by cows' hooves in one generation. The snails breed, and there will now be a slightly greater percentage of yellow snails in the next generation than there had been. But in the next generation, the yellow ones may suffer more trampling, purely by chance. The proportion of yellow offspring will then be lower again. These random events cause fluctuations in the percentage of the two types. Wright proved mathematically that eventually, if no other factors intervene, these fluctuations will bring the population either to 100 percent yellow or 100 percent brown, purely by chance. The population will have evolved, then, but not by natural selection; and there is no improvement of adaptation.

During the period of the New Synthesis, though, genetic drift was emphasized less than natural selection, for which abundant evidence was discovered. Sergei Chetverikov in Russia, and later Theodosius Dobzhansky working in the United States, showed that wild populations of fruit flies contained an immense amount of genetic variation, including the same kinds of mutations that the geneticists had found arising in their laboratories. Dobzhansky and other workers went on to show that these variations affected survival and reproduction: that natural selection was a reality. They showed, moreover, that the genetic differences among related species were indeed compounded of the same kinds of slight genetic variations that they found within species. Thus the taxonomists and the geneticists converged onto a neo-Darwinian theory of evolution: evolution is due not to mutation *or* natural selection, but to both. Random mutations provide abundant genetic variation; natural selection, the antithesis of randomness, sorts out the useful from the deleterious, and transforms the species.

In the following two decades, the paleontologist George Gaylord Simpson showed that this theory was completely adequate to explain the fossil record, and the ornithologists Bernhard Rensch and Ernst Mayr, the botanist G. Ledyard Stebbins, and many other taxonomists showed that the similarities and differences among living species could be fully explained by neo-Darwinism. They also clarified the meaning of "species." Organisms belong to different species if they do not interbreed when the opportunity presents itself, thus remaining genetically distinct. An ancestral species splits into two descendant species when different populations of the ancestor, living in different geographic regions, become so genetically different from each other that they will not or cannot interbreed when they have the chance to do so. As a result, evolution can happen without the formation of new species: a single species can be genetically transformed without splitting into several descendants. Conversely, new species can be formed without much genetic change. If one population becomes different from the rest of its species in, for example, its mating behavior, it will not interbreed with the other populations. Thus it has become a new species, even though it may be identical to its "sister species" in every respect except its behavior. Such a new species is free to follow a new path of genetic change, since it does not become homogenized with its sister species by interbreeding. With time, therefore, it can diverge and develop different adaptations.

The conflict between the geneticists and the Darwinians that was resolved in the New Synthesis was the last major conflict in evolutionary science. Since that time, an enormous amount of research has confirmed most of the major conclusions of neo-Darwinism. We now know that populations contain very extensive genetic variation that continually arises by mutation of pre-existing genes. We also know what genes are and how they become mutated. Many instances of the reality of natural selection in wild populations have been documented, and there is extensive evidence that many species form by the divergence of different populations of an ancestral species.

The major questions in evolutionary biology now tend to be of the form, "All right, factors x and y both operate in evolution, but how important is x compared to y?" For example, studies of biochemical genetic variation have raised the possibility that nonadaptive, random change (genetic drift) may be the major reason for many biochemical differences among species. How important, then, is genetic drift compared to natural selection? Another major question has to do with rates of evolution: Do species usually diverge very slowly, as Darwin thought, or does evolution consist mostly of rapid spurts, interspersed with long periods of constancy? Still another question is raised by mutations, which range all the way from gross changes of the kind Morgan studied to very slight alterations. Does evolution con-

sist entirely of the substitution of mutations that have very slight effects, or are major mutations sometimes important too? Partisans on each side of all these questions argue vigorously for their interpretation of the evidence, but they don't doubt that the major factors of evolution are known. They simply emphasize one factor or another. Minor battles of precisely this kind go on continually in every field of science; without them there would be very little advancement in our knowledge.

Within a decade or two of *The Origin of Species*, the belief that living organisms had evolved over the ages was firmly entrenched in biology. As of 1982, the historical existence of evolution is viewed as fact by almost all biologists. To explain how the fact of evolution has been brought about, a theory of evolutionary mechanisms—mutation, natural selection, genetic drift, and isolation—has been developed.[8] But exactly what is the evidence for the fact of evolution?

Notes

1. Andrew Dickson White, *A History of the Warfare of Science with Theology in Christendom* vol. I (London: Macmillan, 1896; reprint ed., New York: Dover, 1960).

2. A. O. Lovejoy, *The Great Chain of Being* (Cambridge, Mass.: Harvard University Press, 1936).

3. Much of this history is provided by J. C. Greene, *The Death of Adam: Evolution and its Impact on Western Thought* (Ames: Iowa State University Press, 1959).

4. A detailed history of this and other developments in evolutionary biology is given by Ernst Mayr, *The Growth of Biological Thought: Diversity, Evolution, Inheritance* (Cambridge, Mass.: Harvard University Press, 1982).

5. See D. L. Hull, *Darwin and His Critics* (Cambridge, Mass.: Harvard University Press, 1973).

6. Ibid.

7. E. Mayr and W. B. Provine, *The Evolutionary Synthesis* (Cambridge, Mass.: Harvard University Press, 1980).

8. Our modern understanding of the mechanisms of evolution is described in many books. Elementary textbooks include G. L. Stebbins, *Processes of Organic Evolution*, (Englewood Cliffs, N.J.: Prentice-Hall, 1971), and J. Maynard Smith, *The Theory of Evolution* (New York: Penguin Books, 1975). More advanced textbooks include Th. Dobzhansky, F. J. Ayala, G. L. Stebbins, and J. W. Valentine, *Evolution* (San Francisco: Freeman, 1977), and D. J. Futuyma, *Evolutionary Biology* (Sunderland, Mass.: Sinauer, 1979). Unreferenced facts and theories described in the text are familiar enough to most evolutionary biologists that they will be found in most or all of the references cited above.

James D. Watson, a molecular biologist, shared the Nobel Prize for his work in discovering the structure of DNA.

From *Science on Trial* by Douglas J. Futuyma, pp. 23–43. Published by Pantheon Books, a division of Random House, Inc. © 1982 by Douglas J. Futuyma. Reprinted by permission of the author.

Darwin's Influence on Modern Thought

Great minds shape the thinking of successive historical periods. Luther and Calvin inspired the Reformation; Locke, Leibniz, Voltaire and Rousseau, the Enlightenment. Modern thought is most dependent on the influence of Charles Darwin.

ERNST MAYR

Clearly, our conception of the world and our place in it is, at the beginning of the 21st century, drastically different from the zeitgeist at the beginning of the 19th century. But no consensus exists as to the source of this revolutionary change. Karl Marx is often mentioned; Sigmund Freud has been in and out of favor; Albert Einstein's biographer Abraham Pais made the exuberant claim that Einstein's theories "have profoundly changed the way modern men and women think about the phenomena of inanimate nature." No sooner had Pais said this, though, than he recognized the exaggeration. "It would actually be better to say 'modern scientists' than 'modern men and women,'" he wrote, because one needs schooling in the physicist's style of thought and mathematical techniques to appreciate Einstein's contributions in their fullness. Indeed, this limitation is true for all the extraordinary theories of modern physics, which have had little impact on the way the average person apprehends the world.

The situation differs dramatically with regard to concepts in biology. Many biological ideas proposed during the past 150 years stood in stark conflict with what everybody assumed to be true. The acceptance of these ideas required an ideological revolution. And no biologist has been responsible for more—and for more drastic—modifications of the average person's worldview than Charles Darwin.

Darwin's accomplishments were so many and so diverse that it is useful to distinguish three fields to which he made major contributions: evolutionary biology; the philosophy of science; and the modern zeitgeist. Although I will be focusing on this last domain, for the sake of completeness I will put forth a short overview of his contributions—particularly as they inform his later ideas—to the first two areas.

A Secular View of Life

Darwin founded a new branch of life science, evolutionary biology. Four of his contributions to evolutionary biology are es-

pecially important, as they held considerable sway beyond that discipline. The first is the non-constancy of species, or the modern conception of evolution itself. The second is the notion of branching evolution, implying the common descent of all species of living things on earth from a single unique origin. Up until 1859, all evolutionary proposals, such as that of naturalist Jean-Baptiste Lamarck, instead endorsed linear evolution, a teleological march toward greater perfection that had been in vogue since Aristotle's concept of *Scala Naturae*, the chain of being. Darwin further noted that evolution must be gradual, with no major breaks or discontinuities. Finally, he reasoned that the mechanism of evolution was natural selection.

These four insights served as the foundation for Darwin's founding of a new branch of the philosophy of science, a philosophy of biology. Despite the passing of a century before this new branch of philosophy fully developed, its eventual form is based on Darwinian concepts. For example, Darwin introduced historicity into science. Evolutionary biology, in contrast with physics and chemistry, is a historical science—the evolutionist attempts to explain events and processes that have already taken place. Laws and experiments are inappropriate techniques for the explication of such events and processes. Instead one constructs a historical narrative, consisting of a tentative reconstruction of the particular scenario that led to the events one is trying to explain.

For example, three different scenarios have been proposed for the sudden extinction of the dinosaurs at the end of the Cretaceous: a devastating epidemic; a catastrophic change of climate; and the impact of an asteroid, known as the Alvarez theory. The first two narratives were ultimately refuted by evidence incompatible with them. All the known facts, however, fit the Alvarez theory, which is now widely accepted. The testing of historical narratives implies that the wide gap between science and the humanities that so troubled physicist C. P. Snow is actually nonexistent—by virtue of its methodology and its acceptance of the time factor that makes change possible, evolutionary biology serves as a bridge.

The discovery of natural selection, by Darwin and Alfred Russell Wallace, must itself be counted as an extraordinary philosophical advance. The principle remained unknown throughout the more than 2,000-year history of philosophy ranging from the Greeks to Hume, Kant and the Victorian era. The concept of natural selection had remarkable power for explaining directional and adaptive changes. Its nature is simplicity itself. It is not a force like the forces described in the laws of physics; its mechanism is simply the elimination of inferior individuals. This process of nonrandom elimination impelled Darwin's contemporary, philosopher Herbert Spencer, to describe evolution with the now familiar term "survival of the fittest." (This description was long ridiculed as circular reasoning: "Who are the fittest? Those who survive." In reality, a careful analysis can usually determine why certain individuals fail to thrive in a given set of conditions.)

The truly outstanding achievement of the principle of natural selection is that it makes unnecessary the invocation of "final causes"—that is, any teleological forces leading to a particular end. In fact, nothing is predetermined. Furthermore, the objective of selection even may change from one generation to the next, as environmental circumstances vary.

A diverse population is a necessity for the proper working of natural selection. (Darwin's success meant that typologists, for whom all members of a class are essentially identical, were left with an untenable viewpoint.) Because of the importance of variation, natural selection should be considered a two-step process: the production of abundant variation is followed by the elimination of inferior individuals. This latter step is directional. By adopting natural selection, Darwin settled the several-thousand-year-old argument among philosophers over chance or necessity. Change on the earth is the result of both, the first step being dominated by randomness, the second by necessity.

Darwin was a holist: for him the object, or target, of selection was primarily the individual as a whole. The geneticists, almost from 1900 on, in a rather reductionist spirit preferred to consider the gene the target of evolution. In the past 25 years, however, they have largely returned to the Darwinian view that the individual is the principal target.

For 80 years after 1859, bitter controversy raged as to which of four competing evolutionary theories was valid. "Transmutation" was the establishment of a new species or new type through a single mutation, or saltation. "Orthogenesis" held that intrinsic teleological tendencies led to transformation. Lamarckian evolution relied on the inheritance of acquired characteristics. And now there was Darwin's variational evolution, through natural selection. Darwin's theory clearly emerged as the victor during the evolutionary synthesis of the 1940s, when the new discoveries in genetics were married with taxonomic observations concerning systematics, the classification of organisms by their relationships. Darwinism is now almost unanimously accepted by knowledgeable evolutionists. In addition, it has become the basic component of the new philosophy of biology.

A most important principle of the new biological philosophy, undiscovered for almost a century after the publication of *On the Origin of Species*, is the dual nature of biological processes. These activities are governed both by the universal laws of physics and chemistry and by a genetic program, itself the result of natural selection, which has molded the genotype for millions of generations. The causal factor of the possession of a genetic program is unique to living organisms, and it is totally absent in the inanimate world. Because of the backward state of molecular and genetic knowledge in his time, Darwin was unaware of this vital factor.

Another aspect of the new philosophy of biology concerns the role of laws. Laws give way to concepts in Darwinism. In the physical sciences, as a rule, theories are based on laws; for example, the laws of motion led to the theory of gravitation. In evolutionary biology, however, theories are largely based on concepts such as competition, female choice, selection, succession and dominance. These biological concepts, and the theories based on them, cannot be reduced to the laws and theories of the physical sciences. Darwin himself never stated this idea plainly. My assertion of Darwin's importance to modern thought is the result of an analysis of Darwinian theory over the past century. During this period, a pronounced change in the methodology of biology took place. This transformation was not caused exclusively by Darwin, but it was greatly strengthened by developments in evolutionary biology. Observation, comparison and classification, as well as the testing of competing historical narratives, became the methods of evolutionary biology, outweighing experimentation.

I do not claim that Darwin was single-handedly responsible for all the intellectual developments in this period. Much of it, like the refutation of French mathematician and physicist Pierre-Simon Laplace's determinism, was "in the air." But Darwin in most cases either had priority or promoted the new views most vigorously.

The Darwinian Zeitgeist

A 21st-century person looks at the world quite differently than a citizen of the Victorian era did. This shift had multiple sources, particularly the incredible advances in technology. But what is not at all appreciated is the great extent to which this shift in thinking indeed resulted from Darwin's ideas.

Remember that in 1850 virtually all leading scientists and philosophers were Christian men. The world they inhabited had been created by God, and as the natural theologians claimed, He had instituted wise laws that brought about the perfect adaptation of all organisms to one another and to their environment. At the same time, the architects of the scientific revolution had constructed a worldview based on physicalism (a reduction to spatiotemporal things or events or their properties), teleology, determinism and other basic principles. Such was the thinking of Western man prior to the 1859 publication of *On the Origin of Species*. The basic principles proposed by Darwin would stand in total conflict with these prevailing ideas.

First, Darwinism rejects all supernatural phenomena and causations. The theory of evolution by natural selection explains the adaptedness and diversity of the world solely materialistically. It no longer requires God as creator or designer (although one is certainly still free to believe in God even if one accepts evolution). Darwin pointed out that creation, as described in the Bible and the origin accounts of other cultures, was contradicted by almost any aspect of the natural world. Every aspect of the "wonderful design" so admired by the natural theologians could be explained by natural selection. (A closer look also reveals that design is often not so wonderful—see "Evolution and the Origins of Disease," by Randolph M. Nesse and George C. Williams; SCIENTIFIC AMERICAN, November 1998). Eliminating God from science made room for strictly scientific explanations of all natural phenomena; it gave rise to positivism; it produced a powerful intellectual and spiritual revolution, the effects of which have lasted to this day.

Second, Darwinism refutes typology. From the time of the Pythagoreans and Plato, the general concept of the diversity of the world emphasized its invariance and stability. This viewpoint is called typology, or essentialism. The seeming variety, it was said, consisted of a limited number of natural kinds (essences or types), each one forming a class. The members of each class were thought to be identical, constant, and sharply separated from the members of other essences.

Variation, in contrast, is nonessential and accidental. A triangle illustrates essentialism: all triangles have the same fundamental characteristics and are sharply delimited against quadrangles or any other geometric figures. An intermediate between a triangle and a quadrangle is inconceivable. Typological thinking, therefore, is unable to accommodate variation and gives rise to a misleading conception of human races. For the typologist, Caucasians, Africans, Asians or Inuits are types that conspicuously differ from other human ethnic groups. This mode of thinking leads to racism. (Although the ignorant misapplication of evolutionary theory known as "social Darwinism" often gets blamed for justifications of racism, adherence to the disproved essentialism preceding Darwin in fact can lead to a racist viewpoint.)

Darwin completely rejected typological thinking and introduced instead the entirely different concept now called population thinking. All groupings of living organisms, including humanity, are populations that consist of uniquely different individuals. No two of the six billion humans are the same. Populations vary not by their essences but only by mean statistical differences. By rejecting the constancy of populations, Darwin helped to introduce history into scientific thinking and to promote a distinctly new approach to explanatory interpretation in science.

Third, Darwin's theory of natural selection made any invocation of teleology unnecessary. From the Greeks onward, there existed a universal belief in the existence of a teleological force in the world that led to ever greater perfection. This "final cause" was one of the causes specified by Aristotle. After Kant, in the *Critique of Judgment*, had unsuccessfully attempted to describe biological phenomena with the help of a physicalist Newtonian explanation, he then invoked teleological forces. Even after 1859, teleological explanations (orthogenesis) continued to be quite popular in evolutionary biology. The acceptance of the *Scala Naturae* and the explanations of natural

theology were other manifestations of the popularity of teleology. Darwinism swept such considerations away.

(The designation "teleological" actually applied to various different phenomena. Many seemingly end-directed processes in inorganic nature are the simple consequence of natural laws—a stone falls or a heated piece of metal cools because of laws of physics, not some end-directed process. Processes in living organisms owe their apparent goal-directedness to the operation of an inborn genetic or acquired program. Adapted systems, such as the heart or kidneys, may engage in activities that can be considered goal seeking, but the systems themselves were acquired during evolution and are continuously fine-tuned by natural selection. Finally, there was a belief in cosmic teleology, with a purpose and predetermined goal ascribed to everything in nature. Modern science, however, is unable to substantiate the existence of any such cosmic teleology.)

Fourth, Darwin does away with determinism. Laplace notoriously boasted that a complete knowledge of the current world and all its processes would enable him to predict the future to infinity. Darwin, by comparison, accepted the universality of randomness and chance throughout the process of natural selection. (Astronomer and philosopher John Herschel referred to natural selection contemptuously as "the law of the higgledy-piggledy.") That chance should play an important role in natural processes has been an unpalatable thought for many physicists. Einstein expressed this distaste in his statement, "God does not play dice." Of course, as previously mentioned, only the first step in natural selection, the production of variation, is a matter of chance. The character of the second step, the actual selection, is to be directional.

Despite the initial resistance by physicists and philosophers, the role of contingency and chance in natural processes is now almost universally acknowledged. Many biologists and philosophers deny the existence of universal laws in biology and suggest that all regularities be stated in probabilistic terms, as nearly all so-called biological laws have exceptions. Philosopher of science Karl Popper's famous test of falsification therefore cannot be applied in these cases.

Fifth, Darwin developed a new view of humanity and, in turn, a new anthropocentrism. Of all of Darwin's proposals, the one his contemporaries found most difficult to accept was that the theory of common descent applied to Man. For the theologians and philosophers alike, Man was a creature above and apart from other living beings. Aristotle, Descartes and Kant agreed on this sentiment, no matter how else their thinking diverged. But biologists Thomas Huxley and Ernst Haeckel revealed through rigorous comparative anatomical study that humans and living apes clearly had common ancestry, an assessment that has never again been seriously questioned in science. The application of the theory of common descent to Man deprived man of his former unique position.

Ironically, though, these events did not lead to an end to anthropocentrism. The study of man showed that, in spite of his descent, he is indeed unique among all organisms. Human intelligence is unmatched by that of any other creature. Humans are the only animals with true language, including grammar and syntax. Only humanity, as Darwin emphasized, has developed genuine ethical systems. In addition, through high intelligence, language and long parental care, humans are the only creatures to have created a rich culture. And by these means, humanity has attained, for better or worse, an unprecedented dominance over the entire globe.

Sixth, Darwin provided a scientific foundation for ethics. The question is frequently raised—and usually rebuffed—as to whether evolution adequately explains healthy human ethics. Many wonder how, if selection rewards the individual only for behavior that enhances his own survival and reproductive success, such pure selfishness can lead to any sound ethics. The widespread thesis of social Darwinism, promoted at the end of the 19th century by Spencer, was that evolutionary explanations were at odds with the development of ethics.

We now know, however, that in a social species not only the individual must be considered—an entire social group can be the target of selection. Darwin applied this reasoning to the human species in 1871 in *The Descent of Man*. The survival and prosperity of a social group depends to a large extent on the harmonious cooperation of the members of the group, and this behavior must be based on altruism. Such altruism, by furthering the survival and prosperity of the group, also indirectly benefits the fitness of the group's individuals. The result amounts to selection favoring altruistic behavior.

Kin selection and reciprocal helpfulness in particular will be greatly favored in a social group. Such selection for altruism has been demonstrated in recent years to be widespread among many other social animals. One can then perhaps encapsulate the relation between ethics and evolution by saying that a propensity for altruism and harmonious cooperation in social groups *is* favored by natural selection. The old thesis of social Darwinism—strict selfishness—was based on an incomplete understanding of animals, particularly social species.

The Influence of New Concepts

Let me now try to summarize my major findings. No educated person any longer questions the validity of the so-called theory of evolution, which we now know to be a simple fact. Likewise, most of Darwin's particular theses have been fully confirmed, such as that of common descent, the gradualism of evolution, and his explanatory theory of natural selection.

I hope I have successfully illustrated the wide reach of Darwin's ideas. Yes, he established a philosophy of biology by introducing the time factor, by demonstrating the importance of

chance and contingency, and by showing that theories in evolutionary biology are based on concepts rather than laws. But furthermore—and this is perhaps Darwin's greatest contribution—he developed a set of new principles that influence the thinking of every person: the living world, through evolution, can be explained without recourse to supernaturalism; essentialism or typology is invalid, and we must adopt population thinking, in which all individuals are unique (vital for education and the refutation of racism); natural selection, applied to social groups, is indeed sufficient to account for the origin and maintenance of altruistic ethical systems; cosmic teleology, an intrinsic process leading life automatically to ever greater perfection, is fallacious, with all seemingly teleological phenomena explicable by purely material processes; and determinism is thus repudiated, which places our fate squarely in our own evolved hands.

To borrow Darwin's phrase, there is grandeur in this view of life. New modes of thinking have been, and are being, evolved. Almost every component in modern man's belief system is somehow affected by Darwinian principles.

This article is based on the September 23, 1999, lecture that Mayr delivered in Stockholm on receiving the Crafoord Prize from the Royal Swedish Academy of Science.

Further Information

Darwin on Man: A Psychological Study of Scientific Creativity. Second edition. Howard E. Gruber. University of Chicago Press, 1981.

One Long Argument: Charles Darwin and The Genesis of Modern Evolutionary Thought. Ernst Mayr. Harvard University Press, 1993.

Charles Darwin: Voyaging: A Biography. Janet Browne. Princeton University Press, 1996.

The Descent of Man. Charles Darwin. Popular current edition. Prometheus Books, 1997.

The Origin of Species. Charles Darwin. Popular current edition. Bantam Classic, 1999.

ERNST MAYR is one of the towering figures in the history of evolutionary biology. Following his graduation from the University of Berlin in 1926, ornithological expeditions to New Guinea fueled his interest in theoretical evolutionary biology. Mayr emigrated to the U.S. in 1931 and in 1953 joined the faculty of Harvard University, where he is now Alexander Agassiz Professor of Zoology, Emeritus. His conception of rapid speciation of isolated populations formed the basis for the well-known neoevolutionary concept of punctuated equilibrium. The author of some of the 20th century's most influential volumes on evolution, Mayr is the recipient of numerous awards, including the National Medal of Science.

From *Scientific American*, July 2000, pp. 79–83. © 2000 by Ernst Mayr. Reprinted by permission of the author.

Evolution in Action

Finches, monkeyflowers, sockeye salmon, and bacteria are changing before our eyes.

JONATHAN WEINER

Charles Darwin's wife, Emma, was terrified that they would be separated for eternity, because she would go to heaven and he would not. Emma confessed her fears in a letter that Charles kept and treasured, with his reply to her scribbled in the margin: "When I am dead, know that many times, I have kissed and cryed over this."

Close as they were, the two could hardly bear to talk about Darwin's view of life. And today, those of us who live in the United States, by many measures the world's leading scientific nation, find ourselves in a house divided. Half of us accept Darwin's theory, half of us reject it, and many people are convinced that Darwin burns in hell. I find that old debate particularly strange, because I've spent some of the best years of my life as a science writer peering over the shoulders of biologists who actually watch Darwin's process in action. What they can see casts the whole debate in a new light—or it should.

Darwin himself never tried to watch evolution happen. "It may metaphorically be said," he wrote in the *Origin of Species*,

> that natural selection is daily and hourly scrutinizing, throughout the world, the slightest variations; rejecting those that are bad, preserving and adding up all that are good; silently and insensibly working, whenever and wherever opportunity offers.... We see nothing of these slow changes in progress, until the hand of time has marked the lapse of ages.

Darwin was a modest man who thought of himself as a plodder (one of his favorite mottoes was, "It's dogged as does it"). He thought evolution plodded too. If so, it would be more boring to watch evolution than to watch drying paint. As a result, for several generations after Darwin's death, almost nobody tried. For most of the twentieth century the only well-known example of evolution in action was the case of peppered moths in industrial England. The moth had its picture in all the textbooks, as a kind of special case.

Then, in 1973, a married pair of evolutionary biologists, Peter and Rosemary Grant, now at Princeton University, began a study of Darwin's process in Darwin's islands, the Galápagos, watching Darwin's finches. At first, they assumed that they would have to infer the history of evolution in the islands from the distribution of the various finch species, varieties, and populations across the archipelago. That is pretty much what Darwin had done, in broad strokes, after the *Beagle's* five-week survey of the islands in 1835. But the Grants soon discovered that at their main study site, a tiny desert island called Daphne Major, near the center of the archipelago, the finches were evolving rapidly. Conditions on the island swung wildly back and forth from wet years to dry years, and finches on Daphne adapted to each swing, from generation to generation. With the help of a series of graduate students, the Grants began to spend a good part of every year on Daphne, watching evolution in action as it shaped and reshaped the finches' beaks.

At the same time, a few biologists began making similar discoveries elsewhere in the world. One of them was John A. Endler, an evolutionary biologist at the University of California, Santa Barbara, who studied Trinidadian guppies. In 1986 Endler published a little book called *Natural Selection in the Wild*, in which he collected and reviewed all of the studies of evolution in action that had been published to that date. Dozens of new field projects were in progress. Biologists finally began to realize that Darwin had been too modest. Evolution by natural selection can happen rapidly enough to watch.

Now the field is exploding. More than 250 people around the world are observing and documenting evolution, not only in finches and guppies, but also in aphids, flies, grayling, monkeyflowers, salmon, and sticklebacks. Some workers are even documenting pairs of species—symbiotic insects and plants—that have recently found each other, and observing the pairs as they drift off into their own world together like lovers in a novel by D.H. Lawrence.

The Grants' own study gets more sophisticated every year. A few years ago, a group of molecular biologists working with the Grants nailed down a gene that plays a key role in shaping the beaks of the finches. The gene codes for a signaling molecule called bone morphogenic protein 4 (BMP4). Finches with bigger beaks tend to have more BMP4, and finches with smaller beaks have less. In the laboratory, the biologists demonstrated that they could sculpt the beaks themselves by adding or subtracting BMP4. The same gene that shapes the beak of the finch in the egg also shapes the human face in the womb.

Some of the most dramatic stories of evolution in action result from the pressures that human beings are imposing on the planet. As Stephen Palumbi, an evolutionary biologist at Stanford University, points out, we are changing the course of evolution for virtually every living species everywhere, with consequences that are sometimes the opposite of what we might have predicted, or desired.

Take trophy hunting. Wild populations of bighorn mountain sheep are carefully managed in North America for hunters who want a chance to shoot a ram with a trophy set of horns. Hunting permits can cost well into the six figures. On Ram Mountain, in Alberta, Canada, hunters have shot the biggest of the bighorn rams for more than thirty years. And the result? Evolution has made the hunters' quarry scarce. The runts have had a better chance than the giants of passing on their genes. So on Ram Mountain the rams have gotten smaller, and their horns are proportionately smaller yet.

Or take fishing, which is economically much more consequential. The populations of Atlantic cod that swam for centuries off the coasts of Labrador and Newfoundland began a terrible crash in the late 1980s. In the years leading up to the crash, the cod had been evolving much like the sheep on Ram Mountain. Fish that matured relatively fast and reproduced relatively young had the better chance of passing on their genes; so did the fish that stayed small. So even before the population crashed, the average cod had been shrinking.

We often seem to lose out wherever we fight hardest to control nature. Antibiotics drive the evolution of drug-resistant bacteria at a frightening pace. Sulfonamides were introduced in the 1930s, and resistance to them was first observed a decade later. Penicillin was deployed in 1943, and the first penicillin resistance was observed in 1946. In the same way, pesticides and herbicides create resistant bugs and weeds.

Palumbi estimates that the annual bill for such unintended human-induced evolution runs to more than $100 billion in the U.S. alone. Worldwide, the pressure of global warming, fragmented habitats, heightened levels of carbon dioxide, acid rain, and the other myriad perturbations people impose on the chemistry and climate of the planet—all change the terms of the struggle for existence in the air, in the water, and on land. Biologists have begun to worry about those perturbations, but global change may be racing ahead of them.

To me, the most interesting news in the global evolution watch concerns what Darwin called "that mystery of mysteries, the origin of species."

The process whereby a population acquires small, inherited changes through natural selection is known as microevolution. Finches get bigger, fish gets smaller, but a finch is still a finch and a fish is still a fish. For people who reject Darwin's theory, that's the end of the story: no matter how many small, inherited changes accumulate, they believe, natural selection can never make a new kind of living thing. The kinds, the species, are eternal.

Darwin argued otherwise. He thought that many small changes could cause two lines of life to diverge. Whenever animals and plants find their way to a new home, for instance, they suffer, like emigres in new countries. Some individuals fail, others adapt and prosper. As the more successful individuals reproduce, Darwin maintained, the new population begins to differ from the ancestral one. If the two populations diverge widely enough, they become separate species. Change on that scale is known as macroevolution.

In *Origin*, Darwin estimated that a new species might take between ten thousand and fourteen thousand generations to arise. Until recently, most biologists assumed it would take at least that many, or maybe even millions, of generations, before microevolutionary changes led to the origin of new species. So they assumed they could watch evolution by natural selection, but not the divergence of one species into separate, reproductively isolated species. Now that view is changing too.

Not long ago, a young evolution-watcher named Andrew Hendry, a biologist at McGill University in Montreal, reported the results of a striking study of sockeye salmon. Sockeye tend to reproduce either in streams or along lake beaches. When the glaciers of the last ice age melted and retreated, about ten thousand years ago, they left behind thousands of new lakes. Salmon from streams swam into the lakes and stayed. Today their descendants tend to breed among themselves rather than with sockeyes that live in the streams. The fish in the lakes and streams are reproductively isolated from each other. So how fast did that happen?

In the 1930s and 1940s, sockeye salmon were introduced into Lake Washington, in Washington State. Hundreds of thousands of their descendants now live and breed in Cedar River, which feeds the lake. By 1957 some of the introduced sockeye also colonized a beach along the lake called Pleasure Point, about four miles from the mouth of Cedar River.

Hendry could tell whether a full-grown, breeding salmon had been born in the river or at the beach by examining the rings on its otoliths, or ear stones. Otolith rings reflect variations in water temperature while a fish embryo is developing. Water temperatures at the beach are relatively constant compared with the river temperatures. Hendry and his colleagues checked the otoliths and collected DNA samples from the fish—and found that more than a third of the sockeye breeding at Pleasure Point had grown up in the river. They were immigrants.

With such a large number of immigrants, the two populations at Pleasure Point should have blended back together. But they hadn't. So at breeding time many of the river sockeye that swam over to the beach must have been relatively unsuccessful at passing on their genes.

Hendry could also tell the stream fish and the beach fish apart just by looking at them. Where the sockeye's breeding waters are swift-flowing, such as in Cedar River, the males tend to be slender. Their courtship ritual and competition with other males requires them to turn sideways in strong current—an awkward maneuver for a male with a deep, roundish body. So in strong current, slender males have the better chance of passing on their genes. But in still waters, males with the deepest bodies have the best chance of getting mates. So beach males tend to be rounder—their dimensions greater from the top of the back to the bottom of the belly—than river males.

What about females? In the river, where currents and floods are forever shifting and swirling the gravel, females have to dig deep nests for their eggs. So the females in the river tend to be bigger than their lake-dwelling counterparts, because bigger females can dig deeper nests. Where the water is calmer, the gravel stays put, and shallower nests will do.

So all of the beachgoers, male and female, have adapted to life at Pleasure Point. Their adaptations are strong enough that reproductive isolation has evolved. How long did the evolution take? Hendry began studying the salmon's reproductive isolation in 1992. At that time, the sockeyes in the stream and the ones at Pleasure Point had been breeding in their respective habitats for at most thirteen generations. That is so fast that, as Hendry and his colleagues point out, it may be possible someday soon to catch the next step, the origin of a new species.

And it's not just the sockeye salmon. Consider the three-spined stickleback. After the glaciers melted at the end of the last ice age, many sticklebacks swam out of the sea and into new glacial lakes—just as the salmon did. In the sea, sticklebacks wear heavy, bony body armor. In a lake they wear light armor. In a certain new pond in Bergen, Norway, during the past century, sticklebacks evolved toward the lighter armor in just thirty-one years. In Loberg Lake, Alaska, the same kind of change took only a dozen years. A generation for sticklebacks is two years. So that dramatic evolution took just six generations.

Dolph Schluter, a former finch-watcher from the Galapagos and currently a biologist at the University of British Columbia in Vancouver, has shown that, along with the evolution of new body types, sticklebacks also evolve a taste for mates with the new traits. In other words, the adaptive push of sexual selection is going hand-in-hand with natural selection. Schluter has built experimental ponds in Vancouver to observe the phenomenon under controlled conditions, and the same patterns he found in isolated lakes repeat themselves in his ponds. So adaptation can sometimes drive sexual selection and accelerate reproductive isolation.

There are other developments in the evolution watch, too, many to mention in this small space. Some of the fastest action is microscopic. Richard Lenski, a biologist at Michigan State University in East Lansing, watches the evolution of *Escherichia coli*. Because one generation takes only twenty minutes, and billions of *E. coli* can fit in a petri dish, the bacteria make ideal subjects for experimental evolution. Throw some *E. coli* into a new dish, for instance, with food they haven't encountered before, and they will evolve and adapt—quickly at first and then more slowly, as they refine their fit with their new environment.

And then there are the controversies. Science progresses and evolves by controversy, by internal debate and revision. In the United States these days one almost hates to mention that there are arguments among evolutionists. So often, they are taken out of context and hyperamplified to suggest that nothing about Darwinism is solid—that Darwin is dead. But research is messy because nature is messy, and fieldwork is some of the messiest research of all. It is precisely here at its jagged cutting edge that Darwinism is most vigorously alive.

Not long ago, one of the most famous icons of the evolution watch toppled over: the story of the peppered moths, familiar to anyone who remembers biology 101. About half a century ago, the British evolutionist Bernard Kettlewell noted that certain moths in the British Isles had evolved into darker forms when the trunks of trees darkened with industrial pollution. When the trees lightened again, after clean air acts were passed, the moths had evolved into light forms again. Kettlewell claimed that dark moths resting on dark tree trunks were harder for birds to see; in each decade, moths of the right color were safer.

But in the past few years, workers have shown that Kettlewell's explanation was too simplistic. For one thing, the moths don't normally rest on tree trunks. In forty years of observation, only twice have moths been seen resting there. Nobody knows where they do rest. The moths did evolve rapidly, but no one can be certain why.

To me what remains most interesting is the light that studies such as Hendry's, or the Grants', may throw on the origin of species. It's extraordinary that scientists are now examining the very beginnings of the process, at the level of beaks and fins, at the level of the genes. The explosion of evolution-watchers is a remarkable development in Darwin's science. Even as the popular debate about evolution in America is reaching its most heated moment since the trial of John Scopes, evolutionary biologists are pursuing one of the most significant and surprising voyages of discovery since the young Darwin sailed into the Galápagos Archipelago aboard Her Majesty's ship *Beagle*.

Not long ago I asked Hendry if his studies have changed the way he thinks about the origin of species. "Yes," he replied without hesitation, "I think it's occurring all over the place."

JONATHAN WEINER began writing about evolution in 1990, when he met Peter and Rosemary Grant, who observe evolution firsthand in finch populations in the Galápagos. Weiner's book *The Beak of the Finch* (Alfred A. Knopf) won a Pulitzer Prize in 1994. He is a professor in the Graduate School of Journalism at Columbia University, in New York City. He is also working on a book about human longevity for Ecco Press.

From *Natural History*, November 2005, pp. 47–52. Copyright © 2005 by Natural History Magazine. Reprinted by permission.

15 Answers to Creationist Nonsense

Opponents of evolution want to make a place for creationism by tearing down real science, but their arguments don't hold up

JOHN RENNIE

When Charles Darwin introduced the theory of evolution through natural selection 143 years ago, the scientists of the day argued over it fiercely, but the massing evidence from paleontology, genetics, zoology, molecular biology and other fields gradually established evolution's truth beyond reasonable doubt. Today that battle has been won everywhere—except in the public imagination.

Embarrassingly, in the 21st century, in the most scientifically advanced nation the world has ever known, creationists can still persuade politicians, judges and ordinary citizens that evolution is a flawed, poorly supported fantasy. They lobby for creationist ideas such as "intelligent design" to be taught as alternatives to evolution in science classrooms. As this article goes to press, the Ohio Board of Education is debating whether to mandate such a change. Some antievolutionists, such as Philip E. Johnson, a law professor at the University of California at Berkeley and author of *Darwin on Trial*, admit that they intend for intelligent-design theory to serve as a "wedge" for reopening science classrooms to discussions of God.

Besieged teachers and others may increasingly find themselves on the spot to defend evolution and refute creationism. The arguments that creationists use are typically specious and based on misunderstandings of (or outright lies about) evolution, but the number and diversity of the objections can put even well-informed people at a disadvantage.

To help with answering them, the following list rebuts some of the most common "scientific" arguments raised against evolution. It also directs readers to further sources for information and explains why creation science has no place in the classroom.

1. Evolution is only a theory. It is not a fact or a scientific law.

Many people learned in elementary school that a theory falls in the middle of a hierarchy of certainty—above a mere hypothesis but below a law. Scientists do not use the terms that way, however. According to the National Academy of Sciences (NAS), a scientific theory is "a well-substantiated explanation of some aspect of the natural world that can incorporate facts, laws, in-

ferences, and tested hypotheses." No amount of validation changes a theory into a law, which is a descriptive generalization about nature. So when scientists talk about the theory of evolution—or the atomic theory or the theory of relativity, for that matter—they are not expressing reservations about its truth.

In addition to the *theory* of evolution, meaning the idea of descent with modification, one may also speak of the *fact* of evolution. The NAS defines a fact as "an observation that has been repeatedly confirmed and for all practical purposes is accepted as 'true.'" The fossil record and abundant other evidence testify that organisms have evolved through time. Although no one observed those transformations, the indirect evidence is clear, unambiguous and compelling.

All sciences frequently rely on indirect evidence. Physicists cannot see subatomic particles directly, for instance, so they verify their existence by watching for tell-tale tracks that the particles leave in cloud chambers. The absence of direct observation does not make physicists' conclusions less certain.

2. Natural selection is based on circular reasoning: the fittest are those who survive, and those who survive are deemed fittest.

"Survival of the fittest" is a conversational way to describe natural selection, but a more technical description speaks of differential rates of survival and reproduction. That is, rather than labeling species as more or less fit, one can describe how many offspring they are likely to leave under given circumstances. Drop a fast-breeding pair of small-beaked finches and a slower-breeding pair of large-beaked finches onto an island full of food seeds. Within a few generations the fast breeders may control more of the food resources. Yet if large beaks more easily crush seeds, the advantage may tip to the slow breeders. In a pioneering study of finches on the Galápagos Islands, Peter R. Grant of Princeton University observed these kinds of population shifts in the wild [see his article "Natural Selection and Darwin's Finches"; SCIENTIFIC AMERICAN, October 1991].

The key is that adaptive fitness can be defined without reference to survival: large beaks are better adapted for crushing

PATRICIA J. WYNNE

GALÁPAGOS FINCHES show adaptive beak shapes

seeds, irrespective of whether that trait has survival value under the circumstances.

3. Evolution is unscientific, because it is not testable or falsifiable. It makes claims about events that were not observed and can never be re-created.

This blanket dismissal of evolution ignores important distinctions that divide the field into at least two broad areas: microevolution and macroevolution. Microevolution looks at changes within species over time—changes that may be preludes to speciation, the origin of new species. Macroevolution studies how taxonomic groups above the level of species change. Its evidence draws frequently from the fossil record and DNA comparisons to reconstruct how various organisms may be related.

These days even most creationists acknowledge that microevolution has been upheld by tests in the laboratory (as in studies of cells, plants and fruit flies) and in the field (as in Grant's studies of evolving beak shapes among Galapagos finches). Natural selection and other mechanisms—such as chromosomal changes, symbiosis and hybridization—can drive profound changes in populations over time.

The historical nature of macroevolutionary study involves inference from fossils and DNA rather than direct observation. Yet in the historical sciences (which include astronomy, geology and archaeology, as well as evolutionary biology), hypotheses can still be tested by checking whether they accord with physical evidence and whether they lead to verifiable predictions about future discoveries. For instance, evolution implies that between the earliest-known ancestors of humans (roughly five million years old) and the appearance of anatomically modern humans (about 100,000 years ago), one should find a succession of hominid creatures with features progressively less apelike and more modern, which is indeed what the fossil record shows. But one should not—and does not—find modern human fossils embedded in strata from the Jurassic period (65 million years ago). Evolutionary biology routinely makes predictions far more refined and precise than this, and researchers test them constantly.

Evolution could be disproved in other ways, too. If we could document the spontaneous generation of just one complex life-form from inanimate matter, then at least a few creatures seen in the fossil record might have originated this way. If superintelligent aliens appeared and claimed credit for creating life on earth (or even particular species), the purely evo-

lutionary explanation would be cast in doubt. But no one has yet produced such evidence.

It should be noted that the idea of falsifiability as the defining characteristic of science originated with philosopher Karl Popper in the 1930s. More recent elaborations on his thinking have expanded the narrowest interpretation of his principle precisely because it would eliminate too many branches of clearly scientific endeavor.

4. Increasingly, scientists doubt the truth of evolution.

No evidence suggests that evolution is losing adherents. Pick up any issue of a peer-reviewed biological journal, and you will find articles that support and extend evolutionary studies or that embrace evolution as a fundamental concept.

Conversely, serious scientific publications disputing evolution are all but nonexistent. In the mid-1990s George W. Gilchrist of the University of Washington surveyed thousands of journals in the primary literature, seeking articles on intelligent design or creation science. Among those hundreds of thousands of scientific reports, he found none. In the past two years, surveys done independently by Barbara Forrest of Southeastern Louisiana University and Lawrence M. Krauss of Case Western Reserve University have been similarly fruitless.

Creationists retort that a closed-minded scientific community rejects their evidence. Yet according to the editors of *Nature, Science* and other leading journals, few antievolution manuscripts are even submitted. Some antievolution authors have published papers in serious journals. Those papers, however, rarely attack evolution directly or advance creationist arguments; at best, they identify certain evolutionary problems as unsolved and difficult (which no one disputes). In short, creationists are not giving the scientific world good reason to take them seriously.

5. The disagreements among even evolutionary biologists show how little solid science supports evolution.

Evolutionary biologists passionately debate diverse topics: how speciation happens, the rates of evolutionary change, the ancestral relationships of birds and dinosaurs, whether Neandertals were a species apart from modern humans, and much more. These disputes are like those found in all other branches of science. Acceptance of evolution as a factual occurrence and a guiding principle is nonetheless universal in biology.

Unfortunately, dishonest creationists have shown a willingness to take scientists' comments out of context to exaggerate and distort the disagreements. Anyone acquainted with the works of paleontologist Stephen Jay Gould of Harvard University knows that in addition to co-authoring the punctuated-equilibrium model, Gould was one of the most eloquent defenders and articulators of evolution. (Punctuated equilibrium explains patterns in the fossil record by suggesting that most evolutionary changes occur within geologically brief intervals—which may nonetheless amount to hundreds of generations.) Yet creationists delight in dissecting out phrases from Gould's voluminous prose to make him sound as though he had doubted evolution, and they present punctuated equilibrium as though it allows new species to materialize overnight or birds to be born from reptile eggs.

When confronted with a quotation from a scientific authority that seems to question evolution, insist on seeing the statement in context. Almost invariably, the attack on evolution will prove illusory.

6. If humans descended from monkeys, why are there still monkeys?

This surprisingly common argument reflects several levels of ignorance about evolution. The first mistake is that evolution does not teach that humans descended from monkeys; it states that both have a common ancestor.

The deeper error is that this objection is tantamount to asking, "If children descended from adults, why are there still adults?" New species evolve by splintering off from established ones, when populations of organisms become isolated from the main branch of their family and acquire sufficient differences to remain forever distinct. The parent species may survive indefinitely thereafter, or it may become extinct.

7. Evolution cannot explain how life first appeared on earth.

The origin of life remains very much a mystery, but biochemists have learned about how primitive nucleic acids, amino acids and other building blocks of life could have formed and organized themselves into self-replicating, self-sustaining units, laying the foundation for cellular biochemistry. Astrochemical analyses hint that quantities of these compounds might have originated in space and fallen to earth in comets, a scenario that may solve the problem of how those constituents arose under the conditions that prevailed when our planet was young.

Creationists sometimes try to invalidate all of evolution by pointing to science's current inability to explain the origin of life. But even if life on earth turned out to have a non-evolutionary origin (for instance, if aliens introduced the first cells billions of years ago), evolution since then would be robustly confirmed by countless microevolutionary and macroevolutionary studies.

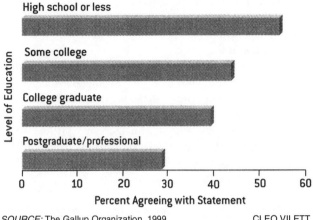

"GOD CREATED HUMANS IN THEIR PRESENT FORM WITHIN THE PAST 10,000 YEARS OR SO."

SOURCE: The Gallup Organization, 1999,   CLEO VILETT

8. Mathematically, it is inconceivable that anything as complex as a protein, let alone a living cell or a human, could spring up by chance.

Chance plays a part in evolution (for example, in the random mutations that can give rise to new traits), but evolution does not depend on chance to create organisms, proteins or other entities. Quite the opposite: natural selection, the principal known mechanism of evolution, harnesses nonrandom change by preserving "desirable" (adaptive) features and eliminating "undesirable" (non-adaptive) ones. As long as the forces of selection stay constant, natural selection can push evolution in one direction and produce sophisticated structures in surprisingly short times.

As an analogy, consider the 13-letter sequence "TOBEOR-NOTTOBE." Those hypothetical million monkeys, each pecking out one phrase a second, could take as long as 78,800 years to find it among the 26^{13} sequences of that length. But in the 1980s Richard Hardison of Glendale College wrote a computer program that generated phrases randomly while preserving the positions of individual letters that happened to be correctly placed (in effect, selecting for phrases more like Hamlet's). On average, the program re-created the phrase in just 336 iterations, less than 90 seconds. Even more amazing, it could reconstruct Shakespeare's entire play in just four and a half days.

9. The Second Law of Thermodynamics says that systems must become more disordered over time. Living cells therefore could not have evolved from inanimate chemicals, and multicellular life could not have evolved from protozoa.

This argument derives from a misunderstanding of the Second Law. If it were valid, mineral crystals and snowflakes would

21

also be impossible, because they, too, are complex structures that form spontaneously from disordered parts.

The Second Law actually states that the total entropy of a closed system (one that no energy or matter leaves or enters) cannot decrease. Entropy is a physical concept often casually described as disorder, but it differs significantly from the conversational use of the word.

More important, however, the Second Law permits parts of a system to decrease in entropy as long as other parts experience an offsetting increase. Thus, our planet as a whole can grow more complex because the sun pours heat and light onto it, and the greater entropy associated with the sun's nuclear fusion more than rebalances the scales. Simple organisms can fuel their rise toward complexity by consuming other forms of life and nonliving materials.

10. Mutations are essential to evolution theory, but mutations can only eliminate traits. They cannot produce new features.

On the contrary, biology has catalogued many traits produced by point mutations (changes at precise positions in an organism's DNA)—bacterial resistance to antibiotics, for example.

Mutations that arise in the homeobox (*Hox*) family of development-regulating genes in animals can also have complex effects. *Hox* genes direct where legs, wings, antennae and body segments should grow. In fruit flies, for instance, the mutation called *Antennapedia* causes legs to sprout where antennae should grow. These abnormal limbs are not functional, but their existence demonstrates that genetic mistakes can produce complex structures, which natural selection can then test for possible uses.

Moreover, molecular biology has discovered mechanisms for genetic change that go beyond point mutations, and these expand the ways in which new traits can appear. Functional modules within genes can be spliced together in novel ways. Whole genes can be accidentally duplicated in an organism's DNA, and the duplicates are free to mutate into genes for new, complex features. Comparisons of the DNA from a wide variety of organisms indicate that this is how the globin family of blood proteins evolved over millions of years.

11. Natural selection might explain microevolution, but it cannot explain the origin of new species and higher orders of life.

Evolutionary biologists have written extensively about how natural selection could produce new species. For instance, in the model called allopatry, developed by Ernst Mayr of Harvard University, if a population of organisms were isolated from the rest of its species by geographical boundaries, it might be subjected to different selective pressures. Changes would accumulate in the isolated population. If those changes became so significant that the splinter group could not or routinely would

not breed with the original stock, then the splinter group would be *reproductively isolated* and on its way toward becoming a new species.

Natural selection is the best studied of the evolutionary mechanisms, but biologists are open to other possibilities as well. Biologists are constantly assessing the potential of unusual genetic mechanisms for causing speciation or for producing complex features in organisms. Lynn Margulis of the University of Massachusetts at Amherst and others have persuasively argued that some cellular organelles, such as the energy-generating mitochondria, evolved through the symbiotic merger of ancient organisms. Thus, science welcomes the possibility of evolution resulting from forces beyond natural selection. Yet those forces must be natural; they cannot be attributed to the actions of mysterious creative intelligences whose existence, in scientific terms, is unproved.

12. Nobody has ever seen a new species evolve.

Speciation is probably fairly rare and in many cases might take centuries. Furthermore, recognizing a new species during a formative stage can be difficult, because biologists sometimes disagree about how best to define a species. The most widely used definition, Mayr's Biological Species Concept, recognizes a species as a distinct community of reproductively isolated populations—sets of organisms that normally do not or cannot breed outside their community. In practice, this standard can be difficult to apply to organisms isolated by distance or terrain or to plants (and, of course, fossils do not breed). Biologists therefore usually use organisms' physical and behavioral traits as clues to their species membership.

Nevertheless, the scientific literature does contain reports of apparent speciation events in plants, insects and worms. In most of these experiments, researchers subjected organisms to various types of selection—for anatomical differences, mating behaviors, habitat preferences and other traits—and found that they had created populations of organisms that did not breed with outsiders. For example, William R. Rice of the University of New Mexico and George W. Salt of the University of California at Davis demonstrated that if they sorted a group of fruit flies by their preference for certain environments and bred those flies separately over 35 generations, the resulting flies would refuse to breed with those from a very different environment.

13. Evolutionists cannot point to any transitional fossils—creatures that are half reptile and half bird, for instance.

Actually, paleontologists know of many detailed examples of fossils intermediate in form between various taxonomic groups. One of the most famous fossils of all time is *Archaeopteryx*, which combines feathers and skeletal structures peculiar to birds with features of dinosaurs. A flock's worth of other feathered fossil species, some more avian and some less, has also been found. A sequence of fossils spans the evolution of modern horses from the tiny *Eohippus*. Whales had four-legged

OTHER RESOURCES FOR DEFENDING EVOLUTION

How to Debate a Creationist: 25 Creationists' Arguments and 25 Evolutionists' Answers. Michael Shermer. Skeptics Society, 1997. This well-researched refutation of creationist claims deals in more depth with many of the same scientific arguments raised here, as well as other philosophical problems. *Skeptic* magazine routinely covers creation/evolution debates and is a solid, thoughtful source on the subject: **www.skeptic.com**

Defending Evolution in the Classroom: A Guide to the Creation/Evolution Controversy. Brian J. Alters and Sandra M. Alters. Jones and Bartlett Publishers, 2001. This up-to-date overview of the creation/evolution controversy explores the issues clearly and readably, with a full appreciation of the cultural and religious influences that create resistance to teaching evolution. It, too, uses a question-and-answer format that should be particularly valuable for teachers.

Science and Creationism: A View from the National Academy of Sciences. Second edition. National Academy Press, 1999. This concise booklet has the backing of the country's top scientific authorities. Although its goal of making a clear, brief statement necessarily limits the detail with which it can pursue its arguments, the publication serves as handy proof that the scientific establishment unwaveringly supports evolution. It is also available at **www7.nationalacademies.org/evalution/**

The Triumph of Evolution and the Failure of Creationism. Niles Eldredge. W. H. Freeman and Company, 2000. The author, a leading contributor to evolution theory and a curator at the American Museum of Natural History in New York City, offers a scathing critique of evolution's opponents.

Intelligent Design Creationism and Its Critics. Edited by Robert T. Pennock. Bradford Books/MIT Press, 2001. For anyone who wishes to understand the "intelligent design" controversy in detail, this book is a terrific one-volume summary of the scientific, philosophical and theological issues. Philip E. Johnson, Michael J. Behe and William A. Dembski make the case for intelligent design in their chapters and are rebutted by evolutionists, including Pennock, Stephen Jay Gould and Richard Dawkins.

Talk.Origins archive (www.talkorigins.org). This wonderfully thorough online resource compiles useful essays and commentaries that have appeared in Usenet discussions about creationism and evolution. It offers detailed discussions (some of which may be too sophisticated for casual readers) and bibliographies relating to virtually any objection to evolution that creationists might raise.

National Center for Science Education Web site (www.ncseweb.org). The center is the only national organization that specializes in defending the teaching of evolution against creationist attacks. Offering resources for combating misinformation and monitoring antievolution legislation, it is ideal for staying current with the ongoing public debate.

PBS Web site for evolution (www.pbs.org/wgbh/evolution/). Produced as a companion to the seven-part television series *Evolution*, this site is an enjoyable guide to evolutionary science. It features multimedia tools for teaching evolution. The accompanying book, *Evolution*, by Carl Zimmer (HarperCollins, 2001), is also useful for explaining evolution to doubters.

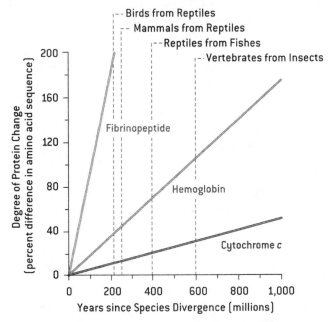

PROTEIN EVOLUTION REFLECTS SPECIES DIVERGENCE

CLEO VILETT

ancestors that walked on land, and creatures known as *Ambulocetus* and *Rodhocetus* helped to make that transition [see "The Mammals That Conquered the Seas," by Kate Wong; SCIENTIFIC AMERICAN, May]. Fossil seashells trace the evolution of various mollusks through millions of years. Perhaps 20 or more hominids (not all of them our ancestors) fill the gap between Lucy the australopithecine and modern humans.

Creationists, though, dismiss these fossil studies. They argue that *Archaeopteryx* is not a missing link between reptiles and birds—it is just an extinct bird with reptilian features. They want evolutionists to produce a weird, chimeric monster that cannot be classified as belonging to any known group. Even if a creationist does accept a fossil as transitional between two species, he or she may then insist on seeing other fossils intermediate between it and the first two. These frustrating requests can proceed ad infinitum and place an unreasonable burden on the always incomplete fossil record.

Nevertheless, evolutionists can cite further supportive evidence from molecular biology. All organisms share most of the same genes, but as evolution predicts, the structures of these genes and their products diverge among species, in keeping with their evolutionary relationships. Geneticists speak of the "molecular clock" that records the passage of time. These molecular data also show how various organisms are transitional within evolution.

14. Living things have fantastically intricate features—at the anatomical, cellular and molecular levels—that could not function if they were any less complex or sophisticated. The only prudent conclusion is that they are the products of intelligent design, not evolution.

This "argument from design" is the backbone of most recent attacks on evolution, but it is also one of the oldest. In 1802 theologian William Paley wrote that if one finds a pocket watch in a field, the most reasonable conclusion is that someone dropped it, not that natural forces created it there. By analogy, Paley argued, the complex structures of living things must be the handiwork of direct, divine invention. Darwin wrote *On the Origin of Species* as an answer to Paley: he explained how natural forces of selection, acting on inherited features, could gradually shape the evolution of ornate organic structures.

Generations of creationists have tried to counter Darwin by citing the example of the eye as a structure that could not have evolved. The eye's ability to provide vision depends on the perfect arrangement of its parts, these critics say. Natural selection could thus never favor the transitional forms needed during the eye's evolution—what good is half an eye? Anticipating this criticism, Darwin suggested that even "incomplete" eyes might confer benefits (such as helping creatures orient toward light) and thereby survive for further evolutionary refinement. Biology has vindicated Darwin: researchers have identified primitive eyes and light-sensing organs throughout the animal kingdom and have even tracked the evolutionary history of eyes through comparative genetics. (It now appears that in various families of organisms, eyes have evolved independently.)

Today's intelligent-design advocates are more sophisticated than their predecessors, but their arguments and goals are not fundamentally different. They criticize evolution by trying to demonstrate that it could nor account for life as we know it and then insist that the only tenable alternative is that life was designed by an unidentified intelligence.

15. Recent discoveries prove that even at the microscopic level, life has a quality of complexity that could not have come about through evolution.

"Irreducible complexity" is the battle cry of Michael J. Behe of Lehigh University, author of *Darwin's Black Box: The Biochemical Challenge to Evolution*. As a household example of irreducible complexity, Behe chooses the mousetrap—a machine that could not function if any of its pieces were missing and whose pieces have no value except as parts of the whole. What is true of the mousetrap, he says, is even truer of the bacterial flagellum, a whiplike cellular organelle used for propulsion that operates like an outboard motor. The proteins that make up a flagellum are uncannily arranged into motor components, a universal joint and other structures like those that a human engineer might specify. The possibility that this intricate array could have arisen through evolutionary modification is virtually nil, Behe argues, and that bespeaks intelligent design. He makes similar points about the blood's clotting mechanism and other molecular systems.

Yet evolutionary biologists have answers to these objections. First, there exist flagellae with forms simpler than the one that Behe cites, so it is not necessary for all those components to be present for a flagellum to work. The sophisticated components of this flagellum all have precedents elsewhere in nature, as described by Kenneth R. Miller of Brown University and others. In fact, the entire flagellum assembly is extremely similar to an organelle that *Yersinia pestis*, the bubonic plague bacterium, uses to inject toxins into cells.

The key is that the flagellum's component structures, which Behe suggests have no value apart from their role in propulsion, can serve multiple functions that would have helped favor their evolution. The final evolution of the flagellum might then have involved only the novel recombination of sophisticated parts that initially evolved for other purposes. Similarly, the blood-clotting system seems to involve the modification and elaboration of proteins that were originally used in digestion, according to studies by Russell F. Doolittle of the University of California at San Diego. So some of the complexity that Behe calls proof of intelligent design is not irreducible at all.

Complexity of a different kind—"specified complexity"—is the cornerstone of the intelligent-design arguments of William A. Dembski of Baylor University in his books *The Design Inference* and *No Free Lunch*. Essentially his argument is that living things are complex in a way that undirected, random processes could never produce. The only logical conclusion, Dembski asserts, in an echo of Paley 200 years ago, is that some superhuman intelligence created and shaped life.

Dembski's argument contains several holes. It is wrong to insinuate that the field of explanations consists only of random processes or designing intelligences. Researchers into nonlinear systems and cellular automata at the Santa Fe Institute and elsewhere have demonstrated that simple, undirected processes can yield extraordinarily complex patterns. Some of the complexity seen in organisms may therefore emerge through natural phenomena that we as yet barely understand. But that is far different from saying that the complexity could not have arisen naturally.

"CREATION SCIENCE" IS A CONTRADICTION IN TERMS. A central tenet of modern science is methodological naturalism—it seeks to explain the universe purely in terms of observed or testable natural mechanisms. Thus, physics describes the atomic nucleus with specific concepts governing matter and energy, and it tests those descriptions experimentally. Physicists introduce new particles, such as quarks, to flesh out their theories only when data show that the previous descriptions cannot adequately explain observed phenomena. The new particles do not have arbitrary properties, moreover—their definitions are tightly constrained, because the new particles must fit within the existing framework of physics.

In contrast, intelligent-design theorists invoke shadowy entities that conveniently have whatever unconstrained abilities are needed to solve the mystery at hand. Rather than expanding scientific inquiry, such answers shut it down. (How does one disprove the existence of omnipotent intelligences?)

Intelligent design offers few answers. For instance, when and how did a designing intelligence intervene in life's history? By creating the first DNA? The first cell? The first human? Was every species designed, or just a few early ones? Proponents of intelligent-design theory frequently decline to be pinned down on these points. They do not even make real attempts to reconcile their disparate ideas about intelligent design. Instead they pursue argument by exclusion—that is, they belittle evolutionary explanations as far-fetched or incomplete and then imply that only design-based alternatives remain.

Logically, this is misleading: even if one naturalistic explanation is flawed, it does not mean that all are. Moreover, it does not make one intelligent-design theory more reasonable than another. Listeners are essentially left to fill in the blanks for themselves, and some will undoubtedly do so by substituting their religious beliefs for scientific ideas.

Time and again, science has shown that methodological naturalism can push back ignorance, finding increasingly detailed and informative answers to mysteries that once seemed impenetrable: the nature of light, the causes of disease, how the brain works. Evolution is doing the same with the riddle of how the living world took shape. Creationism, by any name, adds nothing of intellectual value to the effort.

JOHN RENNIE is editor in chief of *Scientific American*.

Why Should Students Learn Evolution?

Brian J. Alters and Sandra Alters

"When you combine the lack of emphasis on evolution in kindergarten through 12th grade, with the immense popularity of creationism among the public, and the industry discrediting evolution, it's easy to see why half of the population believes humans were created 10,000 years ago and lived with dinosaurs. It is by far the biggest failure of science education from top to bottom."

—Randy Moore, Editor, The American Biology Teacher

"This is an important area of science, with particular significance for a developmental psychologist like me. Unless one has some understanding of the key notions of species, variation, natural selection, adaptation, and the like (and how these "have been discovered), unless one appreciates the perennial struggle among individuals (and populations) for survival in a particular ecological niche, one cannot understand the living world of which we are a part."

—Howard Gardner, Professor,
Harvard Graduate School of Education

With all of the controversy over the teaching of evolution reported in the media, with parents confronting their children's science teachers on this issue, and with students themselves confronting their instructors in high schools and colleges, would it be best—and easiest—to just delete the teaching of evolution in the classroom? Can't students attain a well-rounded background in science without learning this controversial topic? The overwhelming consensus of biologists in the scientific community is "no." Why, then, should science students learn about evolution?

A simple answer is that evolution is the basic context of all the biological sciences. Take away this context, and all that is left is disparate facts without the thread that ties them all together. Put another way, evolution is the explanatory framework, the unifying theory. It is indispensable to the study of biology, just as the atomic theory is indispensable to the study of chemistry. The characteristics and behavior of atoms and their subatomic particles form the basis of this physical science. So, too, biology can be understood fully only in an evolutionary context. In explaining how the organisms of today got to be the way they are, evolution helps make sense out of the history of life and explains relationships among species, It, is a useful and often essential framework within which scientists organize and interpret observations, and make predictions about the living world.

But this simple answer is not the entire reason why students should learn evolution. There are other considerations as well. Evolutionary explanations answer key questions in the biological sciences such as why organisms across species have so many striking similarities yet are tremendously diverse. These key questions are the *why* questions of biology. Much of biology explains *how* organisms work ... how we breathe, how fish swim, or how leopard frogs produce thousands of eggs at one time ... but it is up to evolution to explain the why behind these mechanisms. In answering the key *why* questions of biology, evolutionary explanations become an important lens through which scientists interpret data, whether they are developmental biologists, plant physiologists, or biochemists, to mention just a few of the many foci of those who study life.

Understanding evolution also has practical considerations that affect day-to-day life. Without an understanding of natural selection, students cannot recognize and understand problems based on this process, such as insect resistance to pesticides or microbial resistance to antibiotics. In a report released in June, 2000, Dr. Gro Harlem Brunddand, Director-General of the World Health Organization, stated that the world is at risk of losing drugs that control many infectious diseases because of increasing antimicrobial resistance. The report goes on to give examples, stating that 98% of strains of gonorrhea in Southeast Asia are now resistant to penicillin. Additionally, 14,000 people die each year from drug-resistant infections acquired in hospitals in the United States. And in New Delhi, India, typhoid drugs are no longer effective against this disease. Such problems face every person on our planet, and an understanding of natural selection will help students realize how important their behavior is in either contributing to or helping stem this crisis in medical progress.

Evolution not only enriches and provides a conceptual foundation for biological sciences such as ecology, genetics, developmental biology, and systematics, it provides a framework for scientific disciplines with historical aspects, such as anthropology, astronomy, geology, and paleontology. Evolution is therefore a unifying theme among many sciences, providing students with a framework by which to understand the natural world from many perspectives.

As scientists search for evolutionary explanations to the many questions of life, they develop methods and formulate concepts that are being applied in other fields, such as molecular biology, medicine, and statistics. For example, scientists studying molecular evolutionary change have developed methods to distinguish variations in gene sequences within and among species. These methods not only add to the toolbox of the molecular biologist but also will have likely applications in medicine by helping to identify variations that cause genetic diseases. In characterizing and analyzing variation, evolutionary biologists have also developed statistical methods, such as analysis of variance and path analysis, which are widely used in other fields. Thus, methods and concepts developed by evolutionary biologists have wide relevance in other fields and influence us all daily in ways we cannot realize without an understanding of this important and central idea.

Evolution is not only a powerful and wide-reaching concept among the pure and applied sciences, it also permeates other disciplines such as philosophy, psychology, literature, and the arts. Evolution by means of natural selection, articulated amidst controversy in the mid-nineteenth century, has reached the twenty-first century having had an extensive and expansive impact on human thought. An important intellectual development in the history of ideas, evolution should hold a central place in science teaching and learning.

Why Is Evolution the Context of the Biological Sciences— a Unifying Theory?

First, how does evolution take place? A key idea is that some of the individuals within a population of organisms possess measurable changes in inheritable characteristics that favor their survival. (These characteristics can be morphological, physiological, behavioral, or biochemical.) These individuals are more likely to live to reproductive age than are individuals not possessing the favorable characteristics. These reproductively advantageous traits (called *adaptive traits* or *adaptations*) are passed on from surviving individuals to their offspring. Over time, the individuals carrying these traits will increase in numbers within the population, and the nature of the population as a whole will gradually change. This process of survival of the most. reproductively fit organisms is called *natural selection*.

The process of evolutionary change explains that the organisms of today got to be the way they are, at least in part, as the result of natural selection over billions of years and even billions more generations. Organisms are related to one another, some more distantly, branching from a common ancestor long ago, and some more recently, branching from a common ancestor closer to the present day. The fact that diverse organisms have descended from common ancestors accounts for the similarities exhibited among species. Since biology is the story of life, then evolution is the story of biology and the relatedness of all life.

How Do Evolutionary Explanations Answer Key Questions in the Biological Sciences?

Evolution answers the question of the unity and similarity of life by its relatedness and shared history. But what about its diversity? And how does evolution answer other key questions in the biological sciences? What are these questions and how does evolution answer the *why* question inherent in each?

Evolution explains the diversity of life in the same way that it explains its unity. As mentioned in the preceding paragraphs, some individuals within a population of organisms possess measurable changes in inheritable characteristics that favor their survival. These adaptive traits are passed on from surviving individuals to their offspring. Over time, as populations inhabit different ecological niches, the individuals carrying adaptive traits in each population increase in numbers, and the nature of each population gradually changes. Such divergent evolution, the splitting of single species into multiple, descendant species, accounts for variation. There are different modes, or patterns, of divergence, and various reproductive isolating mechanisms that contribute to divergent evolution. However, the result is the same: Populations split from common ancestral populations and their genetic differences accumulate.

What are some other key questions in biology that are answered by evolution? One key question asks why form is adapted to function. Evolutionary theory tells us that more organisms that have parts of their anatomy (a long, slender beak, for instance) better adapted to certain functions (such as capturing food that lives deep within holes in rotting tree trunks) will live to reproductive age in greater numbers than those with less-well-adapted beaks. Therefore, the organisms with better-adapted beaks will pass on the genes for these features to greater numbers of offspring. Eventually, after numerous generations, natural selection will result in a population that has long slender beaks adapted to procuring food. Thus, anatomical, behavioral, or biochemical traits (the "forms") fit their functions because form fitting function is adaptive. But this idea leads us to yet another important question: Why do organisms have a variety of nonadaptive features that coexist amidst those that are adaptive?

During the course of evolution, traits that no longer confer a reproductive advantage do not disappear in the population unless they are reproductively disadvantageous. A population of beige beach birds that escaped predation because of protective coloration will not change coloration if this population becomes geographically isolated to a grasslands environment, unless the now useless beige coloration allows the birds to be hunted and killed more easily. In other words, if beige coloration is not a liability in the new environment, the genes that code for this trait will be passed on by all surviving birds in this grasslands niche. Even as the population of birds changes over generations, the genes for beige feathers will be retained in the population as long as this trait confers no reproductive disadvantage (and as long as mutation and genetic drift do not result in such a change).

These preceding examples do not cover all the key questions of biology (of course), but do show that such key questions are really questions about evolution and its mechanisms. Only evolutionary theory can answer the *why* questions inherent in these themes of life.

How Does Understanding Evolution Help Us Understand Processes That Affect Our Health and Our Day-to-day Life? And How Are Evolutionary Methods Applied to Other Fields?

As mentioned earlier, without an understanding of natural selection, students cannot recognize and understand problems based on this process, such as insect resistance to pesticides or microbial resistance to antibiotics. Additionally, it is only through such understanding that scientists can hope. to find solutions to these serious situations. Scientists know that the underlying cause of microbial resistance to antibiotics is improper use of these drugs. As explained in the World Health Organization report *Overcoming Antimicrobial Resistance*, in poor countries antibiotics are often used in ways that encourage the development of resistance. Unable to afford the full course of treatment, patients often take antibiotics only until their symptoms go away—killing the most susceptible microbes while allowing those more resistant to survive and reproduce. When these most resistant pathogens infect another host, antibiotics are less effective against the more resistant strains. In wealthy countries such as the United States, antibiotics are overused, being prescribed for viral diseases for which they are ineffective and being used in agriculture to treat sick animals and promote the growth of those that are well. Such misuse and overuse of antibiotics speeds the process whereby less resistant strains of bacteria are wiped out and more resistant strains flourish.

In addition to developing resistance to antibiotics and other therapies, pathogens can evolve resistance to the body's natural defenses. The virulence of pathogens (the ease with which they cause disease) can also evolve rapidly. Understanding the co-evolution of the human immune system and the pathogens that attack it help scientists track and predict disease outbreaks.

Understanding evolution also helps researchers understand the frequency, nature, and distribution of genetic disease. Gene frequencies in populations are affected by selection pressures, mutation, migration, and random genetic drift. Studying genetic diseases from an evolutionary standpoint helps us see that even lethal genes can remain in a population if there is a reproductive advantage in the heterozygote, as in the case of sickle-cell anemia and malaria.

Sickle-cell anemia is one of the most common genetic disorders among African Americans, having arisen in their African ancestors. It has been observed in persons whose ancestors came from the Mediterranean basin, the Indian subcontinent, the Caribbean, and parts of Central and South America (particularly Brazil). The sickle-cell gene has persisted in these populations, even though the disease eventually kills its victims, because carriers who inherit a single defective gene are resistant to malaria. Those with the sickle-cell gene have a survival advantage in regions of the world in which malaria is prevalent, which are the regions of the ancestral populations listed previously. Although many of these peoples have since migrated from these areas, this ancestral gene still persists within their populations.

Scientists are also working to identify gene variations that cause genetic diseases. Molecular evolutionary biologists have developed methods to distinguish between variations in gene sequences that affect reproductive fitness and variations that do not. To do this, scientists analyze human DNA sequences and DNA sequences among closely related species. The Human Genome Project, a worldwide effort to map the positions of all the genes and to sequence the over 3 billion DNA base pairs of the human genome, is providing much of the data for this effort and also is allowing scientists to study the relationships between the structure of genes and the proteins they produce. (On June 26, 2000, scientists announced the completion of the "working draft", of the human genome. The working draft covers 85% of the genome's coding regions in rough form.)

Some diseases are caused by interaction between genes and environment (lifestyle) factors. Genetic factors may predispose a person to a disease. For example, America's number one killers, cardiovascular disease and cancer, have both genetic and environmental causes. However, the complex interplay between genes and environmental factors in the development of these diseases makes it difficult for scientists to study the genetics of these diseases. Nevertheless, using evolutionary principles and approaches, scientists have developed a technique called *gene tree analysis* to discover genetic markers that are predictive of certain diseases. (Genetic markers are pairs of alleles whose inheritance can be traced through a pedigree [family tree].) Analyses of gene trees can help medical researchers identify the mutations in genes that cause certain diseases. This knowledge helps medical researchers understand the cause of the diseases to which these genes are linked and can help them develop treatments for such illnesses.

How Is Evolution Indispensable to the Subdisciplines of Biology and How Does It Enrich Them?

Organizing life, for example, a process on which Linnaeus worked as he grouped organisms by morphological characteristics, continues today with processes that reflect evolutionary relationships. Systematics, the branch of biology that studies the classification of life, does so in the context of evolutionary relationships. Cladistics, the predominant method used in systematics today, classifies organisms with respect to their phylogenetic relationships—those based on their evolutionary history. Therefore, students who do not understand evolution cannot understand modern methods of classification.

Developmental biology is another example of a biological subdiscipline enriched by an evolutionary perspective. In fact, some embryological phenomena can be understood only in the light of evolutionary history. For example, why terrestrial salamanders go through a larval stage with gills and fins that are never used is a question answered by evolution. During evolution, as new species (e.g., terrestrial salamanders) evolve from ancestral forms (e.g., aquatic ancestors), their new developmental instructions are often added to developmental instructions already in place. Thus, patterns of development in groups of organisms were built over the evolutionary history of those groups, thus retaining ancestral instructions. This process results in the embryonic stages of particular vertebrates reflecting the embryonic stages of those vertebrates' ancestors.

The study of animal behavior is enriched by an evolutionary perspective as well. Behavioral traits also evolve, and like morphological traits they are often most similar among closely related species. Phylogenetic studies of behavior have provided examples of how complex behaviors such as the courtship displays of some birds have evolved from simpler ancestral behaviors. Likewise, the study of human behavior can be enhanced by an evolutionary perspective. Evolutionary psychologists seek to uncover evolutionary reasons for many human behaviors, searching through our ancestral programming to determine how natural selection has resulted in a species that behaves as it does.

There are many sciences with significant historical aspects, such as anthropology, astronomy, geology, and paleontology. Geology, for example, is the study of the history of the earth, especially as recorded in the rocks. Paleontology is the study of fossils. Inherent in the work of the geologist and the paleontologist are questions about the relationships of modern animals and plants to ancestral forms, and about the chronology of the history of the earth. Evolution provides the framework within which these questions can be answered.

What Do Science and Education Societies Say about the Study of Evolution?

Instructors often look to scientific societies for answers to many questions regarding their teaching. There is one aspect of teaching on which the scientific societies agree and are emphatic. Evolution is key to scientific study, and should be taught in the science classroom. The National Research Council, part of the National Academy of Science, identified evolution as a major unifying idea in science that transcends disciplinary boundaries. Its publication *National Science Education Standards* lists biological evolution as one of the six content areas in the life sciences that are important for all high school students to study. Likewise, the American Association for the Advancement of Science identified the evolution of life as one of six major areas of study in the life sciences in its publication *Benchmarks for Scientific Literacy*. The National Science Teachers Association, the largest organization in the world committed to promoting excellence and innovation "in science teaching and learning, published a position statement on the teaching of evolution in 1997, which states that "evolution is a major unifying concept of science and should be included as part of K–College science frameworks and curricula." The National Association of Biology Teachers, a leading organization in life science education, also issued a position statement on the teaching of evolution in 1997, which states that evolution has a "central, unifying role … in nature, and therefore in biology. Teaching biology in an effective and scientifically honest manner requires classroom discussions and laboratory experiences on evolution." Evolution has been identified as the unifying theme of biology by almost all science organizations that focus on the biological sciences.

So why should students learn evolution? Eliminating evolution from the education of students removes the context and unifying theory that underpins and permeates the biological sciences. Students thus learn disparate facts in the science classroom without the thread that ties them together, and they miss the answers to its underlying *why* questions. Without an understanding of evolution, they cannot understand processes based on this science, such as insect resistance to pesticides and microbial resistance to antibiotics. Students will not come to understand evolutionary connections to other scientific fields, nor will they fully understand the world of which we are a part. Evolution is, in fact, one of the most important concepts in attaining scientific literacy.

The Illusion of Design

Richard Dawkins

The world is divided into things that look as though somebody designed them (wings and wagon-wheels, hearts and televisions), and things that just happened through the unintended workings of physics (mountains and rivers, sand dunes, and solar systems). Mount Rushmore belonged firmly in the second category until the sculptor Gutzon Borglum carved it into the first. Charles Darwin moved in the other direction. He discovered a way in which the unaided laws of physics—the laws according to which things "just happen"—could, in the fullness of geologic time, come to mimic deliberate design. The illusion of design is so successful that to this day most Americans (including, significantly, many influential and rich Americans) stubbornly refuse to believe it is an illusion. To such people, if a heart (or an eye or a bacterial flagellum) looks designed, that's proof enough that it is designed.

No wonder Thomas Henry Huxley, "Darwin's bulldog" was moved to chide himself on reading the *Origin of Species*: "How extremely stupid not to have thought of that" And Huxley was the least stupid of men. The breathtaking power and reach of Darwin's idea—extensively documented in the field, as Jonathan Weiner reports—is matched by its audacious simplicity. You can write it out in a phrase: nonrandom survival of randomly varying hereditary instructions for building embryos. Yet, given the opportunities afforded by deep time, this simple little algorithm generates prodigies of complexity, elegance, and diversity of apparent design. True design, the kind we see in a knapped flint, a jet plane, or a personal computer, turns out to be a manifestation of an entity—the human brain—that itself was never designed, but is an evolved product of Darwin's mill.

Paradoxically, the extreme simplicity of what the philosopher Daniel C. Dennett called Darwin's dangerous idea may be its greatest barrier to acceptance. People have a hard time believing that so simple a mechanism could deliver such powerful results.

The arguments of creationists, including those creationists who cloak their pretensions under the politically devious phrase "intelligent-design theory," repeatedly return to the same big fallacy. Such-and-such looks designed. Therefore it was designed. To pursue my paradox, there is a sense in which the skepticism that often greets Darwin's idea is a measure of its greatness. Paraphrasing the twentieth-century population geneticist Ronald A. Fisher, natural selection is a mechanism for generating improbability on an enormous scale. Improbable is pretty much a synonym for unbelievable. Any theory that explains the highly improbable is asking to be disbelieved by those who don't understand it.

Yet the highly improbable does exist in the real world, and it must be explained. Adaptive improbability—complexity—is precisely the problem that any theory of life must solve and that natural selection, uniquely as far as science knows, does solve. In truth, it is intelligent design that is the biggest victim of the argument from improbability. Any entity capable of deliberately designing a living creature, to say nothing of a universe, would have to be hugely complex in its own right.

If, as the maverick astronomer Fred Hoyle mistakenly thought, the spontaneous origin of life is as improbable as a hurricane blowing through a junkyard and having the luck to assemble a Boeing 747, then a divine designer is the ultimate Boeing 747. The designer's spontaneous origin *ex nihilo* would have to be even more improbable than the most complex of his alleged creations. Unless, of course, he relied on natural selection to do his work for him! And in that case, one might pardonably wonder (though this is not the place to pursue the question), does he need to exist at all?

The achievement of nonrandom natural selection is to tame chance. By smearing out the luck, breaking down the improbability into a large number of small steps—each one somewhat improbable but not ridiculously so—natural selection ratchets up the improbability. As the generations unfold, ratcheting takes the cumulative improbability up to levels that—in the absence of the ratcheting—would exceed all sensible credence.

Many people don't understand such nonrandom cumulative ratcheting. They think natural selection is a theory of chance, so no wonder they don't believe it! The battle that we biologists face, in our struggle to convince the public and their elected representatives that evolution is a fact, amounts to the battle to convey to them the power of Darwin's ratchet—the blind watchmaker—to propel lineages up the gentle slopes of Mount Improbable.

The misapplied argument from improbability is not the only one deployed by creationists. They are quite fond of gaps, both literal gaps in the fossil record and gaps in their understanding of what Darwinism is all about. In both cases the (lack of) logic in the argument is the same. They allege a gap or deficiency in the Darwinian account. Then, without even inquiring whether intelligent design suffers from the same deficiency, they award victory to the rival "theory" by default. Such reasoning is no way to do science. But science is precisely not what creation "scientists," despite the ambitions of their intelligent-design bullyboys, are doing.

In the case of fossils, today's biologists are more fortunate than Darwin was in having access to beautiful series of transitional stages: almost cinematic records of evolutionary changes in action. Not all transitions are so attested, of course—hence the vaunted gaps. Some small animals just don't fossilize; their phyla are known only from modern specimens: their history is one big gap. The equivalent gaps for any creationist or intelligent-design theory would be the absence of a cinematic record of God's every move on the morning that he created, for example, the bacterial flagellar motor. Not only is there no such divine videotape: there is a complete absence of evidence of any kind for intelligent design.

Absence of evidence for is not positive evidence against, of course. Positive evidence against evolution could easily be found—if it exists. Fisher's contemporary and rival J.B.S. Haldane was asked by a Popperian zealot what would falsify evolution. Haldane quipped, "Fossil rabbits in the Precambrian." No such fossil has ever been found, of course, despite numerous searches for anachronistic species.

There are other barriers to accepting the truth of Darwinism. Many people cannot bear to think that they are cousins not just of chimpanzees and monkeys, but of tapeworms, spiders, and bacteria. The unpalatability of a proposition, however, has no bearing on its truth. I personally find the idea of cousinship to all living species positively agreeable, but neither my warmth toward it, nor the cringing of a creationist, has the slightest bearing on its truth.

The same could be said of political or moral objections to Darwinism. "Tell children they are nothing more than animals and they will behave like animals." I do not for a moment accept that the conclusion follows from the premise. But even if it did, once again, a disagreeable consequence cannot undermine the truth of a premise. Some have said that Hitler founded his political philosophy on Darwinism. This is nonsense: doctrines of racial superiority in no way follow from natural selection, properly understood. Nevertheless, a good case can be made that a society run on Darwinian lines would be a very disagreeable society in which to live. But, yet again, the unpleasantness of a proposition has no bearing on its truth.

Huxley, George C. Williams, and other evolutionists have opposed Darwinism as a political and moral doctrine just as passionately as they have advocated its scientific truth. I count myself in that company. Science needs to understand natural selection as a force in nature, the better to oppose it as a normative force in politics. Darwin himself expressed dismay at the callousness of natural selection: "What a book a Devil's Chaplain might write on the clumsy, wasteful, blundering low & horridly cruel works of nature!"

In spite of the success and admiration that he earned, and despite his large and loving family, Darwin's life was not an especially happy one. Troubled about genetic deterioration in general and the possible effects of inbreeding closer to home, as James Moore documents, and tormented by illness and bereavement, as Richard Milner's interview with the psychiatrist Ralph Colp Jr. shows, Darwin's achievements seem all the more remarkable. He even found the time to excel as an experimenter, particularly with plants. David Kohn's and Sheila Ann Dean's essays lead me to think that, even without his major theoretical achievements, Darwin would have won lasting recognition as an experimenter, albeit an experimenter with the style of a gentlemanly amateur, which might not find favor with modern journal referees.

As for his major theoretical achievements, of course, the details of our understanding have moved on since Darwin's time. That was particularly the case during the synthesis of Darwinism with Mendelian digital genetics. And beyond the synthesis, as Douglas J. Futuyma explains and Sean B. Carroll details further for the exciting new field of "evo-devo," Darwinism proves to be a flourishing population of theories, itself undergoing rapid evolutionary change.

In any developing science there are disagreements. But scientists—and here is what separates real scientists from the pseudoscientists of the school of intelligent design—always know what evidence it would take to change their minds. One thing all real scientists agree upon is the fact of evolution itself. It is a fact that we are cousins of gorillas, kangaroos, starfish, and bacteria. Evolution is as much a fact as the heat of the sun. It is not a theory, and for pity's sake, let's stop confusing the philosophically naive by calling it so. Evolution is a fact.

RICHARD DAWKINS, a world-renowned explicator of Darwinian evolution, is the Charles Simonyi Professor of the Public Understanding of Science at the University of Oxford, where he was educated. Dawkins's popular books about revolution and science include *The Selfish Gene,* Oxford University Press, 1976, *The Blind Watchmaker*, (W.W. Norton, 1986), *Climbing Mount Improbable* (W.W. Norton, 1996), and most recently, *The Ancestor's Tale* (Houghton Mifflin, 2004), which retells the sage of evolution in a Chaucerian mode.

From *Natural History,* November 2005, pp. 35–38. Copyright © 2005 by Natural History Magazine. Reprinted by permission.

Designer Thinking

Mark S. Blumberg

I was home from college for the winter holidays. One evening I had settled into a comfortable recliner to read an old paperback on logic that I had discovered among my father's collection of books. It was cold outside and the house was noticeably still. I was alone, or so I thought. As with most books on logic, the topic of human rationality had to be broached, which this one did by noting that "we are all familiar with the definition of *man* as a rational being." At that moment, I heard a sound and looked up. I was staring into the barrels of two snub-nosed revolvers aimed directly at my head. The rational beings holding the guns were wearing ski masks. One of them ordered me to the ground, whereupon he bound my hands behind my back and placed my head face-down on a soft pillow. Any comfort that I may have felt was quickly dissipated when one of the gunmen suddenly turned up the volume of the television, releasing an explosion of sound that convinced me that I had but a few seconds left to live. Thankfully, my guests were thieves, not murderers. Nonetheless, at that moment I forever lost interest in what logicians have to say about human behavior.

Our commitment to the notion that humans are rational is as old as philosophy itself, inextricably bound to an equally ancient need to associate ourselves with a higher being and distinguish ourselves from mere animals. Humans, according to this still-popular perspective, possess souls, which are the vessels that transport us from this life to the next. Deserving transport to salvation, however, requires all of us to choose correctly between good and evil, and this choice is made through the exercise of reason, a capacity that only we and God possess. Animals, on the other hand, are not members of this elite club, lacking souls and, therefore, the potential for an afterlife; without souls, it is argued, animals have no need for reason.

It has long been recognized, however, that animals cannot be so easily dismissed from the realm of rationality; indeed, they possess behavioral capacities that are reminiscent of the faculty of reason. How, then, many have wondered, can we account for these capacities without elevating animals to our level? The easy and comforting solution has long been to say that animals behave out of instinct.

Long before Darwin, philosophers, naturalists, and theologians noted and struggled to explain the complex behaviors of animals that, seemingly in the absence of learning, allow them to survive the challenges of the natural environment as if designed to do so. When Galen, the second-century Greek physician, extracted an infant goat from its mother's womb and placed before it several bowls containing such liquids as milk, wine, and water, he tells us that the kid chose the milk. Galen did not conclude that the kid was behaving rationally, but that it behaved as *if* it were rational. This form of rationality, however, was thought by Galen—and many others before and after him— to be the gift of a beneficent creator. For how else were such thinkers to explain the natural wisdom expressed by such lowly creatures?

The concept that we call *instinct* has had a long and convoluted history. Part of the difficulty in tracing the history of such a complicated idea is that language, culture, religion, philosophy, and science interact with such turbulence that one can rarely trace with confidence its trajectory over time. Knowing the etymology of a word—for example, that *instinct* is derived from the Latin word *instinguo,* meaning "to excite or urge," and that it is related to the word *stimulus,* a contraction of *stig-mu-lus,* an "object that was used to prod mules"—provides some sense of place but, ultimately, does not satisfy. Etymology may inform us that the originators of the word *instinct* used it to denote the urge to act, but it cannot tell us whether that urge in fact originates in the mind of God or the mind of an animal. What we can say with some confidence, however, is that instinct has been repeatedly employed throughout history as a means of erecting an unbreachable wall between rational man and unthinking brute.

Reason and instinct. For many, these two terms are complete opposites, one denoting the freedom that empowers the human mind and the other the shackles that doom animals to a life of automatism. Despite their differences, however, reason and instinct reduce to a single concept that, when fully appreciated, provides a foundation for understanding many features of biological history across many dimensions of time and space. That concept is design. As we will see, designer thinking has permeated, and continues to permeate, topics as diverse as religion, evolution, mind, and human invention. Understanding the attractions and pitfalls of this form of thinking is essential if we are to understand the nature and origins of instincts.

The Argument from Design

A few years ago I attended a public lecture by a scientist who had written a book proclaiming that Darwin was wrong. The author of this book, Michael Behe, was promoting what many in the audience believed to be a new idea: that animals are the product of *intelligent* design and not, as Darwin argued, the product of *apparent* design. That Behe is a biochemist seemed to lend credence to his argument, as he shrewdly spoke over the heads of his non-scientific and predominantly religious-minded audience and wowed them with the complexities of life in its most miniature forms. Consider the bacterium, he preached, with its flagellum designed so magnificently for forward propulsion, like an outboard motor with all of its parts working interdependently to accomplish its function. Such interdependency, he continued, could no more be the product of gradual, blind evolution than a mousetrap. Each part of the mousetrap has a function only within the context of the complete device; remove the spring and the mousetrap is rendered completely—not only partly—useless. Having deftly maneuvered his audience to this point, he was ready to complete the bait and switch: If an evolutionary explanation fails for a mousetrap, then how could it possibly not fail for a flagellum? To bring home his point, Behe projected onto the screen a mechanical drawing of the flagellum and its associated apparatus that looked like it had been removed from the desk of the mechanical engineer who had designed it. Nice show. The audience bought it.

And why wouldn't they. After all, this argument—the argument from design—has proven its rhetorical effectiveness for centuries. Although little more than an appeal to analogies between human and natural contrivances, the argument from design provided an important intellectual foundation for the religious faith of scientists and theologians alike, with Plato, Thomas Aquinas, and Isaac Newton among its many advocates. Three hundred years ago, Newton invoked the argument in his masterwork *Principia,* thus providing the imprimatur of the greatest scientist of his age. With such an ancient and esteemed pedigree, it is not surprising that it took an iconoclast like the English philosopher David Hume to write what many consider to be the definitive refutation of the argument from design, the *Dialogues Concerning Natural Religion.* Hume knew that he was addressing a touchy subject so, despite his iconoclasm, he arranged for his work to be published posthumously. As it turned out, Hume completed both the *Dialogues* and his life in the same year—1776.

Hume wrote the *Dialogues* from the perspective of three primary characters: Cleanthes, the scientific believer; Demea, the orthodox believer; and Philo, the skeptic (presumed by many to be the voice of Hume himself). It is left to Cleanthes to enunciate the argument from design, which he does with great eloquence:

> Consider, anatomize the eye; survey its structure and contrivance, and tell me, from your own feeling, if the idea of a contriver does not immediately flow in upon you with a force like that of sensation. The most obvious conclusion, surely, is in favor of design; and it re-

quires time, reflection, and study, to summon up those frivolous though abstruse objections which can support infidelity. Who can behold the male and female of each species, the correspondence of their parts and instincts, their passions and whole course of life before and after generation, but must be sensible that the propagation of the species is intended by nature? Millions and millions of such instances present themselves through every part of the universe, and no language can convey a more intelligible, irresistible meaning, than the curious adjustment of final causes. To what degree, therefore, of blind dogmatism must one have attained to reject such natural and such convincing arguments?

Cleanthes's choosing of the eye as definitive evidence for the creative hand of God was an obvious one. Indeed, explaining the origin of the eye's seemingly extreme perfection would prove to be a challenge to Darwin's theory of natural selection and, ultimately, one of that theory's greatest successes. As Darwin acknowledged in *The Origin of Species,* "To suppose that the eye with all its inimitable contrivances for adjusting the focus to different distances, for admitting different amounts of light, and for the correction of spherical and chromatic aberration, could have been formed by natural selection, seems, I freely confess; absurd in the highest degree."

Consistent with his rhetorical style, Darwin erects a seemingly impassable barrier and then easily bounds over it by outlining the process by which "numerous gradations from a simple and imperfect eye to one complex and perfect can be shown to exist, each grade being useful to its possessor, as is certainly the case." For example, we know of animals with mere patches of light-sensitive cells on the skin surface; animals with indentations on the skin surface that contain light-sensitive cells; animals in which these indentations are enlarged to form eye cups that direct light to the light-sensitive cells; animals in which the eye cup has closed to form a pinhole so that light can be focused; animals with eye cups containing a gelatinous substance that acts as a crude lens; animals, like us, with a more refined lens, as well as the ability to adjust the amount of light entering the eye. Thus, eyes did not evolve through the insertion of ready-made parts, but rather evolved such that even the most primitive eyes found throughout nature benefit the organisms that possess them. Better to see a little than not at all.

Although Darwin judges the human eye an organ of extreme perfection, we know that it is not. For example, what we call the blind spot can be reasonably described as the result of a design flaw, produced by a developmental wiring problem that any thoughtful designer, working from scratch, would have avoided. Hume may not have had such detailed biological evidence at his disposal, but he was acutely aware of the "inaccurate workmanship ... of the great machine of nature." Taking this argument to its logical extreme, Philo ridicules the argument from design by noting the many imperfections to be found in the world. Thus, for all we know, our world "was only the first rude essay of some infant deity who afterwards abandoned it, ashamed of his lame performance." In time, this argument from imperfection would provide perhaps the most convincing evidence for evolution and against intelligent design. As the late evolutionary biologist Steven Jay Gould once noted, "Odd arrangements and funny solutions are the proof of evolution—paths that a sensible God would never tread but that a natural process, constrained by history, follows perforce."

Hume's *Dialogues* seriously wounded the argument from design by revealing its logical flaws. Then, also in 1776, the Scottish philosopher Adam Smith published *The Wealth of Nations* and dealt the argument a further blow. Smith's contribution was the notion that economic order can emerge when each individual is free to behave without constraint, as if an invisible hand (to use Smith's metaphor) were molding the order according to a grand design. Inspired in part by Smith's ideas, Darwin provided the knockout punch to the argument from design in *The Origin of Species* by providing a mechanism—natural selection—by which order can arise without thought, mentation, intelligence, or design.

Organic evolution plays out on a grand temporal scale—thousands and millions of years—and it was the cloak provided by the vastness of time that obscured the mechanisms of evolutionary change and enhanced the illusion of intelligent design by an unseen creator, that is, an invisible hand guided by an invisible mind. The direct link between God and mind was apparent to Hume who, in the words of Philo, asks, "And if we are not contented with calling the first and supreme cause a GOD or DEITY, but desire to vary the expression, what can we call him but MIND or THOUGHT, to which he is justly supposed to bear a considerable resemblance?"

If the contrivances of nature, once imagined to be the product of a heavenly god, could be successfully moved to the realm of blind mechanism—of evolutionary trial and error playing out across the immensity of time—what about the contrivances of human beings, imagined to be the product of a creative, rational, purposeful, but more earthly mind? Is the man-made mousetrap evidence of the flagellum's intelligent design, or is a flagellum evidence of the evolution of the mousetrap? Before making the leap to humans, however, let's consider the role of trial and error in animal behavior.

Cat and Mouse

A rational mind lurking behind human contrivances is the foundation upon which the argument from design rests. It is a given. Two centuries ago, William Paley famously invaded the argument from design in his *Natural Theology, Or, Evidences of the Existence and Attributes of the Deity, Collected from the Appearances of Nature*. In that book, he contrasted the obvious simplicity of a stone with the equally obvious complexity of a watch to make the point that all complex objects must have a designer, whether it be a watchmaker or God.

Paley's error is simple yet profound. When we can't directly observe the developmental origins of any historical process, we naturally gravitate toward explanations that make the complex seem simple. Such explanations, however, are mere illusions. To avoid Paley's error, the first step is to ask: Where did *that* come from? The next step is to never stop asking that question.

As already mentioned, Darwin's theory of natural selection broke the spell of the argument from design by providing a natural path to the origins of organismal diversity on our planet. This radically new perspective of life and time removed thought and planning from a process that had been assumed to be the product of a divine mind. But does Darwin's perspective translate to other time scales? For example, what about animal behavior, which unfolds in seconds and minutes?

Toward the end of his career, Darwin eyed a successor, George Romanes, whom he hoped would apply the elder scientist's evolutionary ideas to animal behavior. The study of animal behavior was in its infancy at the end of the nineteenth century, relying to a large degree on the informal observations of amateur naturalists living and traveling throughout the world. Romanes collected and disseminated these often fanciful anecdotes as facts. For example, one anecdote recounted by Romanes gained credence by its being reported by several independent observers in Iceland. It was reported that mice work together to load provisions of food onto cow paddies, whereupon they launch the paddies, hop aboard, and navigate across the river using their tails as rudders. How could mice develop such extraordinary nautical skills? Perhaps, Romanes thought, the mice had observed humans loading provisions onto boats and steering those boats with a rudder. All that was required from the mice, then, was the ability to imitate human behavior. Of course, mice have no such ability, and this fanciful story of marine-going murines has become a cautionary tale about the dangers of anecdotes; nonetheless, the story buttressed Romanes's conception of imitation as a primary source of novel behaviors.

Romanes's belief in the power of imitation is illustrated most famously in an anecdote concerning a cat that belonged to his coachman. Romanes had noticed the cat using its front paws to unlatch the lock on the gate at the front of the house. But how could a cat perform such a feat? Romanes arrived at a simple conclusion. He imagined that the cat reasoned, "If a hand can do it, why not a paw?"

While the anecdote of the Icelandic mice illustrates the unreliability of amateur observations, the ability of the coachman's cat to unlatch a gate has been observed countless times by countless cat owners. Romanes's explanation for both anecdotes, however, is the same: imitation. Romanes was attracted to imitation because it seemed to him like a more simple and reasonable explanation than the two possible alternative explanations, namely, instinct and learning.

The reliance of Romanes on anecdote and unbounded conjecture fueled one young experimental psychologist, Edward Thorndike, to perform an experiment that simultaneously ridiculed Romanes and helped to found the scientific study of animal learning. For his experiments, published in 1899, Thorndike placed hungry cats inside small "puzzle boxes" of his own design, each version of the box requiring a unique set of actions that, when performed, opened the box and freed the cat to exit and receive food and water. After escaping, the cat was placed back in the box and tested again.

Thorndike made two important discoveries from this simple experiment. First, he found that the time required to escape from the box decreased with each successive test, indicating that the cats were learning. But second, he found that cats learned to escape through a process that had nothing to do with reflection and imitation. Naïve cats met confinement in the box with agitated and random behavior that occasionally, by happenstance, produced actions that opened the box. Over time, the cats' random agitation diminished and their behavior became increasingly directed toward the latch; eventually, the cats' behavior was so focused and efficient that, like Romanes's coachman's cat, the final form of the behavior *appeared* purposeful and human-like. Thorndike's point, of course, was that the purposefulness of a behavior can cloak the aimlessness of its origins.

The trial-and-error learning process introduced by Thorndike has been compared to natural selection. In trial-and-error learning, however, behavioral variability within an individual is reduced by weeding out the useless behaviors so that what remains, after many trials, is the elegant solution. In other words, the unfit behaviors are not reproduced in successive learning trials. And just as with natural selection, when we view the final product without having witnessed the process that led up to it, we are easily fooled into thinking that intelligent design is at work.

But does a trial-and-error explanation of some aspects of cat behavior provide any insight into the rational and purposeful behaviors of human beings? Do the behaviors of a trapped cat teach us anything about the invention of a mousetrap—about human creativity and ingenuity?

Failing to Succeed

Sitting in that audience a few years ago, listening to that modern purveyor of the argument from design, I was struck by Behe's confident assertion that human artifacts—clocks, computers, CD players—are created through a rational, thoughtful process. Of course this assertion is hardly new: recall William Paley's analogy between watchmaking and worldmaking. So, at the conclusion of his talk, I raised my hand and asked the speaker if he was aware of the possibility that even human artifacts evolve through a process of trial and error? The blank look on his face answered my question.

Henry Petroski is an engineer who has written extensively about the evolution of human artifacts. By examining in detail the history of such common, low-tech objects as pencils, paperclips, forks, and zippers, he provides a perspective that runs counter to the romantic image of the lonely inventor creating novel devices de novo using little more than reason and inspiration. On the contrary, according to Petroski, invention often is a trial-and-error process in which each successive development of an artifact is achieved through the removal of those features that don't work—those irritants that prevent an artifact from being as useful as it can be. Like evolution, design entails the removal of the unfit, producing in the end an artifact that appears to be the product of forethought, of intelligent design. As Thomas Edison famously expressed it, inventors fail their way to success. As Petroski expresses it, form follows failure.

Petroski demonstrates this evolutionary view of human invention through numerous examples. For instance, one may marvel at the beauty of the Brooklyn Bridge and see it as a stand-alone creation, but the reality is that, from the moment that the first man chopped down a tree and laid it across a creek, bridges have evolved through a trial-and-error process that has included numerous spectacular failures. Indeed, these failures are essential for future innovation. The same can be said for any of the great structural, architectural, or mechanical achievements, from the pyramids to the great medieval cathedrals to the space shuttle. One need not look to the great engineering achievements, however, to gain a sense of the evolutionary forces at play.

Consider a dining room table set for a formal function, replete with a variety of eating utensils: dinner forks, salad forks, dessert forks, carving knife, cheese knife, steak knives, fish knives, butter knives, dinner spoons, soup spoons, etc. For the most part, each utensil seems well-suited, perhaps even ideally suited, to the food for which it is intended. Upon observing such a fit between form and function, one might imagine that these utensils were designed for their purpose by a small group of particularly insightful food fanciers. Not so, says Petroski.

The diversity of eating utensils can be traced to a co-evolutionary relationship between the knife and the fork (the spoon has a separate, but no less interesting, history). This history is convoluted, but it follows a logical progression and is shaped by a few fundamental needs. The first need is basic: getting food from the plate to the mouth without having to touch the food with one's fingers. Because many non-Western cultures consider it acceptable to use fingers when eating (to compensate, Africans and Arabs have developed more rigorous cultural norms concerning hand-washing before and after meals), we are discussing here a tradition that arose primarily in Europe and was transported later to other parts of the world.

We begin with the knife, originally adapted from sharp pieces of flint. Over time, knife-making skills improved and, eventually, bronze and iron replaced stone. Knives were used for many things, including personal defense and, eventually among the more refined, eating. But how does one eat a piece of meat with nothing but a knife? Early on, bread was used to steady the meat for cutting, whereupon the meat was jabbed with the knife and conveyed to the mouth. But this was hardly an easy way to eat. So with the Middle Ages came the advent of the two-knife solution, one knife to steady the food and the other to cut, jab, and convey. Still, as a pointy utensil, a knife does not do a very good job of steadying food on its way to the mouth. For this job, at least two tines are necessary. The fork was born.

Initially, the two-tine fork was used primarily in the kitchen for carving and serving. These were large forks with widely separated tines, and they were useful in this role because the meat could be carved effectively and the fork could be easily removed from the meat. By the seventeenth century, forks began appearing at the dining table in England, but a new problem greeted this transition. Specifically, although a fork with two large, widely separated tines was effective for preventing rotation when carving meat, such a design was not ideal for individual dining because the two large tines were not useful for spearing bite-sized portions of food and, moreover, the wide spacing of the tines was ineffective for scooping. These problems were addressed by the advent of the three-tined fork which continued to solve the rotation problem. In time, a four-tined-fork evolved, followed by brief flirtations with five- and six-tined forks; these flirtations were brief because the width of the mouth sets an effective limit to the width of a fork. Ultimately, the four-tined fork won out, becoming the standard in England by the end of the 1800s.

The fork's impact on the design of the knife has not been trivial. First, the left-hand knife was replaced by the fork as a solution to the rotation problem. Then, the availability of a pointed fork diminished the necessity of a pointed knife. By the end of the seventeenth century, dinner knives were now blunted and, because two-tined forks were the norm at that time and were not ideal for scooping, knives were then given a broad surface to serve a scooping function. Thus has arisen the standard dinner fork and knife. The remaining standard utensils—and the many non-standard ones—have similar histories.

The evolution of eating utensils has shaped the cultural transmission of eating habits in surprising ways. For example, the European style of eating is derived from the history just reviewed, with the knife in the right hand cutting the food and pushing it onto the fork in the left hand for conveyance to the mouth. For some reason, forks were rare in colonial America but spoons were not. Thus, these colonists would use the knife in the European tradition but use the base of the spoon in their left hand to steady the food while cutting, after which the right (and typically more dexterous) hand would lay down the knife and pick up and flip over the spoon to scoop up the food for eating. This crisscross eating method became an ingrained American cultural tradition and therefore survived the introduction of the fork. These European and American, behavioral traditions continue today despite the fact that forks, knives, and spoons are placed on the table in identical positions in Europe and America; in other words, our history with eating utensils is a hidden force that continues to shape our culture and behavior.

Thus, human artifacts are no more the independent creations of any single individual than is the evolutionary invention of sonar in dolphins or echolocation in bats. In Petroski's words, artifacts

> do not spring fully formed from the mind of some maker but, rather, become shaped and reshaped through the (principally negative) experiences of their users within the social, cultural, and technological contexts in which they are embedded.... Imagining how the form of things as seemingly simple as eating utensils might have evolved demonstrates the inadequacy of a 'form follows function' argument to serve as a guiding principle for understanding how artifacts have come to look the way they do. . . . If not in tableware, does form follow function in the genesis and de-

velopment of our more high-tech designs, or is the alliterative phrase just an alluring consonance that lulls the mind to sleep?

None of this is to downplay the significance of the individual's ability to identify and solve problems related to the design of low- and high-tech devices. Petroski does, however, describe the process of invention in a way that is more satisfying than are facile appeals to human consciousness and other indefinable qualities. Humans are clever, ingenious, inventive, resourceful, and persistent. But we do not invent complex, or even simple, devices from a standing start through the mere application of our mind to a problem. We are not that smart.

The argument from design is more than a fallacious argument: it is a reflection of how the human mind works. When we are confronted with complexity and see no path to how that complexity originated, the appeal of the argument from design is immense:

What explains the ability of cats to open gates? *Design, through the action of imitation.*

What explains the ingenuity of human invention? *Design, through the action of reason.*

What explains the complexity of the human eye? *Design, through the action of divine creation.*

Of course, nothing is explained by these answers but, even worse, they numb the very impulse that might lead us to ask deeper and more profound questions. We are lulled to sleep.

Logic is one of the crowning achievements of the human mind, testimony to the promise of human reason. But logic does not guide human thought and action so much as it describes it. Yes, we can apply logic to a problem but, as we know, it takes training and practice to do so effectively. Moreover, we should be cautious about singing the praises of our logical and reasoning abilities when so many of our species' greatest thinkers have declared their allegiance to the argument from design even after its flaws were exposed.

Although my aim here is not to diminish the estimation of our capacity for complex thought, there is little doubt that our intellectual conceits have erected a wall between humans—guided by reason—and animals—guided by instinct. Darwinian thinking, however, does not abide such artificial and arbitrary barriers. So it is no surprise that from Darwin's time to the present day, many who wish to tear down this wall have sought to do so by demonstrating human *instincts* and animal *reason*. Whether one views such attempts as successful or not is related, in no small degree, to one's comfort with such terms as *reason* and *instinct* in the first place.

We have seen in this chapter the pervasiveness of designer thinking and, just as important, we have seen the many benefits to our understanding of the world around us by moving beyond the attractions of designer thinking and asking the next question about origins. When we ask questions about origins, we defeat designer thinking.

Of course, defeating designer thinking when the subject is silverware is one thing. Defeating the appeal to genes as the designers of instinctive behaviors is quite another. And there can be no disputing the fact that—by and large—genes have become biological royalty, imbued with divine powers that are denied all other mere molecules. For those enraptured, the complexities of behavior cannot be comprehended without appealing to a genetic blueprint, a genetic controller, or a genetic program.

Sound familiar?

The Perimeter of Ignorance

A boundary where scientists face a choice: invoke a deity or continue the quest for knowledge

NEIL deGRASSE TYSON

Writing in centuries past, many scientists felt compelled to wax poetic about cosmic mysteries and God's handiwork. Perhaps one should not be surprised at this: most scientists back then, as well as many scientists today, identify themselves as spiritually devout.

But a careful reading of older texts, particularly those concerned with the universe itself, shows that the authors invoke divinity only when they reach the boundaries of their understanding. They appeal to a higher power only when staring into the ocean of their own ignorance. They call on God only from the lonely and precarious edge of incomprehension. Where they feel certain about their explanations, however, God gets hardly a mention.

Let's start at the top. Isaac Newton was one of the greatest intellects the world has ever seen. His laws of motion and his universal law of gravitation, conceived in the mid-seventeenth century, account for cosmic phenomena that had eluded philosophers for millennia. Through those laws, one could understand the gravitational attraction of bodies in a system, and thus come to understand orbits.

Newton's law of gravity enables you to calculate the force of attraction between any two objects. If you introduce a third object, then each one attracts the other two, and the orbits they trace become much harder to compute. Add another object, and another, and another, and soon you have the planets in our solar system. Earth and the Sun pull on each other, but Jupiter also pulls on Earth, Saturn pulls on Earth, Mars pulls on Earth, Jupiter pulls on Saturn, Saturn pulls on Mars, and on and on.

Newton feared that all this pulling would render the orbits in the solar system unstable. His equations indicated that the planets should long ago have either fallen into the Sun or flown the coop—leaving the Sun, in either case, devoid of planets. Yet the solar system, as well as the larger cosmos, appeared to be the very model of order and durability. So Newton, in his greatest work, the *Principia*, concludes that God must occasionally step in and make things right:

> The six primary Planets are revolv'd about the Sun, in circles concentric with the Sun, and with motions directed towards the same parts, and almost in the same plane.... But it is not to be conceived that mere mechan-

ical causes could give birth to so many regular motions.... This most beautiful System of the Sun, Planets, and Comets, could only proceed from the counsel and dominion of an intelligent and powerful Being.

In the *Principia*, Newton distinguishes between hypotheses and experimental philosophy, and declares, "Hypotheses, whether metaphysical or physical, whether of occult qualities or mechanical, have no place in experimental philosophy." What he wants is data, "inferr'd from the phenomena." But in the absence of data, at the border between what he could explain and what he could only honor—the causes he could identify and those he could not—Newton rapturously invokes God:

> Eternal and Infinite, Omnipotent and Omniscient; ... he governs all things, and knows all things that are or can be done.... We know him only by his most wise and excellent contrivances of things, and final causes; we admire him for his perfections; but we reverence and adore him on account of his dominion.

A century later, the French astronomer and mathematician Pierre-Simon de Laplace confronted Newton's dilemma of unstable orbits head-on. Rather than view the mysterious stability of the solar system as the unknowable work of God, Laplace declared it a scientific challenge. In his multipart masterpiece, *Mécanique Céleste*, the first volume of which appeared in 1798, Laplace demonstrates that the solar system is stable over periods of time longer than Newton could predict. To do so, Laplace pioneered a new kind of mathematics called perturbation theory, which enabled him to examine the cumulative effects of many small forces. According to an oft-repeated but probably embellished account, when Laplace gave a copy of *Mécanique Céleste* to his physics-literate friend Napoleon Bonaparte, Napoleon asked him what role God played in the construction and regulation of the heavens. "Sire," Laplace replied, "I have no need of that hypothesis."

Laplace notwithstanding, plenty of scientists besides Newton have called on God—or the gods—wherever their comprehension fades to ignorance. Consider the second-century A.D. Alexandrian astronomer Ptolemy. Armed with a description, but no real understanding, of what the planets were doing up there, he could not contain his religious fervor:

> I know that I am mortal by nature, and ephemeral; but when I trace, at my pleasure, the windings to and fro of the heavenly bodies, I no longer touch Earth with my feet: I stand in the presence of Zeus himself and take my fill of ambrosia.

Or consider the seventeenth-century Dutch astronomer Christiaan Huygens, whose achievements include constructing the first working pendulum clock and discovering the rings of Saturn. In his charming book *The Celestial Worlds Discover'd*, posthumously published in 1696, most of the opening chapter celebrates all that was then known of planetary orbits, shapes, and sizes, as well as the planets' relative brightness and presumed rockiness. The book even includes foldout charts illustrating the structure of the solar system. God is absent from this discussion—even though a mere century earlier, before Newton's achievements, planetary orbits were supreme mysteries.

Celestial Worlds also brims with speculations about life in the solar system, and that's where Huygens raises questions to which he has no answer. That's where he mentions the biological conundrums of the day, such as the origin of life's complexity. And sure enough, because seventeenth-century physics was more advanced than seventeenth-century biology, Huygens invokes the hand of God only when he talks about biology:

> I suppose no body will deny but that there's somewhat more of Contrivance, somewhat more of Miracle in the production and growth of Plants and Animals than in lifeless heaps of inanimate Bodies.... For the finger of God, and the Wisdom of Divine Providence, is in them much more clearly manifested than in the other.

Today secular philosophers call that kind of divine invocation "God of the gaps"—which comes in handy, because there has never been a shortage of gaps in people's knowledge.

As reverent as Newton, Huygens, and other great scientists of earlier centuries may have been, they were also empiricists. They did not retreat from the conclusions their evidence forced them to draw, and when their discoveries conflicted with prevailing articles of faith, they upheld the discoveries. That doesn't mean it was easy: sometimes they met fierce opposition, as did Galileo, who had to defend his telescopic evidence against formidable objections drawn from both scripture and "common" sense.

Galileo clearly distinguished the role of religion from the role of science. To him, religion was the service of God and the salvation of souls, whereas science was the source of exact observations and demonstrated truths. In a long, famous, bristly letter written in the summer of 1615 to the Grand Duchess Christina of Tuscany (but, like so many epistles of the day, circulated among the literati), he quotes, in his own defense, an unnamed yet sympathetic church official saying that the Bible "tells you how to go to heaven, not how the heavens go."

The letter to the duchess leaves no doubt about where Galileo stood on the literal word of the Holy Writ:

> In expounding the Bible if one were always to confine oneself to the unadorned grammatical meaning, one might fall into error....
>
> Nothing physical which ... demonstrations prove to us, ought to be called in question (much less condemned) upon the testimony of biblical passages which may have some different meaning beneath their words....

I do not feel obliged to believe that the same God who has endowed us with senses, reason and intellect has intended us to forgo their use.

A rare exception among scientists, Galileo saw the unknown as a place to explore rather than as an eternal mystery controlled by the hand of God.

As long as the celestial sphere was generally regarded as the domain of the divine, the fact that mere mortals could not explain its workings could safely be cited as proof of the higher wisdom and power of God. But beginning in the sixteenth century, the work of Copernicus, Kepler, Galileo, and Newton—not to mention Maxwell, Heisenberg, Einstein, and everybody else who discovered fundamental laws of physics—provided rational explanations for an increasing range of phenomena. Little by little, the universe was subjected to the methods and tools of science, and became a demonstrably knowable place.

Then, in what amounts to a stunning yet unheralded philosophical inversion, throngs of ecclesiastics and scholars began to declare that it was the laws of physics themselves that served as proof of the wisdom and power of God.

One popular theme of the seventeenth and eighteenth centuries was the "clockwork universe"—an ordered, rational, predictable mechanism fashioned and run by God and his physical laws. The early telescopes, which all relied on visible light, did little to undercut that image of an ordered system. The Moon revolved around Earth. Earth and other planets rotated on their axes and revolved around the Sun. The stars shone. The nebulae floated freely in space.

Not until the nineteenth century was it evident that visible light is just one band of a broad spectrum of electromagnetic radiation—the band that human beings just happen to see. Infrared was discovered in 1800, ultraviolet in 1801, radio waves in 1888, X rays in 1895, and gamma rays in 1900. Decade by decade in the following century, new kinds of telescopes came into use, fitted with detectors that could "see" these formerly invisible parts of the electromagnetic spectrum. Now astrophysicists began to unmask the true character of the universe.

Turns out that some celestial bodies give off more light in the invisible bands of the spectrum than in the visible. And the invisible light picked up by the new telescopes showed that mayhem abounds in the cosmos: monstrous gamma-ray bursts, deadly pulsars, matter-crushing gravitational fields, matter-hungry black holes that flay their bloated stellar neighbors, newborn stars igniting within pockets of collapsing gas. And as our ordinary, optical telescopes got bigger and better, more mayhem emerged: galaxies that collide and cannibalize each other, explosions of supermassive stars, chaotic stellar and planetary orbits. Our own cosmic neighborhood—the inner solar system—turned out to be a shooting gallery, full of rogue asteroids and comets that collide with planets from time to time. Occasionally they've even wiped out stupendous masses of Earth's flora and fauna. The evidence all points to the fact that we occupy not a well-mannered clockwork universe, but a destructive, violent, and hostile zoo.

Of course, Earth can be bad for your health too. On land, grizzly bears want to maul you; in the oceans, sharks want to eat you. Snowdrifts can freeze you, deserts dehydrate you, earthquakes bury you, volcanoes incinerate you. Viruses can infect you, parasites suck your vital fluids, cancers take over your body, congenital diseases force an early death. And even if you have the good luck to be healthy, a swarm of locusts could devour your crops, a tsunami could wash away your family, or a hurricane could blow apart your town.

So the universe wants to kill us all. But let's ignore that complication for the moment.

Many, perhaps countless, questions hover at the front lines of science. In some cases, answers have eluded the best minds of our species for decades or even centuries. And in contemporary America, the notion that a higher intelligence is the single answer to all enigmas has been enjoying a resurgence. This present-day version of God of the gaps goes by a fresh name: "intelligent design." The term suggests that some entity, endowed with a mental capacity far greater than the human mind can muster, created or enabled all the things in the physical world that we cannot explain through scientific methods.

An Interesting Hypothesis

But why confine ourselves to things too wondrous or intricate for us to understand, whose existence and attributes we then credit to a superintelligence? Instead, why not tally all those things whose design is so clunky, goofy, impractical, or unworkable that they reflect the absence of intelligence?

Take the human form. We eat, drink, and breathe through the same hole in the head, and so, despite Henry J. Heimlich's eponymous maneuver, choking is the fourth leading cause of "unintentional injury death" in the United States. How about drowning, the fifth leading cause? Water covers almost three-quarters of Earth's surface, yet we are land creatures—submerge your head for just a few minutes, and you die.

Or take our collection of useless body parts. What good is the pinky toenail? How about the appendix, which stops functioning after childhood and thereafter serves only as the source of appendicitis? Useful parts, too, can be problematic. I happen to like my knees, but nobody ever accused them of being well protected from bumps and bangs. These days, people with problem knees can get them surgically replaced. As for our pain-prone spine, it may be a while before someone finds a way to swap that out.

How about the silent killers? High blood pressure, colon cancer, and diabetes each cause tens of thousands of deaths in the U.S. every year, but it's possible not to know you're afflicted until your coroner tells you so. Wouldn't it be nice if we had built-in biogauges to warn us of such dangers well in advance? Even cheap cars, after all, have engine gauges.

And what comedian designer configured the region between our legs—an entertainment complex built around a sewage system?

The eye is often held up as a marvel of biological engineering. To the astrophysicist, though, it's only a so-so detector. A better one would be much more sensitive to dark things in the sky and to all the invisible parts of the spectrum. How much more breathtaking sunsets would be if we could see ultraviolet and infrared. How useful it would be if, at a glance, we could see every source of microwaves in the environment, or know which radio station transmitters were active. How helpful it would be if we could spot police radar detectors at night.

Think how easy it would be to navigate an unfamiliar city if we, like birds, could always tell which way was north because of the magnetite in our heads. Think how much better off we'd be if we had gills as well as lungs, how much more productive if we had six arms instead of two. And if we had eight, we could safely drive a car while simultaneously talking on a cell phone, changing the radio station, applying makeup, sipping a drink, and scratching our left ear.

Stupid design could fuel a movement unto itself. It may not be nature's default, but it's ubiquitous. Yet people seem to enjoy thinking that our bodies, our minds, and even our universe represent pinnacles of form and reason. Maybe it's a good antidepressant to think so. But it's not science—not now, not in the past, not ever.

Another practice that isn't science is embracing ignorance. Yet it's fundamental to the philosophy of intelligent design: I don't know what this is. I don't know how it works. It's too complicated for me to figure out. It's too complicated for any human being to figure out. So it must be the product of a higher intelligence.

What do you do with that line of reasoning? Do you just cede the solving of problems to someone smarter than you, someone who's not even human? Do you tell students to pursue only questions with easy answers?

There may be a limit to what the human mind can figure out about our universe. But how presumptuous it would be for me to claim that if I can't solve a problem, neither can any other person who has ever lived or who will ever be born. Suppose Galileo and Laplace had felt that way? Better yet, what if Newton had not? He might then have solved Laplace's problem a century earlier, making it possible for Laplace to cross the next frontier of ignorance.

Science is a philosophy of discovery. Intelligent design is a philosophy of ignorance. You cannot build a program of discovery on the assumption that nobody is smart enough to figure out the answer to a problem. Once upon a time, people identified the god Neptune as the source of storms at sea. Today we call these storms hurricanes. We know when and where they start. We know what drives them. We know what mitigates their destructive power. And anyone who has studied global warming can tell you what makes them worse. The only people who still call hurricanes "acts of God" are the people who write insurance forms.

To deny or erase the rich, colorful history of scientists and other thinkers who have invoked divinity in their work would be intellectually dishonest. Surely there's an appropriate place for intelligent design to live in the academic landscape. How about the history of religion? How about philosophy or psychology? The one place it doesn't belong is the science classroom.

If you're not swayed by academic arguments, consider the financial consequences. Allow intelligent design into science textbooks, lecture halls, and laboratories, and the cost to the frontier of scientific discovery—the frontier that drives the economies of the future—would be incalculable. I don't want students who could make the next major breakthrough in renewable energy sources or space travel to have been taught that anything they don't understand, and that nobody yet understands, is divinely constructed and therefore beyond their intellectual capacity. The day that happens, Americans will just sit in awe of what we don't understand, while we watch the rest of the world boldly go where no mortal has gone before.

Astrophysicist **NEIL DEGRASSE TYSON** is the director of the Hayden Planetarium at the American Museum of Natural History. An anthology of his "Universe" columns will be published in 2006 by W.W. Norton.

From *Natural History*, Vol. 114, issue 9, November 2005, pp. 28–32. Copyright © 2005 by Natural History Magazine. Reprinted by permission.

UNIT 2
Primates

Unit Selections

Key Points to Consider

- If we share 98% of our DNA with chimpanzees, why are we so different?

- How is it possible to objectively study and assess emotional and mental states of nonhuman primates?

- What are the implications for human evolution of tool use, social hunting, and food sharing among Ivory Coast chimpanzees?

- Should chimpanzee behavioral patterns be classified as "cultural"?

- Are primates naturally aggressive?

- Under what circumstances have some primates, such as orangutans, become more innovative and intelligent?

- To what extent is cooperation a part of our primate heritage?

- Are we in anthropodenial? Explain your answer.

- How does the human ability to communication compare and contrast with that of our closest living relatives?

Student Website
www.mhcls.com/online

Internet References
Further information regarding these websites may be found in this book's preface or online.

African Primates at Home
  http://www.indiana.edu/~primate/primates.html
Electronic Zoo/NetVet-Primate Page
  http://netvet.wustl.edu/primates.htm
Laboratory Primate Newsletter
  http://www.brown.edu/Research/Primate/other.html

Primates are fun. They are active, intelligent, colorful, emotionally expressive, and unpredictable. Since, in some ways, they are very much like us (see "The 2% Difference"), observing them is like holding up an opaque mirror to ourselves. The image may not be crystal-clear or, indeed, what some would consider flattering, but it is certainly familiar enough to be illuminating.

Primates are, of course, but one of many orders of mammals that adaptively radiated into the variety of ecological niches vacated at the end of the Age of Reptiles about 65 million years ago. Whereas some mammals took to the sea (cetaceans), and some took to the air (chiroptera, or bats), primates are characterized by an arboreal or forested adaptation. While some mammals can be identified by their food-getting habits, such as the meat-eating carnivores, primates have a penchant for eating almost anything and are best described as omnivorous. In taking to the trees, primates did not simply develop a full-blown set of distinguishing characteristics that set them off easily from other orders of mammals, the way the rodent order can be readily identified by its gnawing set of front teeth. Rather, each primate seems to represent degrees of anatomical, biological, and behavioral characteristics on a continuum of change in the direction of the particular traits we humans happen to be interested in.

None of this is meant to imply, of course, that the living primates are our ancestors. Since the prosimians, monkeys, and apes are contemporaries, they are no more our ancestors than we are theirs, and, as living end-products of evolution, we have all descended from a common stock in the distant past. So, if we are interested primarily in our own evolutionary past, why study primates at all? Because, by the criteria we have set up as significant milestones in the evolution of humanity, an inherent reflection of our own bias, primates have not evolved as far as we have. They and their environments, therefore, may represent glimmerings of the evolutionary stages and ecological circumstances through which our own ancestors may have gone. What we stand to gain, for instance, is an educated guess as to how our own ancestors might have appeared and behaved as semi-erect creatures before becoming bipedal. Aside from being a pleasure to observe, then, living primates can teach us something about our past.

This unit demonstrates that the kinds of answers obtained depend upon the kinds of questions asked, and that we have to be very careful in making inferences about human beings from any one particular primate study. We may, if we are not careful, draw conclusions that say more about our own skewed perspectives than about that which we claim to understand. Still another benefit of primate field research is that it provides us with perspectives that the bones and stones of the fossil hunters will never reveal: a sense of the richness and variety of social patterns that must have existed in the primate order for many tens of millions of years. (See Jane Goodall's "The Mind of the Chimpanzee" and Frans B.M. de Waal's "How Animals Do Business.")

Even if we had the physical remains of the earliest hominids in front of us, which we do not have, there is no way such evidence could thoroughly answer the questions that physical anthropologists care most deeply about: How did these creatures move about and get their food? Did they cooperate and share?

On what levels did they think and communicate? Did they have a sense of family, let alone a sense of self? In one way or another, all of the previously mentioned articles on primates relate to these issues, as do some of the subsequent ones on the fossil evidence. But what sets off this unit from the others is how some of the authors attempt to deal with these matters head-on, even in the absence of direct fossil evidence. Christophe Boesch and Hedwige Boesch-Achermann, in "Dim Forest, Bright Chimps," indicate that some aspects of "hominization" (the acquisition of such humanlike qualities as cooperative hunting and food sharing) actually may have begun in the African rain forest rather than in the dry savanna, as has usually been proposed. They base their suggestions on some remarkable first-hand observations of forest-dwelling chimpanzees.

As he shows that chimpanzee behavior may vary according to local circumstances, just as we know human behavior does, Craig Stanford, in "Got Culture?" also makes a strong case for such differences to be classified as cultural.

Recent research has shown some striking resemblances between apes and humans, hinting that such qualities might have been characteristic of our common ancestor. Following this line of reasoning, Frans de Waal (in "Rethinking Primate Aggression" and "Are We in Anthropodenial?") argues that we can make educated guesses as to the mental and physical processes of our hominid predecessors. In "Why Are Some Animals So Smart?" Carel Van Shaik concludes from his study of orangutans that social animals are more innovative and intelligent.

Taken collectively, the articles in this section show how far anthropologists are willing to go to construct theoretical formulations based upon limited data. Although making so much out of so little may be seen as a fault and may generate irreconcilable differences among theorists, a readiness to entertain new ideas should be welcomed for what it is—a stimulus for more intensive and meticulous research

The 2% Difference

Now that scientists have decoded the chimpanzee genome, we know that 98 percent of our DNA is the same. So how can we be so different?

ROBERT SAPOLSKY

If you find yourself sitting close to a chimpanzee, staring face to face and making sustained eye contact, something interesting happens, something that is alternately moving, bewildering, and kind of creepy. When you gaze at this beast, you suddenly realize that the face gazing back is that of a sentient individual, who is recognizably kin. You can't help but wonder, What's the matter with those intelligent design people?

Chimpanzees are close relatives to humans, but they're not identical to us. We are not chimps. Chimps excel at climbing trees, but we beat them hands down at balance-beam routines; they are covered in hair, while we have only the occasional guy with really hairy shoulders. The core differences, however, arise from how we use our brains. Chimps have complex social lives, play power politics, betray and murder each other, make tools, and teach tool use across generations in a way that qualifies as culture. They can even learn to do logic operations with symbols, and they have a relative sense of numbers. Yet those behaviors don't remotely approach the complexity and nuance of human behaviors, and in my opinion there's not the tiniest bit of scientific evidence that chimps have aesthetics, spirituality, or a capacity for irony or poignancy.

What makes the human species brainy are huge numbers of standard-issue neurons

What accounts for those differences? A few years ago, the most ambitious project in the history of biology was carried out: the sequencing of the human genome. Then just four months ago, a team of researchers reported that they had likewise sequenced the complete chimpanzee genome. Scientists have long known that chimps and humans share about 98 percent of their DNA. At last, however, one can sit down with two scrolls of computer printout, march through the two genomes, and see exactly where our 2 percent difference lies.

Given the outward differences, it seems reasonable to expect to find fundamental differences in the portions of the genome that determine chimp and human brains—reasonable, at least, to a brainocentric neurobiologist like me. But as it turns out, the chimp brain and the human brain differ hardly at all in their genetic underpinnings. Indeed, a close look at the chimp genome reveals an important lesson in how genes and evolution work, and it suggests that chimps and humans are a lot more similar than even a neurobiologist might think.

DNA, or deoxyribonucleic acid, is made up of just four molecules, called nucleotides: adenine (A), cytosine (C), guanine (G), and thymine (T). The DNA codebook for every species consists of billions of these letters in a precise order. If, when DNA is being copied in a sperm or an egg, a nucleotide is mistakenly copied wrong, the result is a mutation. If the mutation persists from generation to generation, it becomes a DNA difference—one of the many genetic distinctions that separate one species (chimpanzees) from another (humans). In genomes involving billions of nucleotides, a tiny 2 percent difference translates into tens of millions of ACGT differences. And that 2 percent difference can be very broadly distributed. Humans and chimps each have somewhere between 20,000 and 30,000 genes, so there are likely to be nucleotide differences in every single gene.

To understand what distinguishes the DNA of chimps and humans, one must first ask: What is a gene? A gene is a string of nucleotides that specify how a single distinctive protein should be made. Even if the same gene in chimps and humans differs by an A here and a T there, the result may be of no consequence. Many nucleotide differences are neutral—both the mutation and the normal gene cause the same protein to be made. However, given the right nucleotide difference between the same gene in the two species, the resulting proteins may differ slightly in construction and function.

One might assume that the differences between chimp and human genes boil down to those sorts of typographical errors: one nucleotide being swapped for a different one and altering the gene it sits in. But a close look at the two codebooks reveals very few such instances. And the typos that do occasionally occur follow a compelling pattern. It's important to note that genes don't act alone. Yes, each gene regulates the construction of a specific protein. But what tells that gene *when* and *where* to build that protein? Regulation is everything: It's important not to start up genes related to puberty during, say, infancy, or to activate genes that are related to eye color in the bladder.

In the DNA code list, that critical information is contained in a short stretch of As and Cs and Gs and Ts that lie just before each gene and act as a switch that turns the gene on or off. The switch, in turn, is flicked on by proteins called transcription factors, which activate certain genes in response to certain stimuli. Naturally, every gene is not regulated by its own distinct transcription factor; otherwise, a codebook of as many as 30,000 genes would require 30,000 transcription factors—and 30,000 more genes to code for them. Instead, one transcription factor can flick on an array of functionally related genes. For example, a certain type of injury can activate one transcription factor that turns on a bunch of genes in your white blood cells, triggering inflammation.

Accurate switch flickers are essential. Imagine the consequences if some of those piddly nucleotide changes arose in a protein that happened to be a transcription factor: Suddenly, instead of activating 23 different genes, the protein might charge up 21 or 25 of them—or it might turn on the usual 23 but in different ratios than normal. Suddenly, one minor nucleotide difference would be amplified across a network of gene differences. (And imagine the ramifications if the altered proteins are transcription factors that activate the genes coding for still other transcription factors!) When the chimp and human genomes are compared, some of the clearest cases of nucleotide differences are found in genes coding for transcription factors. Those cases are few, but they have far-ranging implications.

The genomes of chimps and humans reveal a history of other kinds of differences as well. Instead of a simple mutation, in which a single nucleotide is copied incorrectly, consider an insertion mutation, where an extra A, C, G, or T is dropped in, or a deletion mutation, whereby a nucleotide drops out. Insertion or deletion mutations can have major consequences: Imagine the deletion mutation that turns the sentence "I'll have the mousse for dessert" into "I'll have the mouse for dessert," or the insertion mutation implicit in "She turned me down for a date after I asked her to go boweling with me." Sometimes, more than a single nucleotide is involved; whole stretches of a gene may be dropped or added. In extreme cases, entire genes may be deleted or added.

More important than how the genetic changes arise—by insertion, deletion, or straight mutation—is where in the genome they occur. Keep in mind that, for these genetic changes to persist from generation to generation, they must convey some evolutionary advantage. When one examines the 2 percent difference between humans and chimps, the genes in question turn out to be evolutionarily important, if banal. For example, chimps have a great many more genes related to olfaction than we do; they've got a better sense of smell because we've lost many of those genes. The 2 percent distinction also involves an unusually large fraction of genes related to the immune system, parasite vulnerability, and infectious diseases: Chimps are resistant to malaria, and we aren't; we handle tuberculosis better than they do. Another important fraction of that 2 percent involves genes related to reproduction—the sorts of anatomical differences that split a species in two and keep them from interbreeding.

That all makes sense. Still, chimps and humans have very different brains. So which are the brain-specific genes that have evolved in very different directions in the two species? It turns out that there are hardly any that fit that bill. This, too, makes a great deal of sense. Examine a neuron from a human brain under a microscope, then do the same with a neuron from the brain of a chimp, a rat, a frog, or a sea slug. The neurons all look the same: fibrous dendrites at one end, an axonal cable at the other. They all run on the same basic mechanism: channels and pumps that move sodium, potassium, and calcium around, triggering a wave of excitation called an action potential. They all have a similar complement of neurotransmitters: serotonin, dopamine, glutamate, and so on. They're all the same basic building blocks.

The main difference is in the sheer number of neurons. The human brain has 100 million times the number of neurons a sea slug's brain has. Where do those differences in quantity come from? At some point in their development, all embryos—whether human, chimp, rat, frog, or slug—must have a single first cell committed toward generating neurons. That cell divides and gives rise to 2 cells; those divide into 4, then 8, then 16. After a dozen rounds of cell division, you've got roughly enough neurons to run a slug. Go another 25 rounds or so and you've got a human brain. Stop a couple of rounds short of that and, at about one-third the size of a human brain, you've got one for a chimp. Vastly different outcomes, but relatively few genes regulate the number of rounds of cell division in the nervous system before calling a halt. And it's precisely some of those genes, the ones involved in neural development, that appear on the list of differences between the chimp and human genomes.

That's it; that's the 2 percent solution. What's shocking is the simplicity of it. Humans, to be human, don't need to have evolved unique genes that code for entirely novel types of neurons or neurotransmitters, or a more complex hippocampus (with resulting improvements in memory), or a more complex frontal cortex (from which we gain the ability to postpone gratification). Instead, our braininess as a species arises from having humongous numbers of just a few types of off-the-rack neurons and from the exponentially greater number of interactions between them. The difference is sheer quantity: Qualitative distinctions emerge from large numbers. Genes may have something to do with that quantity, and thus with the complexity of the quality that emerges. Yet no gene or genome can ever tell us what sorts of qualities those will be. Remember that when you and the chimp are eyeball to eyeball, trying to make sense of why the other seems vaguely familiar.

The Mind of the Chimpanzee

JANE GOODALL

Often I have gazed into a chimpanzee's eyes and wondered what was going on behind them. I used to look into Flo's, she so old, so wise. What did she remember of her young days? David Greybeard had the most beautiful eyes of them all, large and lustrous, set wide apart. They somehow expressed his whole personality, his serene self-assurance, his inherent dignity—and, from time to time, his utter determination to get his way. For a long time I never liked to look a chimpanzee straight in the eye—I assumed that, as is the case with most primates, this would be interpreted as a threat or at least as a breach of good manners. No so. As long as one looks with gentleness, without arrogance, a chimpanzee will understand, and may even return the look. And then—or such is my fantasy—it is as though the eyes are windows into the mind. Only the glass is opaque so that the mystery can never be fully revealed.

I shall never forget my meeting with Lucy, an eight-year-old home-raised chimpanzee. She came and sat beside me on the sofa and, with her face very close to mine, searched in my eyes—for what? Perhaps she was looking for signs of mistrust, dislike, or fear, since many people must have been somewhat disconcerted when, for the first time, they came face to face with a grown chimpanzee. Whatever Lucy read in my eyes clearly satisfied her for she suddenly put one arm round my neck and gave me a generous and very chimp-like kiss, her mouth wide open and laid over mine. I was accepted.

For a long time after that encounter I was profoundly disturbed. I had been at Gombe for about fifteen years then and I was quite familiar with chimpanzees in the wild. But Lucy, having grown up as a human child, was like a changeling, her essential chimpanzeeness overlaid by the various human behaviours she had acquired over the years. No longer purely chimp yet eons away from humanity, she was man-made, some other kind of being. I watched, amazed, as she opened the refrigerator and various cupboards, found bottles and a glass, then poured herself a gin and tonic. She took the drink to the TV, turned the set on, flipped from one channel to another then, as though in disgust, turned it off again. She selected a glossy magazine from the table and, still carrying her drink, settled in a comfortable chair. Occasionally, as she leafed through the magazine she identified something she saw, using the signs of ASL, the American Sign Language used by the deaf. I, of course, did not understand, but my hostess, Jane Temerlin (who was also

Lucy's 'mother'), translated: 'That dog,' Lucy commented, pausing at a photo of a small white poodle. She turned the page. 'Blue,' she declared, pointing then signing as she gazed at a picture of a lady advertising some kind of soap powder and wearing a brilliant blue dress. And finally, after some vague hand movements—perhaps signed mutterings—'This Lucy's, this mine,' as she closed the magazine and laid it on her lap. She had just been taught, Jane told me, the use of the possessive pronouns during the thrice weekly ASL lessons she was receiving at the time.

The book written by Lucy's human 'father,' Maury Temerlin, was entitled *Lucy, Growing Up Human*. And in fact, the chimpanzee is more like us than is any other living creature. There is close resemblance in the physiology of our two species and genetically, in the structure of the DNA, chimpanzees and humans differ by only just over one per cent. This is why medical research uses chimpanzees as experimental animals when they need substitutes for humans in the testing of some drug or vaccine. Chimpanzees can be infected with just about all known human infectious diseases including those, such as hepatitis B and AIDS, to which other non-human animals (except gorillas, orangutans and gibbons) are immune. There are equally striking similarities between humans and chimpanzees in the anatomy and wiring of the brain and nervous system, and—although many scientists have been reluctant to admit to this—in social behaviour, intellectual ability, and the emotions. The notion of an evolutionary continuity in physical structure from pre-human ape to modern man has long been morally acceptable to most scientists. That the same might hold good for mind was generally considered an absurd hypothesis—particularly by those who used, and often misused, animals in their laboratories. It is, after all, convenient to believe that the creature you are using, while it may react in disturbingly human-like ways, is, in fact, merely a mindless and, above all, unfeeling, 'dumb' animal.

When I began my study at Gombe in 1960 it was not permissible—at least not in ethological circles—to talk about an animal's mind. Only humans had minds. Nor was it quite proper to talk about animal personality. Of course everyone knew that they *did* have their own unique characters—everyone who had ever owned a dog or other pet was aware of that. But ethologists, striving to make theirs a 'hard' science, shied away from the task of trying to explain such things objectively. One respected ethologist, while acknowledging that there was 'variability between individual animals,' wrote that it was best that this fact be 'swept under the carpet.' At that time ethological carpets fairly bulged with all that was hidden beneath them.

How naive I was. As I had not had an undergraduate science education I didn't realize that animals were not supposed to have personalities, or to think, or to feel emotions or pain. I had no idea that it would have been more appropriate to assign each of the chimpanzees a number rather than a name when I got to know him or her. I didn't realize that it was not scientific to discuss behaviour in terms of motivation or purpose. And no one had told me that terms such as *childhood* and *adolescence* were uniquely human phases of the life cycle, culturally determined, not to be used when referring to young chimpanzees. Not knowing, I freely made use of all those forbidden terms and concepts in my initial attempt to describe, to the best of my ability, the amazing things I had observed at Gombe.

I shall never forget the response of a group of ethologists to some remarks I made at an erudite seminar. I described how Figan, as an adolescent, had learned to stay behind in camp after senior males had left, so that we could give him a few bananas for himself. On the first occasion he had, upon seeing the fruits, uttered loud, delighted food calls: whereupon a couple of the older males had charged back, chased after Figan, and taken his bananas. And then, coming to the point of the story, I explained how, on the next occasion, Figan had actually suppressed his calls. We could hear little sounds, in his throat, but so quiet that none of the others could have heard them. Other young chimps, to whom we tried to smuggle fruit without the knowledge of their elders, never learned such self-control. With shrieks of glee they would fall to, only to be robbed of their booty when the big males charged back. I had expected my audience to be as fascinated and impressed as I was. I had hoped for an exchange of views about the chimpanzee's undoubted intelligence. Instead there was a chill silence, after which the chairman hastily changed the subject. Needless to say, after being thus snubbed, I was very reluctant to contribute any comments, at any scientific gatherings, for a very long time. Looking back, I suspect that everyone was interested, but it was, of course, not permissible to present a mere 'anecdote' as evidence for anything.

The editorial comments on the first paper I wrote for publication demanded that every *he* or *she* be replaced with *it*, and every *who* be replaced with *which*. Incensed, I, in my turn, crossed out the *its* and *whichs* and scrawled back the original pronouns. As I had no desire to carve a niche for myself in the world of science, but simply wanted to go on living among and learning about chimpanzees, the possible reaction of the editor of the learned journal did not trouble me. In fact I won that round: the paper when finally published did confer upon the chimpanzees the dignity of their appropriate genders and properly upgraded them from the status of mere 'things' to essential Beingness.

However, despite my somewhat truculent attitude, I did want to learn, and I was sensible of my incredible good fortune in being admitted to Cambridge. I wanted to get my PhD, if only for the sake of Louis Leakey and the other people who had written letters in support of my admission. And how lucky I was to have, as my supervisor, Robert Hinde. Not only because I thereby benefitted from his brilliant mind and clear thinking, but also because I doubt that I could have found a teacher more suited to my particular needs and personality. Gradually he was able to cloak me with at least some of the trappings of a scientist. Thus although I continued to hold to most of my convictions—that animals had personalities; that they could feel happy or sad or fearful; that they could feel pain; that they could strive towards planned goals and achieve greater success if they were highly motivated—I soon realized that these personal convictions were, indeed, difficult to prove. It was best to be circumspect—at least until I had gained some credentials and credibility. And Robert gave me wonderful advice on how best to tie up some of my more rebellious ideas with scientific ribbon. 'You can't *know* that Fifi was jealous,' had admonished on one occasion. We argued a little. And then: 'Why don't you just say *If Fifi were a human child we would say she was jealous.*' I did.

It is not easy to study emotions even when the subjects are human. I know how I feel if I am sad or happy or angry, and if a friend tells me that he is feeling sad, happy or angry, I assume that his feelings are similar to mine. But of course I cannot know. As we try to come to grips with the emotions of beings progressively more different from ourselves the task, obviously, becomes increasingly difficult. If we ascribe human emotions to non-human animals we are accused of being anthropomorphic—a cardinal sin in ethology. But is it so terrible? If we test the effect of drugs on chimpanzees because they are biologically so similar to ourselves, if we accept that there are dramatic similarities in chimpanzee and human brain and nervous system, is it not logical to assume that there will be similarities also in at least the more basic feelings, emotions, moods of the two species?

In fact, all those who have worked long and closely with chimpanzees have no hesitation in asserting that chimps experience emotions similar to those which in ourselves we label pleasure, joy, sorrow, anger, boredom and so on. Some of the emotional states of the chimpanzee are so obviously similar to ours that even an inexperienced observer can understand what is going on. An infant who hurls himself screaming to the ground, face contorted, hitting out with his arms at any nearby object, banging his head, is clearly having a tantrum. Another youngster, who gambols around his mother, turning somersaults, pirouetting and, every so often, rushing up to her and tumbling into her lap, patting her or pulling her hand towards him in a request for tickling, is obviously filled with *joie de vivre*. There are few observers who would not unhesitatingly ascribe his behaviour to a happy, carefree state of well-being. And one cannot watch chimpanzee infants for long without realizing that they have the same emotional need for affection and reassurance as human children. An adult male, reclining in the shade after a good meal, reaching benignly to play with an infant or idly groom an adult female, is clearly in a good mood. When he sits with bristling hair, glaring at his subordinates and threatening them, with irritated gestures, if they come too close, he is clearly feeling cross and grumpy. We make these judgements because the similarity of so much of a chimpanzee's behaviour to our own permits us to empathize.

It is hard to empathize with emotions we have not experienced. I can image, to some extent, the pleasure of a female chimpanzee during the act of procreation. The feelings of her male partner are beyond my knowledge—as are those of the human male in the same context. I have spent countless hours watching mother chimpanzees interacting with their infants. But not until I had an infant of my own did I begin to understand the basic, powerful instinct of mother-love. If someone accidentally did something to frighten Grub, or threaten his well-being in any way, I felt a surge of quite irrational anger. How much more easily could I then understand the feelings of the chimpanzee mother who furiously waves her arm and barks in threat at an individual who approaches her infant too closely, or at a playmate who inadvertently hurts her child. And it was not until I knew the numbing grief that gripped me after the death of my second husband that I could even begin to appreciate the despair and sense of loss that can cause young chimps to pine away and die when they lose their mothers.

Empathy and intuition can be of tremendous value as we attempt to understand certain complex behavioral interactions, provided that the behaviour, as it occurs, is recorded precisely and objectively. Fortunately I have seldom found it difficult to record facts in an orderly manner even during times of powerful emotional involvement. And "knowing" intuitively how a chimpanzee is feeling—after an attack, for example—may help one to understand what happens next. We should not be afraid at least to try to make use of our close evolutionary relationship with the chimpanzees in our attempts to interpret complex behaviour.

Today, as in Darwin's time, it is once again fashionable to speak of and study the animal mind. This change came about gradually, and was, at least in part, due to the information collected during careful studies of animal societies in the field. As these observations became widely known, it was impossible to brush aside the complexities of social behaviour that were revealed in species after species. The untidy clutter under the ethological carpets was brought out and examined, piece by piece. Gradually it was realized that parsimonious explanations of apparently intelligent behaviours were often misleading. This led to a succession of experiments that, taken together, clearly prove that many intellectual abilities that had been thought unique to humans were actually present, though in a less highly developed form, in other, non-human beings. Particularly, of course, in the non-human primates and especially in chimpanzees.

When first I began to read about human evolution, I learned that one of the hallmarks of our own species was that we, and only we, were capable of making tools. *Man the Toolmaker* was an oft-cited definition—and this despite the careful and exhaustive research of Wolfgang Kohler and Robert Yerkes on the tool-using and tool-making abilities of chimpanzees. Those studies, carried out independently in the early twenties, were received with scepticism. Yet both Kohler and Yerkes were respected scientists, and both had a profound understanding of chimpanzee behaviour. Indeed, Kohler's descriptions of the personalities and behaviour of the various individuals in his colony, published in his book *The Mentality of Apes*, remain some of the most vivid and colourful ever written. And his experiments, showing how chimpanzees could stack boxes, then climb the unstable constructions to reach fruit suspended from the ceiling, or join two short sticks to make a pole long enough to rake in fruit otherwise out of reach, have become classic, appearing in almost all textbooks dealing with intelligent behaviour in non-human animals.

By the time systematic observations of tool-using came from Gombe those pioneering studies had been largely forgotten. Moreover, it was one thing to know that humanized chimpanzees in the lab could use implements: it was quite another to find that this was a naturally occurring skill in the wild. I well remember writing to Louis about my first observations, describing how David Greybeard not only used bits of straw to fish for termites but actually stripped leaves from a stem and thus *made* a tool. And I remember too receiving the now oft-quoted telegram he sent in response to my letter: "Now we must redefine *tool*, redefine *Man*, or accept chimpanzees as humans."

There were initially, a few scientists who attempted to write off the termiting observations, even suggesting that I had taught the chimps! By and large, though, people were fascinated by the information and by the subsequent observations of the other contexts in which the Gombe chimpanzees used objects as tools. And there were only a few anthropologists who objected when I suggested that the chimpanzees probably passed their tool-using traditions from one generation to the next, through observations, imitation and practice, so that each population might be expected to have its own unique tool-using culture. Which, incidentally, turns out to be quite true. And when I described how one chimpanzee, Mike, spontaneously solved a new problem by using a tool (he broke off a stick to knock a banana to the ground when he was too nervous to actually take it from my hand) I don't believe there were any raised eyebrows in the scientific community. Certainly I was not attacked viciously, as were Kohler and Yerkes, for suggesting that humans were not the only beings capable of reasoning and insight.

The mid-sixties saw the start of a project that, along with other similar research, was to teach us a great deal about the chimpanzee mind. This was Project Washoe, conceived by Trixie and Allen Gardner. They purchased an infant chimpanzee and began to teach her the signs of ASL, the American Sign Language used by the deaf. Twenty years earlier another husband and wife team, Richard and Cathy Hayes, had tried, with an almost total lack of success, to teach a young chimp, Vikki, to talk. The Hayes's undertaking taught us a lot about the chimpanzee mind, but Vikki, although she did well in IQ tests, and was clearly an intelligent youngster, could not learn human speech. The Gardners, however, achieved spectacular success with their pupil, Washoe. Not only did she learn signs easily, but she quickly began to string them together in meaningful ways. It was clear that each sign evoked, in her mind, a mental image of the object it represented. If, for example, she was asked, in sign language, to fetch an apple, she would go and locate an apple that was out of sight in another room.

Other chimps entered the project, some starting their lives in deaf signing families before joining Washoe. And finally Washoe adopted an infant, Loulis. He came from a lab where no thought of teaching signs had ever penetrated. When he was with Washoe he was given no lessons in language acquisition—not by humans, anyway. Yet by the time he was eight years old he had made fifty-eight signs in their correct contexts. How did he learn them? Mostly, it seems, by imitating the behaviour of Washoe and the other three signing chimps, Dar, Moja and Tatu. Sometimes, though, he received tuition from Washoe herself. One day, for example, she began to swagger about bipedally, hair bristling, signing *food! food! food!* in great excitement. She had seen a human approaching with a bar of chocolate. Loulis, only eighteen months old, watched passively. Suddenly Washoe stopped her swaggering, went over to him, took his hand, and moulded the sign for *food* (fingers pointing towards mouth). Another time, in a similar context, she made the sign for *chewing gum*—but with *her* hand on *his* body. On a third occasion Washoe, apropos of nothing, picked up a small chair, took it over to Loulis, set it down in front of him, and very distinctly made the *chair* sign three times, watching him closely as she did so. The two food signs became incorporated into Loulis's vocabulary but the sign for chair did not. Obviously the priorities of a young chimp are similar to those of a human child!

When news of Washoe's accomplishments first hit the scientific community it immediately provoked a storm of bitter protest. It implied that chimpanzees were capable of mastering a human language, and this, in turn, indicated mental powers of generalization, abstraction and concept-formation as well as an ability to understand and use abstract symbols. And these intellectual skills were surely the prerogatives of *Homo sapiens*. Although there were many who were fascinated and excited by the Gardners' findings, there were many more who denounced the whole project, holding that the data was suspect, the methodology sloppy, and the conclusions not only misleading, but quite preposterous. The controversy inspired all sorts of other language projects. And, whether the investigators were sceptical to start with and hoped to disprove the Gardners' work, or whether they were attempting to demonstrate the same thing in a new way, their research provided additional information about the chimpanzee's mind.

And so, with new incentive, psychologists began to test the mental abilities of chimpanzees in a variety of different ways; again and again the results confirmed that their minds are uncannily like our own. It had long been held that only humans were capable of what is called 'cross-modal transfer of information'—in other words, if you shut your eyes and someone allows you to feel a strangely shaped potato, you will subsequently be able to pick it out from other differently shaped potatoes simply by looking at them. And vice versa. It turned out that chimpanzees can 'know' with their eyes what they 'feel' with their fingers in just the same way. In fact, we now know that some other non-human primates can do the same thing. I expect all kinds of creatures have the same ability.

Then it was proved, experimentally and beyond doubt, that chimpanzees could recognize themselves in mirrors—that they had, therefore, some kind of self-concept. In fact, Washoe, some years previously, had already demonstrated the ability when she spontaneously identified herself in the mirror, staring at her image and making her name sign. But that observation was merely anecdotal. The proof came when chimpanzees who had been allowed to play with mirrors were, while anaesthetized, dabbed with spots of odourless paint in places, such as the ears or the top of the head, that they could see only in the mirror. When they woke they were not only fascinated by their spotted images, but immediately investigated, with their fingers, the dabs of paint.

The fact that chimpanzees have excellent memories surprised no one. Everyone, after all, has been brought up to believe that 'an elephant never forgets' so why should a chimpanzee be any different? The fact that Washoe spontaneously gave the name-sign of Beatrice Gardner, her surrogate mother, when she saw her after a separation of eleven years was no greater an accomplishment than the amazing memory shown by dogs who recognize their owners after separations of almost as long—and the chimpanzee has a much longer life span than a dog. Chimpanzees can plan ahead, too, at least as regards the immediate future. This, in fact, is well illustrated at Gombe, during the termiting season: often an individual prepares a tool for use on a termite mound that is several hundred yards away and absolutely out of sight.

This is not the place to describe in detail the other cognitive abilities that have been studied in laboratory chimpanzees. Among other accomplishments chimpanzees possess pre-mathematical skills: they can, for example, readily differentiate between *more* and *less*. They can classify things into specific categories according to a given criterion—thus they have no difficulty in separating a pile of food into *fruits* and *vegetables* on one occasion, and, on another, dividing the same pile of food into *large* versus *small* items, even though this requires putting some vegetables with some fruits. Chimpanzees who have been taught a language can combine signs creatively in order to describe objects for which they have no symbol. Washoe, for example, puzzled her caretakers by asking, repeatedly, for a *rock berry*. Eventually it transpired that she was referring to Brazil nuts which she had encountered for the first time a while before. Another language-trained chimp described a cucumber as a *green banana*, and another referred to an Alka-Seltzer as a *listen drink*. They can even invent signs. Lucy, as she got older, had to be put on a leash for her outings. One day, eager to set off but having no sign for *leash*, she signalled her wishes by holding a crooked index finger to the ring on her collar. This sign became part of her vocabulary. Some chimpanzees love to draw, and especially to paint. Those who have learned sign language sometimes spontaneously label their works, 'This [is] apple'—or bird, or sweetcorn, or whatever. The fact that the paintings often look, to our eyes, remarkably unlike the objects depicted by the artists either means that the chimpanzees are poor draughtsmen or that we have much to learn regarding ape-style representational art!

People sometimes ask why chimpanzees have evolved such complex intellectual powers when their lives in the wild are so simple. The answer is, of course, that their lives in the wild are not so simple! They use—and need—all their mental skills during normal day-to-day life in their complex society. They are always having to make choices—where to go, or with whom to travel. They need highly developed social skills—particularly those males who are ambitious to attain high positions in the dominance hierarchy. Low-ranking chimpanzees must learn deception—to conceal their intentions or to do things in secret—if they are to get their way in the presence of their superiors. Indeed, the study of chimpanzees in the wild suggests that their intellectual abilities evolved, over the millennia, to help them cope with daily life. And now, the solid core of data concerning chimpanzee intellect collected so carefully in the lab setting provides a background against which to evaluate the many examples of intelligent, rational behaviour that we see in the wild.

It is easier to study intellectual prowess in the lab where, through carefully devised tests and judicious use of rewards, the chimpanzees can be encouraged to exert themselves, to stretch their minds to the limit. It is more meaningful to study the subject in the wild, but much harder. It is more meaningful because we can better understand the environmental pressures that led to the evolution of intellectual skills in chimpanzee societies. It is harder because, in the wild, almost all behaviours are confounded by countless variables; years of observing, recording and analysing take the place of contrived testing; sample size can often be counted on the fingers of one hand; the only experiments are nature's own, and only time— eventually—may replicate them.

In the wild a single observation may prove of utmost significance, providing a clue to some hitherto puzzling aspect of behaviour, a key to the understanding of, for example, a changed relationship. Obviously it is crucial to see as many incidents of this sort as possible. During the early years of my study at Gombe it became apparent that one person alone could never learn more than a fraction of what was going on in a chimpanzee community at any given time. And so, from 1964 onwards, I gradually built up a research team to help in the gathering of information about the behaviour of our closest living relatives.

Got Culture?

CRAIG STANFORD

On my first trip to east africa in the early 1990s, I stood by a dusty, dirt road hitchhiking. I had waited hours in rural Tanzania for an expected lift from a friend who had never shown up, leaving me with few options other than the kindness of strangers. I stood with my thumb out, but the cars and trucks roared by me, leaving me caked in paprika-red dust. I switched to a palm-down gesture I had seen local people using to get lifts. Voilà; on the first try a truck pulled over and I hopped in. A conversation in Kiswahili with the truck driver ensued and I learned my mistake. Hitchhiking with your thumb upturned may work in the United States, but in Africa the gesture can be translated in the way that Americans understand the meaning of an extended, declarative middle finger. Not exactly the best way to persuade a passing vehicle to stop. The universally recognized symbol for needing a lift is not so universal.

Much of culture is the accumulation of thousands of such small differences. Put a suite of traditions together—religion, language, ways of dress, cuisine and a thousand other features—and you have a culture. Of course cultures can be much simpler too. A group of toddlers in a day care center possesses its own culture, as does a multi-national corporation, suburban gardeners, inner-city gang members. Many elements of a culture are functional and hinged to individual survival: thatched roof homes from the tropics would work poorly in Canada, nor would harpoons made for catching seals be very useful in the Sahara. But other features are purely symbolic. Brides in Western culture wear white to symbolize sexual purity. Brides in Hindu weddings wear crimson, to symbolize sexual purity. Whether white or red is more pure is nothing more than a product of the long-term memory and mindset of the two cultures. And the most symbolic of cultural traditions, the one that has always been considered the bailiwick of humanity only, is language. The words "white" and "red" have an entirely arbitrary relationship to the colors themselves. They are simply code names.

Arguing about how to define culture has long been a growth industry among anthropologists. We argue about culture the way the Joint Chiefs of Staff argue about national security: as though our lives depended on it. But given that culture requires symbolism and some linguistic features, can we even talk about culture in other animals?

In 1996 I was attending a conference near Rio de Janeiro when the topic turned to culture.[1] As a biological anthropologist with a decade of field research on African great apes, I offered my perspective on the concept of culture. Chimpanzees, I said with confidence, display a rich cultural diversity. Recent years have shown that each wild chimpanzee population is more than just a gene pool. It is also a distinct culture, comprising a unique assortment of learned traditions in tool use, styles of grooming and hunting, and other features of the sort that can only be seen in the most socially sophisticated primates. Go from one forest to another and you will run into a new culture, just as walking between two human villages may introduce you to tribes who have different ways of building boats or celebrating marriages.

At least that's what I meant to say. But I had barely gotten the word "culture" past my lips when I was made to feel the full weight of my blissful ignorance. The cultural anthropologists practically leaped across the seminar table to berate me for using the words "culture" and "chimpanzee" in the same sentence. I had apparently set off a silent security alarm, and the culture-theory guards came running. How dare you, they said, use a human term like "cultural diversity" to describe what chimpanzees do? Say "behavioral variation," they demanded. "Apes are mere animals, and culture is something that only the human animal can claim. Furthermore, not only can humans alone claim culture, culture alone can explain humanity." It became clear to me that culture, as understood by most anthropologists, is a human concept, and many passionately want it to stay that way. When I asked if this was not just a semantic difference—what are cultural traditions if not learned behavioral variations?—they replied that culture is symbolic, and what animals do lacks symbolism.

When Jane Goodall first watched chimpanzees make simple stick tools to probe into termite mounds, it became clear that tool cultures are not unique to human societies. Of course many animals use tools. Sea otters on the California coast forage for abalones, which they place on their chests and hammer open with stones. Egyptian vultures use stones to break the eggs of ostriches. But these are simple, relatively inflexible lone behaviors. Only among chimpanzees do we see elaborate forms of tools made and used in variable ways, and also see distinct chimp tool cultures across Africa. In Gombe National Park in Tanzania, termite mounds of red earth rise 2 meters high and shelter millions of the almond-colored insects. Chimpanzees pore over the mounds, scratching at plugged tunnels until they find portals into the mound's interior. They will gently insert a twig or blade of grass into a tunnel until the soldier termites latch onto the tools with their powerful mandibles, then they'll withdraw the probe from the mound. With dozens of soldier and worker termites clinging ferociously to the twig, the chimpanzee draws the stick between her lips and reaps a nutritious bounty.

Less than 100 kilometers away from Gombe's termite-fishing apes is another culture. Chimpanzees in Mahale National Park live in a forest that is home to most of the same species of termites, but they practically never use sticks to eat them. If Mahale chimpanzees forage for termites at all, they use their fingers to crumble apart soil and pick out their insect snacks. However, Mahale chimpanzees love to eat ants. They climb up the straight-sided trunks of great trees and poke Gombe-like probes into holes to obtain woodboring species. As adept as Gombe chimpanzees are at fishing for termites, they practically never fish for these ants, even though both the ants and termites occur in both Gombe and Mahale.[2]

Segue 2,000 kilometers westward, to a rainforest in Côte d'Ivoire. In a forest filled with twigs, chimpanzees do not use stick tools. Instead, chimpanzees in Taï National Park and other forests in western Africa use hammers made of rock and wood. Swiss primatologists Christophe and Hedwige Boesch and their colleagues first reported the use of stone tools by chimpanzees twenty years ago.[3] Their subsequent research showed that Taï chimpanzees collect hammers when certain species of nut-bearing trees are in fruit. These hammers are not modified in any way as the stone tools made by early humans were; they are hefted, however, and appraised for weight and smashing value before being carried back to the nut tree. A nut is carefully positioned in a depression in the tree's aboveground root buttresses (the anvil) and struck with precision by the tool-user. The researchers have seen mothers instructing their children on the art of tool use, by assisting them in placing the nut in the anvil in the proper way.

So chimpanzees in East Africa use termite- and ant-fishing tools, and West African counterparts use hammers, but not vice versa. These are subsistence tools; they were almost certainly invented for food-getting. Primatologist William McGrew of Miami University of Ohio has compared the tool technologies of wild chimpanzees with those of traditional human hunter-gatherer societies. He found that in at least some instances, the gap between chimpanzee technology and human technology is not wide. The now-extinct aboriginal Tasmanians, for example, possessed no complex tools or weapons of any kind. Though they are an extreme example, the Tasmanians illustrate that human culture need not be technologically complex.[4]

As McGrew first pointed out, there are three likeliest explanations for the differences we see among the chimpanzee tool industries across Africa.[5] The first is genetic: perhaps there are mutations that arise in one population but not others that govern tool making. This seems extremely unlikely, just as we would never argue that Hindu brides wear red while Western brides wear white due to a genetic difference between Indians and Westerners. The second explanation is ecological: maybe the environment in which the chimpanzee population lives dictates patterns of tool use. Maybe termite-fishing sticks will be in-

vented in places where there are termites and sticks but not rocks and nuts, and hammers invented in the opposite situation. But a consideration of each habitat raises doubts. Gombe is a rugged, rock-strewn place where it is hard to find a spot to sit that is not within arm's reach of a few stones, but Gombe chimpanzees do not use stone tools. The West African chimpanzees who use stone tools live, by contrast, in lowland rainforests that are nearly devoid of rocks. Yet they purposely forage to find them. The tool-use pattern is exactly the opposite of what we would expect if environment and local availability accounted for differences among chimpanzee communities in tool use.

British psychologist Andrew Whiten and his colleagues recently conducted the first systematic survey of cultural differences in tool use among the seven longest-term field studies, representing more than a century and a half of total observation time. They found thirty-nine behaviors that could not be explained by environmental factors at the various sites.[6] Alone with humans in the richness of their behavior repertoire, chimpanzee cultures show variations that can only be ascribed to learned traditions. These traditions, passed from one generation to the next through observation and imitation, are a simple version of human culture.

But wait. I said earlier that human culture must have a symbolic element. Tools that differ in form and function, from sticks to hammers to sponges made of crushed leaves, are all utterly utilitarian. They tell us much about the environment in which they are useful but little about the learned traditions that led to their creation. Human artifacts, on the other hand, nearly always contain some purely symbolic element, be it the designs carved into a piece of ancient pottery or the "Stanley" logo on my new claw hammer. Is there anything truly symbolic in chimpanzee culture, in the human sense of an object or behavior that is completely detached from its use?

Male chimpanzees have various ways of indicating to a female that they would like to mate. At Gombe, one such courtship behavior involves rapidly shaking a small bush or branch several times, after which a female in proximity will usually approach the male and present her swelling to him. But in Mahale, males have learned to use leaves in their courtship gesture. A male plucks a leafy stem from a nearby plant and noisily uses his teeth and fingers to tear off its leaves. Leaf-clipping is done mainly in the context of wanting to mate with a particular female, and appears to function as a purely symbolic signal of sexual desire (it could also be a gesture of frustration). A second leafy symbol is leaf-grooming. Chimpanzees pick leaves and intently groom them with their fingers, as seriously as though they were grooming another chimpanzee. And this may be the function; leaf-grooming may signal a desire for real grooming from a social partner. Since the signal for grooming involves grooming, albeit of another object, this gesture is not symbolic in the sense that leaf-clipping is. But its distribution across Africa is equally spotty; leaf-grooming is commonly practiced in East African chimpanzee cultures but is largely absent in western Africa.[7]

These two cases of potentially symbolic behavior may not seem very impressive. After all, the briefest consideration of human culture turns up a rich array of symbolism, from language to the arts. But are all human cultures highly symbolic? If we use language and other forms of symbolic expression as the criterion for culture, then how about a classroom full of two-year-old toddlers in a day care center? They communicate by a very simple combination of gestures and half-formed sentences. Toddlers have little symbolic communication or appreciation for art and are very little different from chimpanzees in their cultural output. We grant them human qualities because we know they will mature into symbol-using, linguistically expert adults, leaving chimpanzees in the dust. But this is no reason to consider them on a different plane from the apes when both are fifteen months old.

Chimpanzee societies are based on learned traditions passed from mother to child and from adult males to eager wannabe males. These traditions vary from place to place. This is culture. Culture is not limited, however, to those few apes that are genetically 99 percent human. Many primates show traditions. These are usually innovations by younger members of a group, which sweep rapidly through the society and leave it just slightly different than before. Japanese primatologists have long observed such traditions among the macaques native to their island nation. Researchers long ago noticed that a new behavior had arisen in one population of Japanese macaque monkeys living on Koshima Island just offshore the mainland. The monkeys were regularly tossed sweet potatoes, rice and other local treats by the locals. One day Imo, a young female in the group, took her potato and carried it to the sea, where she washed it with salty brine before eating it. This behavior rapidly spread throughout the group, a nice example of innovation happening in real time so that researchers could observe the diffusion. Later, other monkeys invented the practice of scooping up rice offered them with the beach sand it was scattered on, throwing both onto the surf and then scraping up the grains that floated while the sand sank.

At a supremely larger scale, such innovations are what human cultural differences are all about. Of course, only in human cultures do objects such as sweet potatoes take on the kind of symbolic meaning that permits them to stand for other objects and thus become a currency. Chimpanzees lack the top-drawer cognitive capacity needed to invent such a currency. Or do they? Wild chimpanzees hunt for a part of their living. All across equatorial Africa, meat-eating is a regular feature of chimpanzee life, but its style and technique vary from one forest to another. In Taï National Park in western Africa, hunters are highly cooperative; Christophe Boesch has reported specific roles such as ambushers and drivers as part of the apes' effort to corral colobus monkeys in the forest canopy.[8] At Gombe in East Africa, meanwhile, hunting is like a baseball game; a group sport performed on an individual basis. This difference may be environmentally influenced; perhaps the high canopy rain forest at Taï requires cooperation more than the broken, low canopy forest at Gombe. There is a culture of hunting in each forest as well, in which young and eager male wannabes copy the predatory skills of their elders. At Gombe, for instance, chimpanzees relish wild pigs and piglets in addition to monkeys and small antelope. At Taï, wild pigs are ignored even when they stroll in front of a hunting party.

There is also a culture of sharing the kill. Sharing of meat is highly nepotistic at Gombe; sons who make the kill share with their mothers and brothers but snub rival males. They also share preferentially with females who have sexual swellings, and with high-ranking females. At Taï, the captor shares with the other members of the hunting party whether or not they are allies or relatives; a system of reciprocity seems to be in place in which the golden rule works. I have argued that since the energy and time that chimpanzees spend hunting is rarely paid back by the calories, protein and fat gotten from a kill, we should consider hunting a social behavior done at least partly for its own sake.[9] When chimpanzees barter a limited commodity such as meat for other services—alliances, sex, grooming—they are engaging in a very simple and primitive form of a currency exchange. Such an exchange relies on the ability of the participants to remember the web of credits and debts owed one another and to act accordingly. It may be that the two chimpanzee cultures 2,000 kilometers apart have developed their distinct uses of meat as a social currency. In one place meat is used as a reward for cooperation, in the other as a manipulative tool of nepotism. Such systems are commonplace in all human societies, and their roots may be seen in chimpanzees' market economy, too.[10]

I have not yet considered one obvious question. If tool use and other cultural innovations can be so valuable to chimpanzees, why have they not arisen more widely among primates and other big-brained animals? Although chimpanzees are adept tool-users, their very close relatives the bonobos are not. Bonobos do a number of very clever things—dragging their hands beside them as they wade through streams to catch fish is one notable example—but they are not accomplished technicians. Gorillas don't use tools at all, and orangutans have only recently been observed to occasionally use sticks as probing tools in their rainforest canopy world.[11]

Other big-brained animals fare even worse. Wild elephants don't use their wonderfully dexterous trunks to manipulate tools in any major way, although when you're strong enough to uproot trees you may not have much use for a pokey little probe. Dolphins and whales, cognitively gifted though they may be, lack the essential anatomical ingredient for tool manufacture—a pair of nimble hands. Wild bottlenose dolphins have been observed to carry natural sponges about on their snouts to ferret food from the sea bottom, the only known form of cetacean tool use.[12] But that may be the limit of how much a creature that lacks any grasping appendages can manipulate its surroundings.

So to be a cultural animal, it is not enough to be big-brained. You must have the anatomical prerequisites for tool cultures to develop. Even if these are in place, there is no guarantee that a species will generate a subsistence culture in the form of tools. Perhaps environmental necessity dictates which ape species use tools and which don't, except it is hard to imagine that bonobos have much less use for tools than chimpanzees do. There is probably a strong element of chance involved. The chance that a cultural tradition—tool use, hunting style or grooming technique—will develop may be very small in any given century or millennium. Once innovated, the chance that the cultural trait will disappear—perhaps due to the death of the main practitioners from whom everyone learned the behavior—may conversely be great. Instead of a close fit between the environment and the cultural traditions that evolve in it—which many scholars believe explains cultural diversity in human societies—the roots of cultural variation may be much more random. A single influential individual who figures out how to make a better mousetrap, so to speak, can through imitation spread his mousetrap through the group and slowly into other groups.

We tend to think of cultural traditions as highly plastic and unstable compared to biological innovation. It takes hundreds of generations for natural selection to bring about biological change, whereas cultural change can happen in one lifetime, even in a few minutes. Because we live in a culture in which we buy the newest cell phone and the niftiest handheld computer—we fail to appreciate how conservative traditions like tool use can be. *Homo erectus*, with a brain nearly the size of our own, invented a teardrop-shaped stone tool called a hand axe 1.5 million years ago. It was presumably used for butchering carcasses, though some archaeologists think it may have also been a weapon. Whatever its purpose, more than a million years later those same stone axes were still being manufactured and used. Fifty thousand generations passed without a significant change in the major piece of material culture in a very big-brained and intelligent human species. *That's* conservatism and it offers us two lessons. First, if it ain't broke don't fix it: when a traditional way of making a tool works and the environment is not throwing any curves your way, there may be no pressure for a change. Second, we see a human species vastly more intelligent than an ape (*Homo erectus*' neocortical brain volume was a third smaller than a modern human's, but two and a half times larger than a chimpanzee's) whose technology didn't change at all. This tells us that innovations, once made, may last a very long time without being either extinguished or improved upon. It suggests that chimpanzee tool cultures may have been in place for all of the 5 million years since their divergence from our shared ancestor.

The very word *culture*, as William McGrew has pointed out, was invented for humans, and this has long blinded cultural theorists to a more expansive appreciation of the concept. Whether apes have culture or not is not really the issue. The heart of the debate is whether scholars who study culture and consider it their intellectual territory will accept a more expansive definition. In purely academic arguments like this one, the power lies with the party who owns the key concepts of the discipline. They define concepts however they choose, and the choice is usually aimed at fencing off their intellectual turf from all others.

Primatologists are latecomers to the table of culture, and they have had to wait their turn before being allowed to sit. We should be most interested in what the continuum of intelligence tells us about the roots of human behavior, not whether what apes do or don't do fits any particular, rigid definition of culture. When it comes to human practices, from building boats to weddings to choosing mates, we should look at the intersections of our biology and our culture for clues about what has made us who we are.

Notes

1. *Changing Views of Primate Societies: The Role of Gender and Nationality*, June 1996, sponsored by the Wenner-Gren Foundation for Anthropological Research.
2. For an enlightening discussion of cross-cultural differences in chimpanzee tool use, see almost anything William McGrew has written, but especially McGrew (1992).
3. See Boesch and Boesch (1989).
4. Again, see McGrew (1992).
5. McGrew (1979)
6. Whiten *et al.* (1999) combined data from seven long-term chimpanzees studies to produce the most systematic examination of cultural variation in these apes.
7. For further discussion of chimpanzee symbolic behavior in the wild, see Goodall (1986), Wrangham *et al.* (1994), and McGrew et al. (1996).
8. See Boesch and Boesch (1989).
9. See Stanford (1999, 2001).
10. See de Waal (1996) and Stanford (2001).
11. For the first report of systematic tool use by wild orangutans, see van Schaik *et al.* (1996).
12. See Smolker *et al.* (1997).

Dim Forest, Bright Chimps

In the rain forest of Ivory Coast, chimpanzees meet the challenge of life by hunting cooperatively and using crude tools

CHRISTOPHE BOESCH AND
HEDWIGE BOESCH-ACHERMANN

Taï National Park, Ivory Coast, December 3, 1985. Drumming, barking, and screaming, chimps rush through the undergrowth, little more than black shadows. Their goal is to join a group of other chimps noisily clustering around Brutus, the dominant male of this seventy-member chimpanzee community. For a few moments, Brutus, proud and self-confident, stands fairly still, holding a shocked, barely moving red colobus monkey in his hand. Then he begins to move through the group, followed closely by his favorite females and most of the adult males. He seems to savor this moment of uncontested superiority, the culmination of a hunt high up in the canopy. But the victory is not his alone. Cooperation is essential to capturing one of these monkeys, and Brutus will break apart and share this highly prized delicacy with most of the main participants of the hunt and with the females. Recipients of large portions will, in turn, share more or less generously with their offspring, relatives, and friends.

In 1979, we began a long-term study of the previously unknown chimpanzees of Taï National Park, 1,600 square miles of tropical rain forest in the Republic of the Ivory Coast (Côte d'Ivoire). Early on, we were most interested in the chimps' use of natural hammers—branches and stones—to crack open the five species of hard-shelled nuts that are abundant here. A sea otter lying on its back, cracking an abalone shell with a rock, is a familiar picture, but no primate had ever before been observed in the wild using stones as hammers. East Africa's savanna chimps, studied for decades by Jane Goodall in Gombe, Tanzania, use twigs to extract ants and termites from their nests or honey from a bees' nest, but they have never been seen using hammerstones.

As our work progressed, we were surprised by the many ways in which the life of the Taï forest chimpanzees differs from that of their savanna counterparts, and as evidence accumulated, differences in how the two populations hunt proved the most intriguing. Jane Goodall had found that chimpanzees hunt monkeys, antelope, and wild pigs, findings confirmed by Japanese biologist Toshida Nishida, who conducted a long-term study 120 miles south of Gombe, in the Mahale Mountains. So we were not surprised to discover that the Taï chimps eat meat. What intrigued us was the degree to which they hunt cooperatively. In 1953 Raymond Dart proposed that group hunting and cooperation were key ingredients in the evolution of *Homo sapiens*. The argument has been modified considerably since Dart first put it forward, and group hunting has also been observed in some social carnivores (lions and African wild dogs, for instance), and even some birds of prey. Nevertheless, many anthropologists still hold that hunting cooperatively and sharing food played a central role in the drama that enabled early hominids, some 1.8 million years ago, to develop the social systems that are so typically human.

We hoped that what we learned about the behavior of forest chimpanzees would shed new light on prevailing theories of human evolution. Before we could even begin, however, we had to habituate a community of chimps to our presence. Five long years passed before we were able to move with them on their daily trips through the forest, of which "our" group appeared to claim some twelve square miles. Chimpanzees are alert and shy animals, and the limited field of view in the rain forest—about sixty-five feet at best—made finding them more difficult. We had to rely on sound, mostly their vocalizations and drumming on trees. Males often drum regularly while moving through the forest: pant-hooting, they draw near a big buttress tree; then, at full speed they fly over the buttress, hitting it repeatedly with their hands and feet. Such drumming may resound more than half a mile in the forest. In the beginning, our ignorance about how they moved and who was drumming led to failure more often than not, but eventually we learned that the dominant males drummed during the day to let other group members know the direction of travel. On some days, however, intermittent drumming about dawn was the only signal for the whole day. If we were out of earshot at the time, we were often reduced to guessing.

During these difficult early days, one feature of the chimps' routine proved to be our salvation: nut cracking is a noisy business. So noisy, in fact, that in the early days of French colonial rule, one officer apparently even proposed the theory that some unknown tribe was forging iron in the impenetrable and dangerous jungle.

Guided by the sounds made by the chimps as they cracked open nuts, which they often did for hours at a time, we were gradually able to get within sixty feet of the animals. We still seldom saw the chimps themselves (they fled if we came too close), but even so, the evidence left after a session of nut cracking taught us a great deal about what types of nuts they were eating, what sorts of hammer and anvil tools they were using, and—thanks to the very distinctive noise a nut makes when it finally splits open—how many hits were needed to crack a nut and how many nuts could be opened per minute.

After some months, we began catching glimpses of the chimpanzees before they fled, and after a little more time, we were able to draw close enough to watch them at work. The chimps gather nuts from the ground. Some nuts are tougher to crack than others. Nuts of the *Panda oleosa* tree are the most demanding, harder than any of the foods processed by present-day hunter-gatherers and breaking open only when a force of 3,500 pounds is applied. The stone hammers used by the Taï chimps range from stones of ten ounces to granite blocks of four to forty-five pounds. Stones of any size, however, are a rarity in the forest and are seldom conveniently placed near a nut-bearing tree. By observing closely, and in some cases imitating the way the chimps handle hammerstones, we learned that they have an impressive ability to find just the right tool for the job at hand. Taï chimps could remember the positions of many of the stones scattered, often out of sight, around a panda tree. Without having to run around rechecking the stones, they would select one of appropriate size that was closest to the tree. These mental abilities in spatial representation compare with some of those of nine-year-old humans.

To extract the four kernels from inside a panda nut, a chimp must use a hammer with extreme precision. Time and time again, we have been impressed to see a chimpanzee raise a twenty-pound stone above its head, strike a nut with ten or more powerful blows, and then, using the same hammer, switch to delicate little taps from a height of only four inches. To finish the job, the chimps often break off a small piece of twig and use it to extract the last tiny fragments of kernel from the shell. Intriguingly, females crack panda nuts more often than males, a gender difference in tool use that seems to be more pronounced in the forest chimps than in their savanna counterparts.

After five years of fieldwork, we were finally able to follow the chimpanzees at close range, and gradually, we gained insights into their way of hunting. One morning, for example, we followed a group of six male chimps on a three-hour patrol that had taken them into foreign territory to the north. (Our study group is one of five chimpanzee groups more or less evenly distributed in the Taï forest.) As always during these approximately monthly incursions, which seem to be for the purpose of territorial defense, the chimps were totally silent, clearly on edge and on the lookout for trouble. Once the patrol was over, however, and they were back within their own borders, the chimps shifted their attention to hunting. They were after monkeys, the most abundant mammals in the forest. Traveling in large, multi-species groups, some of the forest's ten species of monkeys are more apt than others to wind up as a meal for the chimps. The relatively sluggish and large (almost thirty pounds) red colobus monkeys are the chimps' usual fare. (Antelope also live in the forest, but in our ten years at Taï, we have never seen a chimp catch, or even pursue, one. In contrast, Gombe chimps at times do come across fawns, and when they do, they seize the opportunity—and the fawn.)

The six males moved on silently, peering up into the vegetation and stopping from time to time to listen for the sound of monkeys. None fed or groomed; all focused on the hunt. We followed one old male, Falstaff, closely, for he tolerates us completely and is one of the keenest and most experienced hunters. Even from the rear, Falstaff set the pace; whenever he stopped, the others paused to wait for him. After thirty minutes, we heard the unmistakable noises of monkeys jumping from branch to branch. Silently, the chimps turned in the direction of the sounds, scanning the canopy. Just then, a diana monkey spotted them and gave an alarm call. Dianas are very alert and fast; they are also about half the weight of colobus monkeys. The chimps quickly gave up and continued their search for easier, meatier prey.

Shortly after, we heard the characteristic cough of a red colobus monkey. Suddenly Rousseau and Macho, two twenty-year-olds, burst into action, running toward the cough. Falstaff seemed surprised by their precipitousness, but after a moment's hesitation, he also ran. Now the hunting barks of the chimps mixed with the sharp alarm calls of the monkeys. Hurrying behind Falstaff, we saw him climb up a conveniently situated tree. His position, combined with those of Schubert and Ulysse, two mature chimps in their prime, effectively blocked off three of the monkeys' possible escape routes. But in another tree, nowhere near any escape route and thus useless, waited the last of the hunters, Kendo, eighteen years old and the least experienced of the group. The monkeys, taking advantage of Falstaff's delay and Kendo's error, escaped.

The six males moved on and within five minutes picked up the sounds of another group of red colobus. This time, the chimps approached cautiously, nobody hurrying. They screened the canopy intently to locate the monkeys, which were still unaware of the approaching danger. Macho and Schubert chose two adjacent trees, both full of monkeys, and started climbing very quietly, taking care not to move any branches. Meanwhile, the other four chimps blocked off anticipated escape routes. When Schubert was halfway up, the monkeys finally detected the two chimps. As we watched the colobus monkeys take off in literal panic, the appropriateness of the chimpanzees' scientific name—*Pan* came to mind: with a certain stretch of the imagination, the fleeing monkeys could be shepherds and shepherdesses frightened at the sudden appearance of Pan, the wild Greek god of the woods, shepherds, and their flocks.

Taking off in the expected direction, the monkeys were trailed by Macho and Schubert. The chimps let go with loud hunting barks. Trying to escape, two colobus monkeys jumped into smaller trees lower in the canopy. With this, Rousseau and Kendo, who had been watching from the ground, sped up into the trees and tried to grab them. Only a third of the weight of the chimps, however, the monkeys managed to make it to the next tree along branches too small for their pursuers. But Falstaff had anticipated this move and was waiting for them. In the following confusion, Falstaff seized a juvenile and killed it with a bite to the neck. As the chimps met in a rush on the ground, Falstaff began to eat, sharing with Schubert and Rousseau. A juvenile colobus does not provide much meat, however, and this time, not all the chimps got a share. Frustrated individuals soon started off on another hunt, and relative calm returned fairly quickly: this sort of hunt, by a small band of chimps acting on their own at the edge of their territory, does not generate the kind of high excitement that prevails when more members of the community are involved.

So far we have observed some 200 monkey hunts and have concluded that success requires a minimum of three motivated hunters acting cooperatively. Alone or in pairs, chimps succeed less than 15 percent of the time, but when three or four act as a group, more than half the hunts result in a kill. The chimps seem well aware of the odds; 92 percent of all the hunts we observed were group affairs.

Gombe chimps also hunt red colobus monkeys, but the percentage of group hunts is much lower: only 36 percent. In addition, we learned from Jane Goodall that even when Gombe chimps do hunt in groups, their strategies are different. When Taï chimps arrive under a group of monkeys, the hunters scatter, often silently, usually out of sight of one another but each aware of the others' positions. As the hunt progresses, they gradually close in, encircling the quarry. Such movements require that each chimp coordinate his movements with those of the other hunters, as well as with those of the prey, at all times.

Coordinated hunts account for 63 percent of all those observed at Taï but only 7 percent of those at Gombe. Jane Goodall says that in a Gombe group hunt, the chimpanzees typically travel together until they arrive at a tree with monkeys. Then, as the chimps begin climbing nearby trees, they scatter as each pursues a different target. Goodall gained the impression that Gombe chimps boost their success by hunting independently but simultaneously, thereby disorganizing their prey; our impression is that the Taï chimps owe their success to being organized themselves.

Just why the Gombe and Taï chimps have developed such different hunting strategies is difficult to explain, and we plan to spend some time at Gombe in the hope of finding out. In the meantime, the mere existence of differences is interesting enough and may perhaps force changes in our understanding of human evolution. Most currently accepted theories propose that some three million years ago, a dramatic climate change in Africa east of the Rift Valley turned dense forest into open, drier habitat. Adapting to the difficulties of life under these new conditions, our ancestors supposedly evolved into cooperative hunters and began sharing food they caught. Supporters of this idea point out that plant and animal remains indicative of dry, open environments have been found at all early hominid excavation sites in Tanzania, Kenya, South Africa, and Ethiopia. That the large majority of apes in Africa today live west of the Rift Valley appears to many anthropologists to lend further support to the idea that a change in environment caused the common ancestor of apes and humans to evolve along a different line from those remaining in the forest.

Our observations, however, suggest quite another line of thought. Life in dense, dim forest may require more sophisticated behavior than is commonly assumed: compared with their savanna relatives, Taï chimps show greater complexity in both hunting and tool use. Taï chimps use tools in nineteen different ways and have six different ways of making them, compared with sixteen uses and three methods of manufacture at Gombe.

Anthropologist colleagues of mine have told me that the discovery that some chimpanzees are accomplished users of hammerstones forces them to look with a fresh eye at stone tools turned up at excavation sites. The important role played by female Taï chimps in tool use also raises the possibility that in the course of human evolution, women may have been decisive in the development of many of the sophisticated manipulative skills characteristic of our species. Taï mothers also appear to pass on their skills by actively teaching their offspring. We have observed mothers providing their young with hammers and then stepping in to help when the inexperienced youngsters encounter difficulty. This help may include carefully showing how to position the nut or hold the hammer properly. Such behavior has never been observed at Gombe.

Similarly, food sharing, for a long time said to be unique to humans, seems more general in forest than in savanna chimpanzees. Taï chimp mothers share with their young up to 60 percent of the nuts they open, at least until the latter become sufficiently adept, generally at about six years old. They also share other foods acquired with tools, including honey, ants, and bone marrow. Gombe mothers share such foods much less often, even with their infants. Taï chimps also share meat more frequently than do their Gombe relatives, sometimes dividing a chunk up and giving portions away, sometimes simply allowing beggars to grab pieces.

Any comparison between chimpanzees and our hominid ancestors can only be suggestive, not definitive. But our studies lead us to believe that the process of hominization may have begun independently of the drying of the environment. Savanna life could even have delayed the process; many anthropologists have been struck by how slowly hominid-associated remains, such as the hand ax, changed after their first appearance in the Olduvai age.

Will we have the time to discover more about the hunting strategies or other, perhaps as yet undiscovered abilities of these forest chimpanzees? Africa's tropical rain forests, and their inhabitants, are threatened with extinction by extensive logging, largely to provide the Western world with tropical timber and such products as coffee, cocoa, and rubber. Ivory Coast has lost 90 percent of its original forest, and less than 5 percent of the remainder can be considered pristine. The climate has changed dramatically. The harmattan, a cold, dry wind from the Sahara previously unknown in the forest, has now swept through the Taï forest every year since 1986. Rainfall has diminished; all the rivulets in our study region are now dry for several months of the year.

In addition, the chimpanzee, biologically very close to humans, is in demand for research on AIDS and hepatitis vaccines.

Captive-bred chimps are available, but they cost about twenty times more than wild-caught animals. Chimps taken from the wild for these purposes are generally young, their mothers having been shot during capture. For every chimp arriving at its sad destination, nine others may well have died in the forest or on the way. Such priorities—cheap coffee and cocoa and chimpanzees—do not do the economies of Third World countries any good in the long run, and they bring suffering and death to innocent victims in the forest. Our hope is that Brutus, Falstaff, and their families will survive, and that we and others will have the opportunity to learn about them well into the future. But there is no denying that modern times work against them and us.

Reprinted with permission from *Natural History,* September 1991, pp. 50, 52–56. © 1991 by Natural History Magazine, Inc.

Why Are Some Animals So Smart?

The unusual behavior of orangutans in a Sumatran swamp suggests a surprising answer

CAREL VAN SCHAIK

Even though we humans write the textbooks and may justifiably be suspected of bias, few doubt that we are the smartest creatures on the planet. Many animals have special cognitive abilities that allow them to excel in their particular habitats, but they do not often solve novel problems. Some of course do, and we call them intelligent, but none are as quick-witted as we are.

What favored the evolution of such distinctive brainpower in humans or, more precisely, in our hominid ancestors? One approach to answering this question is to examine the factors that might have shaped other creatures that show high intelligence and to see whether the same forces might have operated in our forebears. Several birds and nonhuman mammals, for instance, are much better problem solvers than others: elephants, dolphins, parrots, crows. But research into our close relatives, the great apes, is surely likely to be illuminating.

Scholars have proposed many explanations for the evolution of intelligence in primates, the lineage to which humans and apes belong (along with monkeys, lemurs, and lorises). Over the past 13 years, though, my group's studies of orangutans have unexpectedly turned up a new explanation that we think goes quite far in answering the question.

Incomplete Theories

One influential attempt at explaining primate intelligence credits the complexity of social life with spurring the development of strong cognitive abilities. This Machiavellian intelligence hypothesis suggests that success in social life relies on cultivating the most profitable relationships and on rapidly reading the social situation—for instance, when deciding whether to come to the aid of an ally attacked by another animal. Hence, the demands of society foster intelligence because the most intelligent beings would be most successful at making self-protective choices and thus would survive to pass their genes to the next generation. Machiavellian traits may not be equally beneficial to other lineages, however, or even to all primates, and so this notion alone is unsatisfying.

One can easily envisage many other forces that would promote the evolution of intelligence, such as the need to work hard for one's food. In that situation, the ability to figure out how to skillfully extract hidden nourishment or the capacity to remember the perennially shifting locations of critical food items would be advantageous, and so such cleverness would be rewarded by passing more genes to the next generation.

My own explanation, which is not incompatible with these other forces, puts the emphasis on social learning. In humans, intelligence develops over time. A child learns primarily from the guidance of patient adults. Without strong social—that is, cultural--inputs, even a potential wunderkind will end up a bungling bumpkin as an adult. We now have evidence that this process of social learning also applies to great apes, and will argue that, by and large, the animals that are intelligent are the ones that are cultural: they learn from one another innovative solutions to ecological or social problems. In short, I suggest that culture promotes intelligence.

Overview/The Orangutan Connection

- The author has discovered extensive tool use among orangutans in a Sumatran swamp. No one has observed orangutans systematically using tools in the wild before.
- This unexpected finding suggests to the author a resolution to a long-standing puzzle: Why are some animals so smart?
- He proposes that culture is the key. Primatologists define culture as the ability to learn—by observation—skills invented by others. Culture can unleash ever increasing accomplishments and can bootstrap a species toward greater and greater intelligence.

I came to this proposition circuitously, by way of the swamps on the western coast of the Indonesian island of Sumatra, where my colleagues and I were observing orangutans. The orangutan is Asia's only great ape, confined to the islands of Borneo and Sumatra and known to be something of a loner. Compared with its more familiar relative, Africa's chimpanzee, the red ape is serene rather than hyperactive and reserved socially rather than convivial. Yet we discovered in them the conditions that allow culture to flourish.

Technology in the Swamp

We were initially attracted to the swamp because it sheltered disproportionately high numbers of orangutans—unlike the islands' dryland forests, the moist swamp habitat supplies abundant food for the apes year-round and can thus support a large population. We worked in an area near Suaq Balimbing in the Kluet swamp, which may have been paradise for orangutans but, with its sticky mud, profusion of biting insects, and oppressive heat and humidity, was hell for researchers.

One of our first finds in this unlikely setting astonished us: the Suaq orangutans created and wielded a variety of tools. Although captive red apes are avid tool users, the most striking feature of tool use among the wild orangutans observed until then was its absence. The animals at Suaq ply their tools for two major purposes. First, they hunt for ants, termites and, especially, honey (mainly that of stingless bees)—more so than all their fellow orangutans elsewhere. They often cast discerning glances at tree trunks, looking for air traffic in and out of small holes. Once discovered, the holes become the focus of visual and then manual inspection by a poking and picking finger. Usually the finger is not long enough, and the orangutan prepares a stick tool. After carefully inserting the tool, the ape delicately moves it back and forth, and then withdraws it, licks it off and sticks it back in. Most of this "manipulation" is done with the tool clenched between the teeth; only the largest tools, used primarily to hammer chunks off termite nests, are handled.

The second context in which the Suaq apes employ tools involves the fruit of the *Neesia*. This tree produces woody, five-angled capsules up to 10 inches long and four inches wide. The capsules are filled with brown seeds the size of lima beans, which, because they contain nearly 50 percent fat, are highly nutritious—a rare and sought-after treat in a natural habitat without fast food. The tree protects its seeds by growing a very tough husk. When the seeds are ripe, however, the husk begins to split open; the cracks gradually widen, exposing neat rows of seeds, which have grown nice red attachments (arils) that contain some 80 percent fat. To discourage seed predators further, a mass of razor-sharp needles fills the husk. The orangutans at Suaq strip the bark off short, straight twigs, which they then hold in their mouths and insert into the cracks. By moving the tool up and down inside the crack, the animal detaches the seeds from their stalks. After this maneuver, it can drop the seeds straight into its mouth. Late in the season, the orangutans eat only the red arils, deploying the same technique to get at them without injury.

Both these methods of fashioning sticks for foraging are ubiquitous at Suaq. In general, "fishing" in tree holes is occasional and lasts only a few minutes, but when *Neesia* fruits ripen, the apes devote most of their waking hours to ferreting out the seeds or arils, and we see them grow fatter and sleeker day by day.

Why the Tool Use Is Cultural

What explains this curious concentration of tool use when wild orangutans elsewhere show so little propensity? We doubt that the animals at Suaq are intrinsically smarter: the observation that most captive members of this species can learn to use tools suggests that the basic brain capacity to do so is present.

So we reasoned that their environment might hold the answer. The orangutans studied before mostly live in dry forest, and the swamp furnishes a uniquely lush habitat. More insects make their nests in the tree holes there than in forests on dry land, and *Neesia* grows only in wet places, usually near flowing water. Tempting as the environmental explanation sounds, however, it does not explain why orangutans in several populations outside Suaq ignore altogether these same rich food sources. Nor does it explain why some populations that do eat the seeds harvest them without tools (which results, of course, in their eating much less than the orangutans at Suaq do). The same holds for tree-hole tools. Occasionally, when the nearby hills—which have dryland forests—show massive fruiting, the Suaq orangutans go there to indulge, and while they are gathering fruit they use tools to exploit the contents of tree holes. The hill habitat is a dime a dozen throughout the orangutan's geographic range, so if tools can be used on the hillsides above Suaq, why not everywhere?

Another suggestion we considered, captured in the old adage that necessity is the mother of invention, is that the Suaq animals, living at such high density, have much more competition for provisions. Consequently, many would be left without food unless they could get at the hard-to-reach supplies—that is, they *need* tools in order to eat. The strongest argument against this possibility is that the sweet or fat foods that the tools make accessible sit very high on the orangutan preference list and should therefore be sought by these animals everywhere. For instance, red apes in all locations are willing to be stung many times by honeybees to get at their honey. So the necessity idea does not hold much water either.

A different possibility is that these behaviors are innovative techniques a couple of clever orangutans invented, which then spread and persisted in the population because other individuals learned by observing these experts. In other words, the tool use is cultural. A major obstacle to studying culture in nature is that, barring experimental introductions, we can never demonstrate convincingly that an animal we observe invents some new trick rather than simply applying a well remembered but rarely practiced habit. Neither can we prove that one individual learned a new skill from another group member rather than figuring out what to do on its own. Although we can show that orangutans in the lab are capable of observing and learning socially, such studies tell us nothing about culture in nature—neither what it is generally about nor how much of it exists. So field-workers have had to develop a system of criteria to demonstrate that a certain behavior has a cultural basis.

First, the behavior must vary geographically, showing that it was invented somewhere, and it must be common where it is found, showing that it spread and persisted in a population. The tool uses at Suaq easily pass these first two tests. The second step is to eliminate simpler explanations that produce the same spatial pattern but without involving social learning. We have already excluded an ecological explanation, in which individuals exposed to a particular habitat independently converge on the same skill. We can also eliminate genetics because of the fact that most captive orangutans can learn to use tools.

The third and most stringent test is that we must be able to find geographic distributions of behavior that can be explained by culture and are not easily explained any other way. One key pattern would be the presence of a behavior in one place and its absence beyond some natural barrier to dispersal. In the case of the tool users at Suaq, the geographic distribution of *Neesia* gave us decisive clues. *Neesia* trees (and orangutans) occur on both sides of the wide Alas River. In the Singkil swamp, however, just south of Suaq and on the same side of the Alas River tools littered the floor, whereas in Batu-Batu swamp across the river they were conspicuously absent, despite our numerous visits in different years. In Batu-Batu, we did find that many of the fruits were ripped apart, showing that these orangutans ate *Neesia* seeds in the same way as their colleagues did at a site called Gunung Palung in distant Borneo but in a way completely different from their cousins right across the river in Singkil.

Batu-Batu is a small swamp area, and it does not contain much of the best swamp forest; thus, it supports a limited number of orangutans. We do not know whether tool use was never invented there or whether it could not be maintained in the smaller population, but we do know that migrants from across the river never brought it in because the Alas is so wide there that it is absolutely impassable for an orangutan. Where it is passable, farther upriver, *Neesia* occasionally grows, but the orangutans in that area ignore it altogether, apparently unaware of its rich offerings. A cultural interpretation, then, most parsimoniously explains the unexpected juxtaposition of knowledgeable tool users and brute-force foragers living practically next door to one another, as well as the presence of ignoramuses farther upriver.

Tolerant Proximity

Why do we see these fancy forms of tool use at Suaq and not elsewhere? To look into this question, we first made detailed comparisons among all the sites at which orangutans have been studied. We found that even when we excluded tool use, Suaq had the largest number of innovations that had spread throughout the population. This finding is probably not an artifact of our own interest in unusual behaviors, because some other sites have seen far more work by researchers eager to discover socially learned behavioral innovations.

We guessed that populations in which individuals had more chances to observe others in action would show a greater diversity of learned skills than would populations offering fewer learning opportunities. And indeed, we were able to confirm that sites in which individuals spend more time with others have greater repertoires of learned innovations—a relation, by the way, that also holds among chimpanzees. This link was strongest for food-related behavior, which makes sense because acquiring feeding skills from somebody else requires more close-range observation than, say, picking up a conspicuous communication signal. Put another way, those animals exposed to the fewest educated individuals have the smallest collection of cultural variants, exactly like the proverbial country bumpkin.

When we looked closely at the contrasts among sites, we noticed something else. Infant orangutans everywhere spend over 20,000 daylight hours in close contact with their mothers, acting as enthusiastic apprentices. Only at Suaq, however, did we also see adults spending considerable time together while foraging. Unlike any other orangutan population studied so far, they even regularly fed on the same food item, usually termite-riddled branches, and shared food—the meat of a slow loris, for example. This unorthodox proximity and tolerance allowed less skilled adults to come close enough to observe foraging methods, which they did as eagerly as kids.

Acquisition of the most cognitively demanding inventions, such as the tool uses found only at Suaq, probably requires face time with proficient individuals, as well as several cycles of observation and practice. The surprising implication of this need is that even though infants learn virtually all their skills from their mothers, a population will be able to perpetuate particular innovations only if tolerant role models other than the mother are around; if mom is not particularly skillful, knowledgeable experts will be close at hand, and a youngster will still be able to learn the fancy techniques that apparently do not come automatically. Thus, the more connected a social network, the more likely it is that the group will retain any skill that is invented, so that in the end tolerant populations support a greater number of such behaviors.

Our work in the wild shows us that most learning in nature, aside from simple conditioning, may have a social component, at least in primates. In contrast, most laboratory experiments that investigate how animals learn are aimed at revealing the subject's ability for individual learning. Indeed, if the lab psychologist's puzzle were presented under natural conditions, where myriad stimuli compete for attention, the subject might never realize that a problem was waiting to be solved. In the wild, the actions of knowledgeable members of the community serve to focus the attention of the naive animal.

The Cultural Roots of Intelligence

Our analyses of orangutans suggest that not only does culture—social learning of special skills—promote intelligence, it favors the evolution of greater and greater intelligence in a population over time. Different species vary greatly in the mechanisms that enable them to learn from others, but formal experiments confirm the strong impression one gets from observing great apes in the wild: they are capable of learning by watching what others do. Thus, when a wild orangutan, or an African great ape for that matter, pulls off a cognitively complex behavior, it has acquired the ability through a mix of observational learning and individual practice, much as a human child has garnered his or her skills. And when an orangutan in Suaq has acquired more of these tricks than its less fortunate cousins elsewhere, it has done so because it had greater opportunities for social learning throughout its life. In brief, social learning may bootstrap an animal's intellectual performance onto a higher plane.

To appreciate the importance of social inputs to the evolution of ever higher intelligence, let us do a thought experiment. Imagine an individual that grows up without any social inputs yet is provided with all the shelter and nutrition it needs. This situation is equivalent to that in which no contact exists between the generations or in which young fend for themselves after they emerge from the nest. Now imagine that some female in this species invents a useful skill—for instance, how to open a nut to extract its nutritious meat. She will do well and perhaps have more offspring than others in the population. Unless the skill gets transferred to the next generation, however, it will disappear when she dies.

Now imagine a situation in which the offspring accompany their mother for a while before they strike out on their own. Most youngsters will learn the new technique from their mother and thus transfer it—and its attendant benefits—to the next generation. This process would generally take place in species with slow development and long association between at least one parent and offspring, but it would get a strong boost if several individuals form socially tolerant groups.

We can go one step further. For slowly developing animals that live in socially tolerant societies, natural selection will tend to reward a slight improvement in the ability to learn through observation more strongly than a similar increase in the ability to innovate, because in such a society, an individual can stand on the shoulders of those in both present and past generations. We will then expect a feed-forward process in which animals can become more innovative and develop better techniques of social learning because both abilities rely on similar cognitive mechanisms. Hence, being cultural predisposes species with some innovative capacities to evolve toward higher intelligence. This, then, brings us to the new explanation for cognitive evolution.

This new hypothesis makes sense of an otherwise puzzling phenomenon. Many times during the past century people reared great ape infants as they would human children. These so-called enculturated apes acquired a surprising set of skills, effortlessly imitating complex behavior—understanding pointing, for example, and even some human language, becoming humorous pranksters and creating drawings. More recently, formal experiments such as those performed by E. Sue Savage-Rumbaugh of Georgia State University, involving the bonobo Kanzi, have revealed startling language abilities. Though often dismissed as lacking in scientific rigor, these consistently replicated cases reveal the astonishing cognitive potential that lies dormant in great apes. We may not fully appreciate the complexity of life in the jungle, but guess that these enculturated apes have truly become overqualified. In a process that encapsulates the story of human evolution, an ape growing up like a human can be bootstrapped to cognitive peaks higher than any of its wild counterparts.

The same line of thinking solves the long-standing puzzle of why many primates in captivity readily use—and sometimes even make—tools, when their counterparts in the wild seem to lack any such urges. The often-heard suggestion that they do not need tools is belied by observations of orangutans, chimpanzees and capuchin monkeys showing that some of this tool use makes available the richest food in the animals' natural habitats or tides the creatures over during lean periods. The conundrum is resolved if we realize that two individuals of the same species can differ dramatically in their intellectual performance, depending on the social environment in which they grew up.

Orangutans epitomize this phenomenon. They are known as the escape artists of the zoo world, cleverly unlocking the doors of their cages. But the available observations from the wild, despite decades of painstaking monitoring by dedicated field-workers, have uncovered precious few technological accomplishments outside Suaq. Wild-caught individuals generally never take to being locked up, always retaining their deeply ingrained shyness and suspicion of humans. But zoo-born apes happily consider their keepers valuable role models and pay attention to their activities and to the objects strewn around the enclosures, learning to learn and thus accumulating numerous skills.

The critical prediction of the intelligence-through-culture theory is that the most intelligent animals are also likely to live in populations in which the entire group routinely adopts innovations introduced by members. This prediction is not easily tested. Animals from different lineages vary so much in their senses and in their ways of life that a single yardstick for intellectual performance has traditionally been hard to find. For now, we can merely ask whether lineages that show incontrovertible signs of intelligence also have innovation-based cultures, and vice versa. Recognizing oneself in a mirror, for example, is a poorly understood but unmistakable sign of self-awareness, which is taken as a sign of high intelligence. So far, despite widespread attempts in numerous lineages, the only mammalian groups to pass this test are great apes and dolphins, the same animals that can learn to understand many arbitrary symbols and that show the best evidence for imitation, the basis for innovation-based culture. Flexible, innovation-based tool use, another expression of intelligence, has a broader distribution in mammals: monkeys and apes, cetaceans, and elephants—all lineages in which social learning is common. Although so far only these very crude tests can be done, they support the intelligence-through-culture hypothesis.

Another important prediction is that the propensities for innovation and social learning must have coevolved. Indeed, Simon Reader, now at Utrecht University in the Netherlands, and Kevin N. Laland, currently at the University of St. Andrews in Scotland, found that primate species that show more evidence of innovation are also those that show the most evidence for social learning. Still more indirect tests rely on correlations among species between the relative size of the brain (after statistically correcting for body size) and social and developmental variables. The well-established correlations between gregariousness and relative brain size in various mammalian groups are also consistent with the idea.

Although this new hypothesis is not enough to explain why our ancestors, alone among great apes, evolved such extreme intelligence, the remarkable bootstrapping ability of the great apes in rich cultural settings makes the gap seem less formidable. The explanation for the historical trajectory of change involves many details that must be painstakingly pieced together from a sparse and confusing fossil and archaeological record. Many researchers suspect that a key change was the invasion of the savanna by tool-wielding, striding early *Homo*. To dig up tubers and deflesh and defend carcasses of large mammals, they had to work collectively and create tools and strategies. These demands fostered ever more innovation and more interdependence, and intelligence snowballed.

Once we were human, cultural history began to interact with innate ability to improve performance. Nearly 150,000 years after the origin of our own species, sophisticated expressions of human symbolism, such as finely worked nonfunctional artifacts (art, musical instruments and burial gifts), were widespread. The explosion of technology in the past 10,000 years shows that cultural inputs can unleash limitless accomplishments, all with Stone Age brains. Culture can indeed build a new mind from an old brain.

CAREL VAN SCHAIK is director of the Anthropological Institute and Museum at the University of Zurich in Switzerland. A native of the Netherlands, he earned his doctorate at Utrecht University in 1985. After a postdoc at Princeton University and another short stint at Utrecht, he went to Duke University, where he was professor of biological anthropology until he returned to the Old World in 2004. His book *Among Orangutans: Red Apes and the Rise of Human Culture* (Harvard University Press, 2004) gives a more detailed treatment of the ideas covered in this article.

How Animals Do Business

Humans and other animals share a heritage of economic tendencies—including cooperation, repayment of favors and resentment at being shortchanged

FRANS B. M. DE WAAL

J ust as my office would not stay empty for long were I to move out, nature's real estate changes hands all the time. Potential homes range from holes drilled by woodpeckers to empty shells on the beach. A typical example of what economists call a "vacancy chain" is the housing market among hermit crabs. To protect its soft abdomen, each crab carries its house around, usually an abandoned gastropod shell. The problem is that the crab grows, whereas its house does not. Hermit crabs are always on the lookout for new accommodations. The moment they upgrade to a roomier shell, other crabs line up for the vacated one.

One can easily see supply and demand at work here, but because it plays itself out on a rather impersonal level, few would view the crab version as related to human economic transactions. The crab interactions would be more interesting if the animals struck deals along the lines of "you can have my house if I can have that dead fish." Hermit crabs are not deal makers, though, and in fact have no qualms about evicting homeowners by force. Other, more social animals do negotiate, however, and their approach to the exchange of resources and services helps us understand how and why human economic behavior may have evolved.

The New Economics

Classical economics views people as profit maximizers driven by pure selfishness. As 17th-century English philosopher Thomas Hobbes put it, "Every man is presumed to seek what is good for himself naturally, and what is just, only for Peaces sake, and accidentally." In this still prevailing view, sociality is but an afterthought, a "social contract" that our ancestors entered into because of its benefits, not because they were attracted to one another. For the biologist, this imaginary history falls as wide off the mark as can be. We descend from a long line of group-living primates, meaning that we are naturally equipped with a strong desire to fit in and find partners to live and work with. This evolutionary explanation for why we interact as we do is gaining influence with the advent of a new

Overview/Evolved Economics

■ The new field of behavioral economics views the way humans conduct business as an evolved heritage of our species.
■ Just as tit for tat and supply and demand influence the trading of goods and services in human economies, they also affect trading activities among animals.
■ Emotional reactions—such as outrage at unfair arrangements—underlie the negotiations of both animals and humans.
■ This shared psychology may explain such curious behaviors as altruism—they are part of our background as cooperative primates.

school, known as behavioral economics, that focuses on actual human behavior rather than on the abstract forces of the marketplace as a guide for understanding economic decision making. In 2002 the school was recognized by a shared Nobel Prize for two of its founders: Daniel Kahneman and Vernon L. Smith.

Animal behavioral economics is a fledgling field that lends support to the new theories by showing that basic human economic tendencies and preoccupations—such as reciprocity, the division of rewards, and cooperation—are not limited to our species. They probably evolved in other animals for the same reasons they evolved in us—to help individuals take optimal advantage of one another without undermining the shared interests that support group life.

Take a recent incident during my research at the Yerkes National Primate Research Center in Atlanta. We had taught capuchin monkeys to reach a cup of food on a tray by pulling on a bar attached to the tray. By making the tray too heavy for a single individual, we gave the monkeys a reason to work together.

On one occasion, the pulling was to be done by two females, Bias and Sammy. Sitting in adjoining cages, they successfully brought a tray bearing two cups of food within reach. Sammy, however, was in such a hurry to collect her reward that she released the bar and grabbed her cup before Bias had a chance to get hers. The tray bounced back, out of Bias's reach. While Sammy munched away, Bias threw a tantrum. She screamed her lungs out for half a minute until Sammy approached her pull bar again. She then helped Bias bring in the tray a second time. Sammy did not do so for her own benefit, because by now the cup accessible to her was empty.

Sammy's corrective behavior appeared to be a response to Bias's protest against the loss of an anticipated reward. Such action comes much closer to human economic transactions than that of the hermit crabs, because it shows cooperation, communication and the fulfillment of an expectation, perhaps even a sense of obligation. Sammy seemed sensitive to the quid pro quo of the situation. This sensitivity is not surprising given that the group life of capuchin monkeys revolves around the same mixture of cooperation and competition that marks our own societies.

The Evolution of Reciprocity

Animals and people occasionally help one another without any obvious benefits for the helper. How could such behavior have evolved? If the aid is directed at a family member, the question is relatively easy to answer. "Blood is thicker than water," we say, and biologists recognize genetic advantages to such assistance: if your kin survive, the odds of your genes making their way into the next generation increase. But cooperation among unrelated individuals suggests no immediate genetic advantages. Pëtr Kropotkin, a Russian prince, offered an early explanation in his book *Mutual Aid*, published in 1902. If helping is communal, he reasoned, all parties stand to gain—everyone's chances for survival go up. We had to wait until 1971, however, for Robert L. Trivers, then at Harvard University, to phrase the issue in modern evolutionary terms with his theory of reciprocal altruism.

Trivers contended that making a sacrifice for another pays off if the other later returns the favor. Reciprocity boils down to "I'll scratch your back, if you scratch mine." Do animals show such tit for tat? Monkeys and apes form coalitions; two or more individuals, for example, gang up on a third. And researchers have found a positive correlation between how often A supports B and how often B supports A. But does this mean that animals actually keep track of given and received favors? They may just divide the world into "buddies," whom they prefer, and "non-buddies," whom they care little about. If such feelings are mutual, relationships will be either mutually helpful or mutually unhelpful. Such symmetries can account for the reciprocity reported for fish, vampire bats (which regurgitate blood to their buddies), dolphins and many monkeys.

Just because these animals may not keep track of favors does not mean they lack reciprocity. The issue rather is how a favor done for another finds its way back to the original altruist. What exactly is the reciprocity mechanism? Mental record keeping is just one way of getting reciprocity to work, and whether animals do this remains to be tested. Thus far chimpanzees are the only exception. In the wild, they hunt in teams to capture colobus monkeys. One hunter usually captures the prey, after which he tears it apart and shares it. Not everyone gets a piece, though,

and even the highest-ranking male, if he did not take part in the hunt, may beg in vain. This by itself suggests reciprocity: hunters seem to enjoy priority during the division of spoils.

To try to find the mechanisms at work here, we exploited the tendency of these apes to share—which they also show in captivity—by handing one of the chimpanzees in our colony a watermelon or some branches with leaves. The owner would be at the center of a sharing cluster, soon to be followed by secondary clusters around individuals who had managed to get a major share, until all the food had trickled down to everyone. Claiming another's food by force is almost unheard of among chimpanzees—a phenomenon known as "respect of possession." Beggars hold out their hand, palm upward, much like human beggars in the street. They whimper and whine, but aggressive confrontations are rare. If these do occur, the possessor almost always initiates them to make someone leave the circle. She whacks the offenders over the head with a sizable branch or barks at them in a shrill voice until they leave her alone. Whatever their rank, possessors control the food flow.

We analyzed nearly 7,000 of these approaches, comparing the possessor's tolerance of specific beggars with previously received services. We had detailed records of grooming on the mornings of days with planned food tests. If the top male, Socko, had groomed May, for example, his chances of obtaining a few branches from her in the afternoon were much improved. This relation between past and present behavior proved general. Symmetrical connections could not explain this outcome, as the pattern varied from day to day. Ours was the first animal study to demonstrate a contingency between favors given and received. Moreover, these food-for-grooming deals were partner-specific—that is, May's tolerance benefited Socko, the one who had groomed her, but no one else.

This reciprocity mechanism requires memory of previous events as well as the coloring of memory such that it induces friendly behavior. In our own species, this coloring process is known as "gratitude," and there is no reason to call it something else in chimpanzees. Whether apes also feel obligations remains unclear, but it is interesting that the tendency to return favors is not the same for all relationships. Between individuals who associate and groom a great deal, a single grooming session carries little weight. All kinds of daily exchanges occur between them, probably without their keeping track. They seem instead to follow the buddy system discussed before. Only in the more distant relationships does grooming stand out as specifically deserving reward. Because Socko and May are not close friends, Socko's grooming was duly noticed.

A similar difference is apparent in human behavior, where we are more inclined to keep track of give-and-take with strangers and colleagues than with our friends and family. In fact, scorekeeping in close relationships, such as between spouses, is a sure sign of distrust.

Biological Markets

Because reciprocity requires partners, partner choice ranks as a central issue in behavioral economics. The hand-me-down housing of hermit crabs is exceedingly simple compared with

What Makes Reciprocity Tick

Humans and other animals exchange benefits in several ways, known technically as reciprocity mechanisms. No matter what the mechanism, the common thread is that benefits find their way back to the original giver.

RECIPROCITY MECHANISM	KEY FEATURES
Symmetry-based "We're buddies"	Mutual affection between two parties prompts similar behavior in both directions without need to keep track of daily give-and-take, so long as the overall relationship remains satisfactory. Possibly the most common mechanism of reciprocity in nature, this kind is typical of humans and chimpanzees in close relationships. **Example:** Chimpanzee friends associate, groom together and support each other in fights.
Attitudinal "If you're nice, I'll be nice"	Parties mirror one another's attitudes, exchanging favors on the spot. Instant attitudinal reciprocity occurs among monkeys, and people often rely on it with strangers. **Example:** Capuchins share food with those who help them pull a treat-laden tray.
Calculated "What have you done for me lately?"	Individuals keep track of the benefits they exchange with particular partners, which helps them decide to whom to return favors. This mechanism is typical of chimpanzees and common among people in distant and professional relationships. **Example:** Chimpanzees can expect food in the afternoon from those they groomed in the morning.

Roberto Osti

the interactions among primates, which involve multiple partners exchanging multiple currencies, such as grooming, sex, support in fights, food, babysitting and so on. This "marketplace of services," as I dubbed it in *Chimpanzee Politics*, means that each individual needs to be on good terms with higher-ups, to foster grooming partnerships and—if ambitious—to strike deals with like-minded others. Chimpanzee males form coalitions to challenge the reigning ruler, a process fraught with risk. After an overthrow, the new ruler needs to keep his supporters contented: an alpha male who tries to monopolize the privileges of power, such as access to females, is unlikely to keep his position for long. And chimps do this without having read Niccoló Machiavelli.

With each individual shopping for the best partners and selling its own services, the framework for reciprocity becomes one of supply and demand, which is precisely what Ronald Noë and Peter Hammerstein, then at the Max Planck Institute for Behavioral Physiology in Seewiesen, Germany, had in mind with their biological market theory. This theory, which applies whenever trading partners can choose with whom to deal, postulates that the value of commodities and partners varies with their availability. Two studies of market forces elaborate this point: one concerns the baby market among baboons, the other the job performance of small fish called cleaner wrasses.

Like all primate females, female baboons are irresistibly attracted to infants-not only their own but also those of others. They give friendly grunts and try to touch them. Mothers are highly protective, however, and reluctant to let anyone handle their precious newborns. To get close, interested females groom the mother while peeking over her shoulder or underneath her arm at the baby. After a relaxing grooming session, a mother may give in to the groomer's desire for a closer look. The other thus buys infant time. Market theory predicts that the value of babies should go up if there are fewer around. In a study of wild chacma baboons in South Africa, Louise Barrett of the University of Liverpool and Peter Henzi of the University of Central Lancashire, both in England, found that, indeed, mothers of rare infants were able to extract a higher price (longer grooming) than mothers in a troop full of babies.

Cleaner wrasses (*Labroides dirnidiatus*) are small marine fish that feed on the external parasites of larger fish. Each cleaner owns a "station" on a reef where clientele come to spread their pectoral fins and adopt postures that offer the cleaner a chance to do its job. The exchange exemplifies a perfect mutualism.

The cleaner nibbles the parasites off the client's body surface, gills and even the inside of its mouth. Sometimes the cleaner is so busy that clients have to wait in line. Client fish come in two varieties: residents and roamers. Residents belong to species with small territories; they have no choice but to go to their local cleaner. Roamers, on the other hand, either hold large territories or travel widely, which means that they have

Roberto Osti

TRAY-PULLING EXPERIMENT demonstrates that capuchin monkeys are more likely to share food with cooperative partners than with those who are not helpful. The test chamber houses two capuchins, separated by mesh. To reach their treat cups, they must use a bar to pull a counterweighted tray; the tray is too heavy for one monkey to handle alone. The "laborer" (*on left*), whose transparent cup is obviously empty, works for the "winner," who has food in its cup. The winner generally shares food with the laborer through the mesh. Failing to do so will cause the laborer to lose interest in the task.

several cleaning stations to choose from. They want short waiting times, excellent service and no cheating. Cheating occurs when a cleaner fish takes a bite out of its client, feeding on healthy mucus. This makes clients jolt and swim away.

Research on cleaner wrasses by Redouan Bshary of the Max Planck institute in Seewiesen consists mainly of observations on the reef but also includes ingenious experiments in the laboratory. His papers read much like a manual for good business practice. Roamers are more likely to change stations if a cleaner has ignored them for too long or cheated them. Cleaners seem to know this and treat roamers better than they do residents. If a roamer and a resident arrive at the same time, the cleaner almost always services the roamer first. Residents have nowhere else to go, and so they can be kept waiting. The only category of fish that cleaners never cheat are predators, who possess a radical counterstrategy, which is to swallow the cleaner. With predators, cleaner fish wisely adopt, in Bshary's words, an "unconditionally cooperative strategy."

Biological market theory offers an elegant solution to the problem of freeloaders, which has occupied biologists for a long time because reciprocity systems are obviously vulnerable to those who take rather than give. Theorists often assume that offenders must be punished, although this has yet to be demonstrated for animals. Instead cheaters can be taken care of in a much simpler way. If there is a choice of partners, animals can simply abandon unsatisfactory relationships and replace them with those offering more benefits. Market mechanisms are all that is needed to sideline profiteers. In our own societies, too, we neither like nor trust those who take more than they give, and we tend to stay away from them.

Fair Is Fair

To reap the benefits of cooperation, an individual must monitor its efforts relative to others and compare its rewards with the effort put in. To explore whether animals actually carry out such monitoring, we turned again to our capuchin monkeys, testing them in a miniature labor market inspired by field observations of capuchins attacking giant squirrels. Squirrel hunting is a group effort, but one in which all rewards end up in the hands of a single individual: the captor. If captors were to keep the prey solely for themselves, one can imagine that others would lose interest in joining them in the future. Capuchins share meat for the same reason chimpanzees (and people) do: there can be no joint hunting without joint payoffs.

We mimicked this situation in the laboratory by making certain that only one monkey (whom we called the winner) of a tray-pulling pair received a cup with apple pieces. Its partner (the laborer) had no food in its cup, which was obvious from the outset because the cups were transparent. Hence, the laborer pulled for the winner's benefit. The monkeys sat side by side, separated by mesh. From previous tests we knew that food possessors might bring food to the partition and permit their neighbor to reach for it through the mesh. On rare occasions, they push pieces to the other.

We contrasted collective pulls with solo pulls. In one condition, both animals had a pull bar and the tray was heavy; in the other, the partner lacked a bar and the winner handled a lighter tray on its own. We counted more acts of food sharing after collective than solo pulls: winners were in effect compensating their partners for the assistance they had received. We also con-

How Humans Do Business

The emotions that Frans de Waal describes in the economic exchanges of social animals have parallels in our own transactions. Such similarities suggest that human economic interactions are controlled at least in part by ancient tendencies and emotions. Indeed, the animal work supports a burgeoning school of research known as behavioral economics. This new discipline is challenging and modifying the "standard model" of economic research, which maintains that humans base economic decisions on rational thought processes. For example, people reject offers that strike them as unfair, whereas classical economics predicts that people take anything they can get. In 2002 the Noble Prize in Economics went to two pioneers of the field: Daniel Kahneman, a psychologist at Princeton University, and Vernon L. Smith, an economist at George Mason University.

Kahneman, with his colleague Amos Tversky, who died in 1996 and thus was not eligible for the prize, analyzed how humans make decisions when confronted by uncertainty and risk. Classical economists had thought of human decisions in terms of expected utility—the sum of the gains people think they will get from some future event multiplied by its probability of occurring. But Kahneman and Tversky demonstrated that people are much more frightened of losses than they are encouraged by potential gains and that people follow that herd. The bursting of the stock-market bubble in 2000 provides a potent example: the desire to stay with the herd may have led people to shell out far more for shares than any purely rational investor would have paid.

Smith's work demonstrated that laboratory experiments would function in economics, which had traditionally been considered a nonexperimental science that relied solely on observation. Among his findings in the lab: emotional decisions are not necessarily unwise.

—The Editors

firmed that sharing affects future cooperation. Because a pair's success rate would drop if the winner failed to share, payment of the laborer was a smart strategy.

Sarah F. Brosnan, one of my colleagues at Yerkes, went further in exploring reactions to the way rewards are divided. She would offer a capuchin monkey a small pebble, then hold up a slice of cucumber as enticement for returning the pebble. The monkeys quickly grasped the principle of exchange. Placed side by side, two monkeys would gladly exchange pebbles for cucumber with the researcher. If one of them got grapes, however, whereas the other stayed on cucumber, things took an unexpected turn. Grapes are much preferred. Monkeys who had been perfectly willing to work for cucumber suddenly went on strike. Not only did they perform reluctantly seeing that the other was getting a better deal, but they became agitated, hurling the pebbles out of the test chamber and sometimes even the cucumber slices. A food normally never refused had become less than desirable.

To reject unequal pay—which people do as well—goes against the assumptions of traditional economics. If maximizing benefits were all that mattered, one should take what one can get and never let resentment or envy interfere. Behavioral economists, on the other hand, assume evolution has led to emotions that preserve the spirit of cooperation and that such emotions powerfully influence behavior. In the short run, caring about what others get may seem irrational, but in the long run it keeps one from being taken advantage of. Discouraging exploitation is critical for continued cooperation.

It is a lot of trouble, though, to always keep a watchful eye on the flow of benefits and favors. This is why humans protect themselves against freeloading and exploitation by forming buddy relationships with partners—such as spouses and good friends—who have withstood the test of time. Once we have determined whom to trust, we relax the rules. Only with more distant partners do we keep mental records and react strongly to imbalances, calling them "unfair."

We found indications for the same effect of social distance in chimpanzees. Straight tit for tat, as we have seen, is rare among friends who routinely do favors for one another. These relationships also seem relatively immune to inequity. Brosnan conducted her exchange task using grapes and cucumbers with chimpanzees as well as capuchins. The strongest reaction among chimpanzees concerned those who had known one another for a relatively short time, whereas the members of a colony that had lived together for more than 30 years hardly reacted at all. Possibly, the greater their familiarity, the longer the time frame over which chimpanzees evaluate their relationships. Only distant relations are sensitive to day-to-day fluctuations.

All economic agents, whether human or animal, need to come to grips with the freeloader problem and the way yields are divided after joint efforts. They do so by sharing most with those who help them most and by displaying strong emotional reactions to violated expectations. A truly evolutionary discipline of economics recognizes this shared psychology and considers the possibility that we embrace the golden rule not accidentally, as Hobbes thought, but as part of our background as cooperative primates.

More To Explore

"The Chimpanzee's Service Economy: Food for Grooming." Frans B. M. de Waal in *Evolution and Human Behavior*, Vol. 18, No. 6, pages 375–386; November 1997.

"Payment for Labour in Monkeys." Frans B. M. de Waal and Michelle L. Berger in *Nature*, Vol. 404, page 563; April 6, 2000.

"Choosy Reef Fish Select Cleaner Fish That Provide High-Quality Service." R. Bshary and D. Schäffer in *Animal Behaviour*, Vol. 63, No. 3, pages 557–564; March 2002.

"Infants as a Commodity in a Baboon Market." S. P. Henzi and L. Barrett in *Animal Behaviour*, Vol. 63, No. 5, pages 915–921; 2002.

"Monkeys Reject Unequal Pay." Sarah F. Brosnan and Frans B. M. de Waal in *Nature*, Vol. 425, pages 297–299; September 18, 2003.

Living Links Center site: www.emory.edu/LIVING-LINKS/
Classic cooperation experiment with chimpanzees:
    www.emory.edu/LIVING-LINKS/crawfordvideo.html

FRANS B. M. DE WAAL is C. H. Candler Professor of Primate Behavior at Emory University and director of the Living Links Center at the university's Yerkes National Primate Research Center. De Waal specializes in the social behavior and cognition of monkeys, chimpanzees and bonobos, especially cooperation, conflict resolution and culture. His books include *Chimpanzee Politics, Peacemaking among Primates, The Ape and the Sushi Master* and the forthcoming *Our Inner Ape*.

Are We in Anthropodenial?

FRANS DE WAAL

When guests arrive at the Yerkes Regional Primate Research Center in Georgia, where I work, they usually pay a visit to the chimpanzees. And often, when she sees them approaching the compound, an adult female chimpanzee named Georgia will hurry to the spigot to collect a mouthful of water. She'll then casually mingle with the rest of the colony behind the mesh fence, and not even the sharpest observer will notice anything unusual. If necessary, Georgia will wait minutes, with her lips closed, until the visitors come near. Then there will be shrieks, laughs, jumps—and sometimes falls—when she suddenly sprays them.

I have known quite a few apes that are good at surprising people, naive and otherwise. Heini Hediger, the great Swiss zoo biologist, recounts how he—being prepared to meet the challenge and paying attention to the ape's every move—got drenched by an experienced chimpanzee. I once found myself in a similar situation with Georgia; she had taken a drink from the spigot and was sneaking up to me. I looked her straight in the eye and pointed my finger at her, warning in Dutch, "I have seen you!" She immediately stepped back, let some of the water dribble from her mouth, and swallowed the rest. I certainly do not wish to claim that she understands Dutch, but she must have sensed that I knew what she was up to, and that I was not going to be an easy target.

To endow animals with human emotions has long been a scientific taboo. But if we do not, we risk missing something fundamental, about both animals and us.

Now, no doubt even a casual reader will have noticed that in describing Georgia's actions, I've implied human qualities such as intentions, the ability to interpret my own awareness, and a tendency toward mischief. Yet scientific tradition says I should avoid such language—I am committing the sin of anthropomorphism, of turning nonhumans into humans. The word comes from the Greek, meaning "human form," and it was the ancient Greeks who first gave the practice a bad reputation. They did not have chimpanzees in mind: the philosopher Xenophanes objected to Homer's poetry because it treated Zeus and the other gods as if they were people. How could we be so arrogant, Xe-

nophanes asked, as to think that the gods should look like us? If horses could draw pictures, he suggested mockingly, they would no doubt make their gods look like horses.

Nowadays the intellectual descendants of Xenophanes warn against perceiving animals to be like ourselves. There are, for example, the behaviorists, who follow psychologist B. F. Skinner in viewing the actions of animals as responses shaped by rewards and punishments rather than the result of internal decision making, emotions, or intentions. They would say that Georgia was not "up to" anything when she sprayed water on her victims. Far from planning and executing a naughty plot, Georgia merely fell for the irresistible reward of human surprise and annoyance. Whereas any person acting like her would be scolded, arrested, or held accountable, Georgia is somehow innocent.

Behaviorists are not the only scientists who have avoided thinking about the inner life of animals. Some sociobiologists—researchers who look for the roots of behavior in evolution—depict animals as "survival machines" and "pre-programmed robots" put on Earth to serve their "selfish" genes. There is a certain metaphorical value to these concepts, but is has been negated by the misunderstanding they've created. Such language can give the impression that only genes are entitled to an inner life. No more delusively anthropomorphizing idea has been put forward since the pet-rock craze of the 1970s. In fact, during evolution, genes—a mere batch of molecules—simply multiply at different rates, depending on the traits they produce in an individual. To say that genes are selfish is like saying a snowball growing in size as it rolls down a hill is greedy for snow.

Logically, these agnostic attitudes toward a mental life in animals can be valid only if they're applied to our own species as well. Yet it's uncommon to find researchers who try to study human behavior as purely a matter of reward and punishment. Describe a person as having intentions, feelings, and thoughts and you most likely won't encounter much resistance. Our own familiarity with our inner lives overrules whatever some school of thought might claim about us. Yet despite this double standard toward behavior in humans and animals, modern biology leaves us no choice other than to conclude that we *are* animals. In terms of anatomy, physiology, and neurology we are really no more exceptional than, say, an elephant or a platypus is in its own way. Even such presumed hallmarks of humanity as warfare, politics, culture, morality, and language may not be completely unprecedented. For example, different groups of wild chimpanzees employ different technologies—some fish for termites with sticks, others crack nuts with stones—that are transmitted from one generation to the next through a process reminiscent of human culture.

Given these discoveries, we must be very careful not to exaggerate the uniqueness of our species. The ancients apparently never gave much thought to this practice, the opposite of anthropomorphism, and so we lack a word for it. I will call it anthropodenial: a blindness to the human-like characteristics of other animals, or the animal-like characteristics of ourselves.

Those who are in anthropodenial try to build a brick wall to separate humans from the rest of the animal kingdom. They carry on the tradition of René Descartes, who declared that while humans possessed souls, animals were mere automatons. This produced a serious dilemma when Charles Darwin came along: If we descended from such automatons, were we not automatons ourselves? If not, how did we get to be so different?

Each time we must ask such a question, another brick is pulled out of the dividing wall, and to me this wall is beginning to look like a slice of Swiss cheese. I work on a daily basis with animals from which it is about as hard to distance yourself as from "Lucy," the famed 3.2-million-year-old fossil australopithecine. If we owe Lucy the respect of an ancestor, does this not force a different look at the apes? After all, as far as we can tell, the most significant difference between Lucy and modern chimpanzees is found in their hips, not their craniums.

A s soon as we admit that animals are far more like our relatives than like machines, then anthropodenial becomes impossible and anthropomorphism becomes inevitable—and scientifically acceptable. But not *all* forms of anthropomorphism, of course. Popular culture bombards us with examples of animals being humanized for all sorts of purposes, ranging from education to entertainment to satire to propaganda. Walt Disney, for example, made us forget that Mickey is a mouse, and Donald a duck. George Orwell laid a cover of human societal ills over a population of livestock. I was once struck by an advertisement for an oil company that claimed its propane saved the environment, in which a grizzly bear enjoying a pristine landscape had his arm around his mate's shoulders. In fact, bears are nearsighted and do not form pair-bonds, so the image says more about our own behavior than theirs.

Perhaps that was the intent. The problem is, we do not always remember that, when used in this way, anthropomorphism can provide insight only into human affairs and not into the affairs of animals. When my book *Chimpanzee Politics* came out in France, in 1987, my publisher decided (unbeknownst to me) to put François Mitterrand and Jacques Chirac on the cover with a chimpanzee between them. I can only assume he wanted to imply that these politicians acted like "mere" apes. Yet by doing so he went completely against the whole point of my book, which was not to ridicule people but to show that chimpanzees live in complex societies full of alliances and power plays that in some ways mirror our own.

You can often hear similar attempts at anthropomorphic humor in the crowds that form around the monkey exhibit at a typical zoo. Isn't it interesting that antelopes, lions, and giraffes rarely elicit hilarity? But people who watch primates end up hooting and yelling, scratching themselves in exaggeration, and pointing at the animals while shouting, "I had to look twice, Larry. I thought it was you!" In my mind, the laughter reflects anthropodenial: it is a nervous reaction caused by an uncomfortable resemblance.

That very resemblance, however, can allow us to make better use of anthropomorphism, but for this we must view it as a means rather than an end. It should not be our goal to find some quality in an animal that is precisely equivalent to an aspect of our own inner lives. Rather, we should use the fact that we are similar to animals to develop ideas we can test. For example, after observing a group of chimpanzees at length, we begin to suspect that some individuals are attempting to "deceive" others— by giving false alarms to distract unwanted attention from the theft of food or from forbidden sexual activity. Once we frame the observation in such terms, we can devise testable predictions. We can figure out just what it would take to demonstrate deception on the part of chimpanzees. In this way, a speculation is turned into a challenge.

Naturally, we must always be on guard. To avoid making silly interpretations based on anthropomorphism, one must always interpret animal behavior in the wider context of a species' habits and natural history. Without experience with primates, one could imagine that a grinning rhesus monkey must be delighted, or that a chimpanzee running toward another with loud grunts must be in an aggressive mood. But primatologists know from many hours of observation that rhesus monkeys bare their teeth when intimidated, and that chimpanzees often grunt when they meet and embrace. In other words, a grinning rhesus monkey signals submission, and a chimpanzee's grunting often serves as a greeting. A careful observer may thus arrive at an informed anthropomorphism that is at odds with extrapolations from human behavior.

One must also always be aware that some animals are more like ourselves than others. The problem of sharing the experiences of organisms that rely on different senses is a profound one. It was expressed most famously by the philosopher Thomas Nagel when he asked, "What is it like to be a bat?" A bat perceives its world in pulses of reflected sound, something we creatures of vision would have a hard time imagining. Perhaps even more alien would be the experience of an animal such as the star-nosed mole. With 22 pink, writhing tentacles around its nostrils, it is able to feel microscopic textures on small objects in the mud with the keenest sense of touch of any animal on Earth.

Humans can barely imagine a star-nosed mole's *Umwelt*—a German term for the environment as perceived by the animal. Obviously, the closer a species is to us, the easier it is to enter its *Umwelt*. This is why anthropomorphism is not only tempting in the case of apes but also hard to reject on the grounds that we cannot know how they perceive the world. Their sensory systems are essentially the same as ours.

Last summer, an ape saved a three-year-old boy. The child, who had fallen 20 feet into the primate exhibit at Chicago's Brookfield Zoo, was scooped up and carried to safety by Binti Jua, an eight-year-old western lowland female gorilla. The gorilla sat down on a log in a stream, cradling the boy in her lap and patting his back, and then carried him to one of the exhibit doorways before laying him down and continuing on her way.

Binti became a celebrity overnight, figuring in the speeches of leading politicians who held her up as an example of much-needed compassion. Some scientists were less lyrical, however. They cautioned that Binti's motives might have been less noble than they appeared, pointing out that this gorilla had been raised by people and had been taught parental skills with a stuffed animal. The whole affair might have been one of a confused maternal instinct, they claimed.

Bonobos have been known to assist companions new to their quarters in zoos, taking them by the hand to guide them through the maze of corridors connecting parts of their building.

The intriguing thing about this flurry of alternative explanations was that nobody would think of raising similar doubts when a person saves a dog hit by a car. The rescuer might have grown up around a kennel, have been praised for being kind to animals, have a nurturing personality, yet we would still see his behavior as an act of caring. Whey then, in Binti's case, was her background held against her? I am not saying that I know what went through Binti's head, but I do know that no one had prepared her for this kind of emergency and that it is unlikely that, with her own 17-month-old infant on her back, she was "maternally confused." How in the world could such a highly intelligent animal mistake a blond boy in sneakers and a red T-shirt for a juvenile gorilla? Actually, the biggest surprise was how surprised most people were. Students of ape behavior did not feel that Binti had done anything unusual. Jörg Hess, a Swiss gorilla expert, put it most bluntly, "The incident can be sensational only for people who don't know a thing about gorillas."

Binti's action made a deep impression mainly because it benefited a member of our own species, but in my work on the evolution of morality and empathy, I have encountered numerous instances of animals caring for one another. For example, a chimpanzee consoles a victim after a violent attack, placing an arm around him and patting his back. And bonobos (or pygmy chimpanzees) have been known to assist companions new to their quarters in zoos, taking them by the hand to guide them through the maze of corridors connecting parts of their building. These kinds of cases don't reach the newspapers but are consistent with Binti's assistance to the unfortunate boy and the idea that apes have a capacity for sympathy.

The traditional bulwark against this sort of cognitive interpretation is the principle of parsimony—that we must make as few assumptions as possible when trying to construct a scientific explanation, and that assuming an ape is capable of something like sympathy is too great a leap. But doesn't that same principle of parsimony argue against assuming a huge cognitive gap when the evolutionary distance between humans and apes is so small? If two closely related species act in the same manner, their underlying mental processes are probably the same, too. The incident at the Brookfield Zoo shows how hard it is to avoid anthropodenial and anthropomorphism at the same time: in trying to avoid thinking of Binti as a human being, we run straight into the realization that Binti's actions make little sense if we refuse to assume intentions and feelings.

In the end we must ask: What kind of risk we are willing to take—the risk of underestimating animal mental life or the risk of overestimating it? There is no simple answer. But from an evolutionary perspective, Binti's kindness, like Georgia's mischief, is most parsimoniously explained in the same way we explain our own behavior—as the result of a complex, and familiar, inner life.

FRANS DE WAAL is a professor of psychology at Emory University and research professor at the Yerkes Regional Primate Research Center in Atlanta. He is the author of several books, including *Chimpanzee Politics* and *Good Natured: The Origins of Right and Wrong in Humans and Other Animals*. His latest book, in collaboration with acclaimed wildlife photographer Frans Lanting, is *Bonobo: The Forgotten Ape*, published by the University of California Press (1997).

A Telling Difference

Animals can communicate, but evidence that any of them can emulate human language remains elusive.

STEPHEN R. ANDERSON

Doctor Dolittle, the fictional hero of Hugh Lofting's novels, was said to be able to talk with animals. Whether he, or any of his would-be imitators in real life today, could actually speak with an animal is an entirely different question.

Clearly animals can communicate; communication is virtually universal among living things. Cats meow; songbirds sing; whales call: A dog wagging its tail can be providing information about how it feels and what it wants. Surely, though, such communication is not of a piece with the languages people use. Describing any of those behaviors with the word "talk" would be nothing more than a (feeble) attempt to be clever.

Animals communicate—without language—for a variety of purposes. The vocal sac of a male frog, ready to mate, expands; a randy male sage grouse's feathers ruffle. Threats provoke communication: a cobra hisses; a porcupine bristles its quills, and a dog responds by barking and baring its teeth. Some animal communication is inadvertent, such as the scents exuded by and covering scavenging ants.

But people do use words such as "talk" to describe certain kinds of animal communication. In fact, since the 1970s many claims have been made about the potential abilities of apes to talk with people in sign languages or via other means. What is the nonspecialist to make of such claims? Are animals simply handicapped by their vocal anatomy? Could they, like Dr. Dolittle's friends, one day be taught to converse with us, via some vocal or nonvocal channel, and perhaps even to pass along their newly acquired linguistic" culture" to their offspring? Or is the evidence presented so far, purportedly in support of such

claims, at best irrelevant to them, and at worst a gross overstatement of what the data really mean?

One evening I returned home to find my wife correcting papers. When I asked her what we were doing for dinner, she said, "I want to go out." Her words left no doubt that she wanted us to go to a restaurant, where we would have dinner.

When I came home the following night, I found my cat Pooh in the kitchen. She looked at me, walked over to an oriental rug in the next room, and began to sharpen her claws on it. Pooh knows I hate that, and as I went to stop her, she ran to the sliding glass door that leads outside. I yelled at her, but my wife said, "Don't get mad; she's just saying, 'I want to go out.'"

Voila! Both my wife and my cat can say, "I want to go out." But do they both have language? Surely that is at best an oversimplification. Each can behave in such a way as to convey information to me. But the means by which they do this are radically different.

One way of approaching the distinction between communication and language is to note that communication is something we—and lots of other animals—do, whereas language is a tool that people can use to do that. People can, of course, communicate without language, though the range of information we can transmit by such means is limited. The same is true of the communication by animals and other organisms: it can transmit some fairly complex information or requests, but it still falls far short of something that is uniquely human: language.

Pychologists and others seeking to establish that such and such an animal either has language or at least has the cognitive ability to acquire language need to be able to say what language is. Their approach is commonly to make a list of characteristics and show that their animal does indeed exhibit all of them, or at least that they can teach it enough to pass a battery of tests.

That strategy poses a number of problems. Any specific set of characteristics is liable to need constant revision, because as linguists learn more about language, its characteristics change.

Nevertheless, at least two of them—the use of arbitrary symbols for nouns and verbs, and the use of syntax—are critical to assessing language ability.

Investigators exploring the cognitive capacities of other species, particularly the people who study the putative language abilities of apes, have often complained about what seems to them a double standard. E. Sue Savage-Rumbaugh, a biologist at Georgia State University in Atlanta, objects that linguists "keep raising the bar."

> First the linguists said we had to get our animals to use signs in a symbolic way if we wanted to say they learned language, OK, we did that, and then they said "No, that's not language, because you don't have syntax." So we proved our apes could produce some combinations of signs, but the linguists said that wasn't enough syntax, or the right syntax. They'll never agree that we've done enough.

But linguists are not being capricious. What gives human language its power and its centrality in our lives is its capacity to articulate a range of novel expressions, thoughts, and ideas, bounded only by imagination. In our native language, you and I can produce and understand sentences we have never encountered before. Human languages have the property of including such a potentially infinite number of distinct sentences with discrete meanings because they are organized in a hierarchical and recursive fashion. Words are not just strung out one after another. Rather, they are organized into phrases, which themselves can be constituents of larger phrases, and so on—in principle, without any limit.

To see why that property is so important, suppose language did not have any syntax. Suppose that knowing how to speak a language were really just knowing a collection of words—lots of words, perhaps, but still a finite collection. In that case, speakers could talk only about a fixed range of things—namely, what they happen to have words for. Imagine you and I were at a baseball game, and our communication were restricted, somehow; to just such a finite collection of words. If the language did not have a specific word for "the third person in from the aisle in the front row of the upper deck," without syntax I could not refer to that specific individual. I might be able to say something like "catchperson!" and point to the spectator I meant. But with the resources of English, even without special words, I can tell you that "the third person in from the aisle in the front row of the upper deck caught Bonds's home run, but the guy behind him grabbed it away from him." You can understand all that instantly and unambiguously, even if neither one of us, or the individuals referred to, are present. And the reason I can put this expression together, and that you can understand it, is that we share the syntax of English.

What gives us the power to talk about an unlimited range of things, even though we only know a fixed set of words at any one time, is our capacity for putting those words together into larger structures, the meanings of which are a function of both the meaning of the individual words and the way the words are put together. Thus we can make up new expressions of arbitrary complexity (such as the preceding sentence!) by putting together known pieces in regular ways.

Kanzi, a bonobo that can communicate via a symbol-laden keyboard, is at the center of the debate about whether nonhumans can acquire language.

Furthermore, the system of combination is recursive. What that means is that language users only need to know how to construct a limited number of different kinds of structures, because those structures can be used repeatedly as building blocks. Recursion enables speakers to build linguistic entities of unlimited complexity from a few basic patterns.

Among animals in the wild, there is simply no evidence that their communication incorporates any of these structures. Instead, communication is limited to a rather small, fixed set of "words." Vervet monkeys, for instance, distinguish among a small number of different predators (eagle, leopard, and snake) and warn their fellow monkeys with a few distinct calls of alarm. Some groups have even adapted certain calls to announce something like "unfamiliar human coming"; others have developed a call for warning of dogs accompanying human hunters. Impressive as those behaviors may be, such an augmentation of the call system happened slowly, and the system itself remains limited. What's more, vervets have no way of saying anything about "the leopard that almost sneaked up on us yesterday."

The most persuasive claims about animal language come not from observing animals in the wild, but from attempting to teach various species of great apes to communicate via sign language or electronic consoles of symbols. Perhaps the best-known such project is Francine (Penny) Patterson's effort to teach American Sign Language (ASL) to a gorilla named Koko. Koko is, by Patterson's account, the ape that "really" learned sign language, using it the way humans do—swearing, using metaphors, telling jokes, making puns. Unfortunately, we have nothing but Patterson's word for any of that. She says she has kept systematic records, but no one else has been able to study them. And without a way to assess Koko's behavior independently, the project is the best illustration imaginable of the adage that "the plural of 'anecdote' is not 'data.'"

Koko was a year old when Patterson began working with her in 1972. Patterson initially trained Koko by molding the gorilla's hands into the desired form, as she exposed the animal to whatever the sign symbolized. Koko caught on after a while and began to imitate. By the age of three and a half, Koko reportedly had acquired about a hundred signs, and by age five, almost 250.

Patterson also spoke aloud while signing, and it is reasonably clear that Koko's input was a kind of pidgin signed English rather than genuine ASL. That circumstance turns out to be a problem affecting several of the ape-language projects. Real ASL is a fully structured natural language, as expressively rich as English is. Hardly any of the investigators pursuing this research, however, have been fluent in ASL. As a result, the apes have not really been exposed to ASL, and so it is not surprising that they have come far short of learning it.

Since 1981, information about Koko has come only in forms such as *NOVA* or *National Geographic* television features, stories in the popular press, children's books, Internet chat sessions with Koko (mediated by Patterson, who acts as both interpreter and translator for the gorilla), and the ongoing public relations activities of Patterson's Gorilla Foundation. Such accounts make bold claims about how clever and articulate Koko is, but in the absence of evidence it is impossible to evaluate those claims.

And the information in the popular accounts does not inspire great confidence. Here is dialogue from a *NOVA* program filmed ten years after the start of the project (with translations for Koko's and Patterson's signing in capital letters):

KOKO: YOU KOKO LOVE DO KNEE YOU

PATTERSON: KOKO LOVE WHAT?

KOKO: LOVE THERE CHASE KNEE DO

OBSERVER: The tree, she wants to play in it!

PATTERSON: No, the girl behind the tree!

Patterson's interpretation, that Koko wanted to chase the girl behind the tree, is not self-evident, to say the least.

Sue Savage-Rumbaugh, whom I quoted above, has undertaken—along with her husband Duane Rumbaugh and others—what has proved to be the most substantial attempt so far to teach language to apes. Their subject, a bonobo named Kanzi, presents the most serious and genuine challenge to those who doubt the linguistic capacities of any nonhuman animal. Kanzi displays fascinating cognitive abilities that have not been documented before in any nonhuman primate. Yet he still falls well short of what an animal would have to do to truly acquire the structural essence of a human language.

Monkeys can say nothing about "the leopard that almost sneaked up on us yesterday."

What sets Kanzi's experience apart is that no one tried to teach him ASL or any other naturally occurring language. Instead, Savage-Rumbaugh and her team taught him a completely artificial symbol system, based on associations between meanings and arbitrary graphic designs called lexigrams. The lexigrams were available to the animal on a computer keyboard. Thus instead of issuing a series of signing gestures with his hands, Kanzi was expected to press the keys corresponding to what he (presumably) meant to say.

Actually, the research did not begin with Kanzi, but with his mother, Matata. At first, Matata was to have been trained to use the lexigram keyboard, but she turned out to be a rather poor student. The experimenters spent many long training sessions pressing lexigram keys on a keyboard connected to a computer, and indicating the intended referent. The computer responded by lighting up the key and uttering the spoken English word, but the training seemed to get nowhere.

Then something remarkable happened. Matata's infant son, Kanzi, was too young to be separated from her during the training sessions. When he was about two-and-a-half years old, however, Matata was removed to another facility for breeding. Suddenly Kanzi emerged from her shadow. Even though he had had no explicit training at all, he had learned to use the lexigram keyboard in a systematic way. He would make the natural bonobo hand-clapping gesture to provoke chasing, for instance, and then immediately hit the CHASE lexigram on the keyboard.

From then on, the focus of study became the abilities Kanzi had developed without direct instruction. In his subsequent training, the keyboard was carried around, and the trainers would press lexigrams as they spoke in English about what they and the animals were doing. While tickling Kanzi, the teacher said, "Liz is tickling Kanzi," and pressed the three keyboard keys LIZ TICKLE KANZI. Kanzi himself used the keyboard freely to express objects he wanted, places he wanted to go, and things he wanted to do. The experimenters also tested Kanzi in more structured interactions, and Kanzi could still identify objects with lexigrams and vice versa.

By the time he was about four years old, Kanzi had roughly forty-four lexigrams in his productive vocabulary, and he could recognize the corresponding spoken English words. He performed almost flawlessly on double-blind tests that required him to match pictures, lexigrams, and spoken words. He also used his lexigrarns in ways that clearly showed an extension from an initial, highly specific reference to a more generalized one. COKE, for instance, came to be used for all dark liquids, and BREAD for all kinds of bread—including taco shells.

Certainly further questions can be (and have been) raised about just what the lexigrams represent for Kanzi. Nearly all the lexigrarns for which his comprehension can be tested are associated with objects, not actions, and so it is hard to assess the richness of his internal representations of meaning. Nevertheless, the lexigrams do appear to function . as symbols, independent of specific exemplars or other contextual conditions. And there is no question that he has learned a collection of "words," in the sense that he has associated arbitrary shapes (the abstract lexigram patterns) with an arbitrary sound (the spoken English equivalent), and he has associated each of those with a meaning of some sort.

Assessing Kanzi's use of syntax is another matter. A major difficulty is that one must evaluate two different systems, those of language production and of language recognition. Kanzi's production centers on the keyboard; his recognition, on spoken English. To be sure not to underestimate

Kanzi's abilities, one must examine both systems for evidence of syntactic understanding.

Kanzi uses his keyboard, but he does not produce enough multi-lexigram sequences to permit a detailed analysis of their structure. That is not to say he does not produce complex utterances. In addition to his keyed-in lexigrams, he expresses himself with a number of natural, highly iconic gestures, with meanings such as "come," "go," and "chase." He also employs pointing gestures to designate people, and he frequently combines a lexigram with a gesture to make a complex utterance.

Such combinations, taken out of context, might look like evidence for internalized rules of syntax. Kanzi does exhibit some reliable tendencies, such as combining words in certain orders: an action word precedes an agent word, a goal precedes an action, an object precedes an agent. But the full data make a semantic analysis of those orderings beside the point, because virtually all Kanzi's complex utterances follow a single rule: lexigram first, then gesture. The combining principle is intriguing, but it is not evidence of syntax, because it has nothing to do with the role that the "words" involved play in the meaning of a communication. It is as if, in English, we wrote the first word of the sentence, spoke the second, and emailed the third. Comparing the way the words were expressed, however, would tell us nothing about their meaning.

Because of such problems in interpreting his productions, arguments for Kanzi's command of syntax rely instead on his comprehension of spoken English. Investigators compared Kanzi's understanding with that of a child named Alia, the daughter of one of Kanzi's trainers. The two were studied at a similar stage of language development, at least in terms of the size of their vocabularies and the average length of their utterances.

Both Kanzi and Alia were quite skilled at responding appropriately to requests such as "put the ball on the pine needles," "put the ice water in the potty," "give the lighter to Rose," and "take the snake outdoors." Many of the actions requested (squeezing hot dogs, washing the TV, and the like) were entirely novel, so the subjects could not succeed simply by doing what one normally does with the object named.

The range of possibilities to which both Kanzi and Alia correctly responded was broad enough to show that each of them could form a conceptual representation of an action involving one, two, or more roles—that is, words could correspond to participants in action or locations of participants or actions. Both were then able to connect information in the utterance with those roles. Kanzi is the first nonhuman to show evidence for such an ability.

Kanzi can also make connections between word order and what the words express about the world. For example, he can distinguish between the sentences "make the doggie bite the snake" and "make the snake bite the doggie." His success, at a minimum, implies he must be sensitive to regularities in word order. Such an ability is unprecedented in studies of animal cognition. Still, it does not in itself prove that Kanzi represents sentences in terms of the kind of structure that characterize human understanding of language.

Remarkably, Kanzi can distinguish "make the doggie bite the snake" from "make the snake bite the doggie."

In contrast, when the understanding of a sentence depends on "grammatical" words, such as prepositions or conjunctions, Kanzi's performance is quite poor. He does not seem to distinguish between putting something *in, on,* or *next to* something else. Sentences in which the word *and* links two nouns (as in "give the peas and the sweet potatoes to Kelly,") or two sentences (as in "go to the refrigerator and get the banana") frequently lead to mistakes that suggest Kanzi cannot interpret such words.

It is also not clear that Kanzi can understand subordinating conjunctions, such as *that* or *which*. It is true that he can correcdy respond to sentences such as: "Go get the carrot that's in the microwave." But appropriate behavior alone does not imply that he has understood the sentence as having a hierarchical structure with an embedded clause, specifying a particular carrot (the one in the microwave) rather than any other (such as one on the counter or on the floor). The content words alone ("go," "get," "carrot," "microwave") are enough to convey the command: "carrot" has to be the object of "go get," but "microwave" has no role to play in that action and can only be interpreted as a property of the carrot (its location). Kanzi needn't understand the grammar to get the right carrot.

Concrete verbs and nouns correspond to observable actions and things in the world, and they can constitute the meanings of symbols for Kanzi. Prepositions and conjunctions, however, are important because they govern how words and phrases relate to one another. Kanzi can associate lexigrams and some spoken words with parts of complex concepts in his mind, but words that are solely grammatical in content can only be ignored, because he has no grammar in which they might play a role.

What then does it mean to use a natural language? Here is a classic example: "The chickens are ready to eat." The sentence has two strikingly different interpretations, but the ambiguity has nothing to do with grammatical words, ambiguous words, or the multiple ways of organizing words into phrases. One interpretation is that someone has chickens that are ready for people to eat. The other interpretation is that some chickens are hungry.

The point is that the syntax of a language involves more than merely combining elements into sequences of words. Sentences incorporate that kind of structure, to be sure, but they also have much more structure, involving abstractions that are not readily apparent in the superficial form of the sentence—abstractions that allow the same group of words to communicate very different pieces of information in systematic ways.

As speakers of a natural language, we manage such abstractions without noticing them. But without them, language would not be the flexible instrument of expression and communication that it is. Perhaps that is the most important "take-home lesson" from the studies of animal communication. Nonhuman animals lack the kind of system that linguists are still hard at work trying to understand, and without such a capacity, animals can communicate only in much more restricted ways. The ability to use language is probably grounded in the biological nature that makes us the particular animal we are.

Reprinted from *Natural History,* November 2004, pp. 38-43, which was adapted from *Doctor Dolittle's Delusion: Animals and the Uniqueness of Human Language,* by Stephen R. Anderson. Copyright © 2004 by Stephen R. Anderson. Reprinted by permission of Yale University Press. www.yale.edu/yup/

UNIT 3
Sex and Society

Unit Selections

Key Points to Consider

- Why is friendship important to olive baboons? What implications does this have for the origins of pair-bonding in hominid evolution?

- What implications does bonobo sexual behavior have for understanding human evolution?

- How do social bonds provide females with protection against abusive males?

Student Website
www.mhcls.com/online

Internet References
Further information regarding these websites may be found in this book's preface or online.

American Anthropologist
http://www.aaanet.org/aa/index.htm

American Scientist
http://www.amsci.org/amsci.html

Bonobo Sex and Society
http://songweaver.com/info/bonobos.html

Any account of hominid evolution would be remiss if it did not at least attempt to explain that most mystifying of all human experiences—our sexuality.

No other aspect of our humanity—whether it be upright posture, tool-making ability, or intelligence in general—seems to elude our intellectual grasp at least as much as it dominates our subjective consciousness. While we are a long way from reaching a consensus as to why it arose and what it is all about, there is widespread agreement that our very preoccupation with sex is in itself one of the hallmarks of being human. Even as we experience it and analyze it, we exalt it and condemn it. Beyond seemingly irrational fixations, however, there is the further tendency to project our own values onto the observations we make and the data we collect.

There are many who argue quite reasonably that the human bias has been more male- than female-oriented and that the recent "feminization" of anthropology has resulted in new kinds of research and refreshingly new theoretical perspectives. Not only should we consider the source when evaluating the old theories, so goes the reasoning, but we should also welcome the source when considering the new.

One reason for studying the sexual and social lives of primates is that they allow us to test certain notions too often taken for granted. For instance, Barbara Smuts, in "What Are Friends For?" reveals that friendship bonds, as illustrated by the olive ba-

boons of East Africa, have little if anything to do with a sexual division of labor or even sexual exclusivity between a pair-bonded male and female. Smuts challenges the traditional male-oriented idea that primate societies are dominated solely by males and for males.

To take another example, traditional theory would have predicted that wherever male primates are larger than females, the size difference will be a major factor in sexual coercion. In actuality, says Barbara Smuts in "Apes of Wrath," such coercion is reduced in those primate species in which females are able to form alliances against male aggression. That this is not so, that making love can be more important than making war, and that females do not necessarily have to live in fear of competitive males, just goes to show that, even among monkeys, nothing can be taken for granted.

Finally, there is the question of the social significance of sexuality in humans. In "What's Love Got to Do With It?" Meredith Small shows that the chimp-like bonobos of Zaire use sex to reduce tensions and cement social relations and, in so doing, have achieved a high degree of equality between the sexes. Whether we see parallels in the human species, says Small, depends on our willingness to interpret bonobo behavior as a "modern version of our own ancestors' sex play," and this, in turn, may depend on our prior theoretical commitments.

What Are Friends For?

Among East African baboons, friendship means companions, health, safety… and, sometimes, sex

BARBARA SMUTS

Virgil, a burly adult male olive baboon, closely followed Zizi, a middle-aged female easily distinguished by her grizzled coat and square muzzle. On her rump Zizi sported a bright pink swelling, indicating that she was sexually receptive and probably fertile. Virgil's extreme attentiveness to Zizi suggested to me—and all rival males in the troop—that he was her current and exclusive mate.

Zizi, however, apparently had something else in mind. She broke away from Virgil, moved rapidly through the troop, and presented her alluring sexual swelling to one male after another. Before Virgil caught up with her, she had managed to announce her receptive condition to several of his rivals. When Virgil tried to grab her, Zizi screamed and dashed into the bushes with Virgil in hot pursuit. I heard sounds of chasing and fighting coming from the thicket. Moments later Zizi emerged from the bushes with an older male named Cyclops. They remained together for several days, copulating often. In Cyclops's presence, Zizi no longer approached or even glanced at other males.

Primatologists describe Zizi and other olive baboons (*Papio cynocephalus anubis*) as promiscuous, meaning that both males and females usually mate with several members of the opposite sex within a short period of time. Promiscuous mating behavior characterizes many of the larger, more familiar primates, including chimpanzees, rhesus macaques, and gray langurs, as well as olive, yellow, and chacma baboons, the three subspecies of savanna baboon. In colloquial usage, promiscuity often connotes wanton and random sex, and several early studies of primates supported this stereotype. However, after years of laboriously recording thousands of copulations under natural conditions, the Peeping Toms of primate fieldwork have shown that, even in promiscuous species, sexual pairings are far from random.

Some adult males, for example, typically copulate much more often than others. Primatologists have explained these differences in terms of competition: the most dominant males monopolize females and prevent lower-ranking rivals from mating. But exceptions are frequent. Among baboons, the exceptions often involve scruffy, older males who mate in full view of younger, more dominant rivals.

A clue to the reason for these puzzling exceptions emerged when primatologists began to question an implicit assumption of the dominance hypothesis—that females were merely passive objects of male competition. But what if females were active arbiters in this system? If females preferred some males over others and were able to express these preferences, then models of mating activity based on male dominance alone would be far too simple.

Once researchers recognized the possibility of female choice, evidence for it turned up in species after species. The story of Zizi, Virgil, and Cyclops is one of hundreds of examples of female primates rejecting the sexual advances of particular males and enthusiastically cooperating with others. But what is the basis for female choice? Why might they prefer some males over others?

This question guided my research on the Eburru Cliffs troop of olive baboons, named after one of their favorite sleeping sites, a sheer rocky outcrop rising several hundred feet above the floor of the Great Rift Valley, about 100 miles northwest of Nairobi, Kenya. The 120 members of Eburru Cliffs spent their days wandering through open grassland studded with occasional acacia thorn trees. Each night they retired to one of a dozen sets of cliffs that provided protection from nocturnal predators such as leopards.

Most previous studies of baboon sexuality had focused on females who, like Zizi, were at the peak of sexual receptivity. A female baboon does not mate when she is pregnant or lactating, a period of abstinence lasting about eighteen months. The female then goes into estrus, and for about two weeks out of every thirty-five-day cycle, she mates. Toward the end of this two-week period she may ovulate, but usually the female undergoes four or five estrous cycles before she conceives. During pregnancy, she once again resumes a chaste existence. As a result, the typical female baboon is sexually active for less than 10 percent of her adult life. I thought that by focusing on the other 90 percent, I might learn something new. In particular, I suspected that routine, day-to-day relationships between males and pregnant or lactating (nonestrous) females might provide clues to female mating preferences.

Nearly every day for sixteen months, I joined the Eburru Cliffs baboons at their sleeping cliffs at dawn and traveled several miles with them while they foraged for roots, seeds, grass, and occasionally, small prey items, such as baby gazelles or hares (see "Predatory Baboons of Kekopey," *Natural History*, March 1976). Like all savanna baboon troops, Eburru Cliffs functioned as a cohesive unit organized around a core of related females, all of whom were born in the troop. Unlike the females, male savanna baboons leave their natal troop to join another where they may remain for many years, so most of the Eburru Cliffs adult males were immigrants. Since membership in the troop remained relatively constant during the period of my study, I learned to identify each individual. I relied on differences in size, posture, gait, and especially, facial features. To the practiced observer, baboons look as different from one another as human beings do.

As soon as I could recognize individuals, I noticed that particular females tended to turn up near particular males again and again. I came to think of these pairs as friends. Friendship among animals is not a well-documented phenomenon, so to convince skeptical colleagues that baboon friendship was real, I needed to develop objective criteria for distinguishing friendly pairs.

I began by investigating grooming, the amiable simian habit of picking through a companion's fur to remove dead skin and ectoparasites (see "Little Things That Tick Off Baboons," *Natural History*, February 1984). Baboons spend much more time grooming than is necessary for hygiene, and previous research had indicated that it is a good measure of social bonds.

Although eighteen adult males lived in the troop, each nonestrous female performed most of her grooming with just one, two, or occasionally, three males. For example, of Zizi's twenty-four grooming bouts with males, Cyclops accounted for thirteen, and a second male, Sherlock, accounted for all the rest. Different females tended to favor different males as grooming partners.

Another measure of social bonds was simply who was observed near whom. When foraging, traveling, or resting, each pregnant or lactating female spent a lot of time near a few males and associated with the others no more often than expected by chance. When I compared the identities of favorite grooming partners and frequent companions, they overlapped almost completely. This enabled me to develop a formal definition of friendship: any male that scored high on both grooming and proximity measures was considered a friend.

Virtually all baboons made friends; only one female and three males who had most recently joined the troop lacked such companions. Out of more than 600 possible adult female-adult male pairs in the troop, however, only about one in ten qualified as friends; these really were special relationships.

Several factors seemed to influence which baboons paired up. In most cases, friends were unrelated to each other, since the male had immigrated from another troop. (Four friendships, however, involved a female and an adolescent son who had not yet emigrated. Unlike other friends, these related pairs never mated.) Older females tended to be friends with older males; younger females with younger males. I witnessed occasional May–December romances, usually involving older females and

young adult males. Adolescent males and females were strongly rule-bound, and with the exception of mother-son pairs, they formed friendships only with one another.

Regardless of age or dominance rank, most females had just one or two male friends. But among males, the number of female friends varied greatly from none to eight. Although high-ranking males enjoyed priority of access to food and sometimes mates, dominant males did not have more female friends than low-ranking males. Instead it was the older males who had lived in the troop for many years who had the most friends. When a male had several female friends, the females were often closely related to one another. Since female baboons spend a lot of time near their kin, it is probably easier for a male to maintain bonds with several related females at once.

When collecting data, I focused on one nonestrous female at a time and kept track of her every movement toward or away from any male; similarly, I noted every male who moved toward or away from her. Whenever the female and male moved close enough to exchange intimacies, I wrote down exactly what happened. When foraging together, friends tended to remain a few yards apart. Males more often wandered away from females than the reverse, and females, more often than males, closed the gap. The female behaved as if she wanted to keep the male within calling distance, in case she needed his protection. The male, however, was more likely to make approaches that brought them within actual touching distance. Often, he would plunk himself down right next to his friend and ask her to groom him by holding a pose with exaggerated stillness. The female sometimes responded by grooming, but more often, she exhibited the most reliable sign of true intimacy: she ignored her friend and simply continued whatever she was doing.

In sharp contrast, when a male who was not a friend moved close to a female, she dared not ignore him. She stopped whatever she was doing and held still, often glancing surreptitiously at the intruder. If he did not move away, she sometimes lifted her tail and presented her rump. When a female is not in estrus, this is a gesture of appeasement, not sexual enticement. Immediately after this respectful acknowledgement of his presence, the female would slip away. But such tense interactions with nonfriend males were rare, because females usually moved away before the males came too close.

These observations suggest that females were afraid of most of the males in their troop, which is not surprising: male baboons are twice the size of females, and their canines are longer and sharper than those of a lion. All Eburru Cliffs males directed both mild and severe aggression toward females. Mild aggression, which usually involved threats and chases but no body contact, occurred most often during feeding competition or when the male redirected aggression toward a female after losing a fight with another male. Females and juveniles showed aggression toward other females and juveniles in similar circumstances and occasionally inflicted superficial wounds. Severe aggression by males, which involved body contact and sometimes biting, was less common and also more puzzling, since there was no apparent cause.

An explanation for at least some of these attacks emerged one day when I was watching Pegasus, a young adult male, and his friend Cicily, sitting together in the middle of a small clearing. Cicily moved to the edge of the clearing to feed, and a higher-ranking female, Zora, suddenly attacked her. Pegasus stood up and looked as if he were about to intervene when both females disappeared into the bushes. He sat back down, and I remained with him. A full ten minutes later, Zora appeared at the edge of the clearing; this was the first time she had come into view since her attack on Cicily. Pegasus instantly pounced on Zora, repeatedly grabbed her neck in his mouth and lifted her off the ground, shook her whole body, and then dropped her. Zora screamed continuously and tried to escape. Each time, Pegasus caught her and continued his brutal attack. When he finally released her five minutes later she had a deep canine gash on the palm of her hand that made her limp for several days.

This attack was similar in form and intensity to those I had seen before and labeled "unprovoked." Certainly, had I come upon the scene after Zora's aggression toward Cicily, I would not have understood why Pegasus attacked Zora. This suggested that some, perhaps many, severe attacks by males actually represented punishment for actions that had occurred some time before.

Whatever the reasons for male attacks on females, they represent a serious threat. Records of fresh injuries indicated that Eburru Cliffs adult females received canine slash wounds from males at the rate of one for every female each year, and during my study, one female died of her injuries. Males probably pose an even greater threat to infants. Although only one infant was killed during my study, observers in Botswana and Tanzania have seen recent male immigrants kill several young infants.

Protection from male aggression, and from the less injurious but more frequent aggression of other females and juveniles, seems to be one of the main advantages of friendship for a female baboon. Seventy times I observed an adult male defend a female or her offspring against aggression by another troop member, not infrequently a high-ranking male. In all but six of these cases, the defender was a friend. Very few of these confrontations involved actual fighting; no male baboon, subordinate or dominant, is anxious to risk injury by the sharp canines of another.

Males are particularly solicitous guardians of their friends' youngest infants. If another male gets too close to an infant or if a juvenile female plays with it too roughly, the friend may intervene. Other troop members soon learn to be cautious when the mother's friend is nearby, and his presence provides the mother with a welcome respite from the annoying pokes and prods of curious females and juveniles obsessed with the new baby. Male baboons at Gombe Park in Tanzania and Amboseli Park in Kenya have also been seen rescuing infants from chimpanzees and lions. These several forms of male protection help to explain why females in Eburru Cliffs stuck closer to their friends in the first few months after giving birth than at any other time.

The male-infant relationship develops out of the male's friendship with the mother, but as the infant matures, this new bond takes on a life of its own. My co-worker Nancy Nicolson found that by about nine months of age, infants actively sought out their male friends when the mother was a few yards away,

suggesting that the male may function as an alternative caregiver. This seemed to be especially true for infants undergoing unusually early or severe weaning. (Weaning is generally a gradual, prolonged process, but there is tremendous variation among mothers in the timing and intensity of weaning. See "Mother Baboons," *Natural History*, September 1980). After being rejected by the mother, the crying infant often approached the male friend and sat huddled against him until its whimpers subsided. Two of the infants in Eburru Cliffs lost their mothers when they were still quite young. In each case, their bond with the mother's friend subsequently intensified, and—perhaps as a result—both infants survived.

A close bond with a male may also improve the infant's nutrition. Larger than all other troop members, adult males monopolize the best feeding sites. In general, the personal space surrounding a feeding male is inviolate, but he usually tolerates intrusions by the infants of his female friends, giving them access to choice feeding spots.

Although infants follow their male friends around rather than the reverse, the males seem genuinely attached to their tiny companions. During feeding, the male and infant express their pleasure in each other's company by sharing spirited, antiphonal grunting duets. If the infant whimpers in distress, the male friend is likely to cease feeding, look at the infant, and grunt softly, as if in sympathy, until the whimpers cease. When the male rests, the infants of his female friends may huddle behind him, one after the other, forming a "train," or, if feeling energetic, they may use his body as a trampoline.

When I returned to Eburru Cliffs four years after my initial study ended, several of the bonds formed between males and the infants of their female friends were still intact (in other cases, either the male or the infant or both had disappeared). When these bonds involved recently matured females, their long-time male associates showed no sexual interest in them, even though the females mated with other adult males. Mothers and sons, and usually maternal siblings, show similar sexual inhibitions in baboons and many other primate species.

The development of an intimate relationship between a male and the infant of his female friend raises an obvious question: Is the male the infant's father? To answer this question definitely we would need to conduct genetic analysis, which was not possible for these baboons. Instead, I estimated paternity probabilities from observations of the temporary (a few hours or days) exclusive mating relationships, or consortships, that estrous females form with a series of different males. These estimates were apt to be fairly accurate, since changes in the female's sexual swelling allow one to pinpoint the timing of conception to within a few days. Most females consorted with only two or three males during this period, and these males were termed likely fathers.

In about half the friendships, the male was indeed likely to be the father of his friend's most recent infant, but in the other half he was not—in fact, he had never been seen mating with the female. Interestingly, males who were friends with the mother but not likely fathers nearly always developed a relationship with her infant, while males who had mated with the female but were not her friend usually did not. Thus friendship with the mother, rather than paternity, seems to mediate the development of male-infant bonds. Recently, a similar pattern was documented for South American capuchin monkeys in a laboratory study in which paternity was determined genetically.

These results fly in the face of a prominent theory that claims males will invest in infants only when they are closely related. If males are not fostering the survival of their own genes by caring for the infant, then why do they do so? I suspected that the key was female choice. If females preferred to mate with males who had already demonstrated friendly behavior, then friendships with mothers and their infants might pay off in the future when the mothers were ready to mate again.

To find out if this was the case, I examined each male's sexual behavior with females he had befriended before they resumed estrus. In most cases, males consorted considerably more often with their friends than with other females. Baboon females typically mate with several different males, including both friends and nonfriends, but prior friendship increased a male's probability of mating with a female above what it would have been otherwise.

This increased probability seemed to reflect female preferences. Females occasionally overtly advertised their disdain for certain males and their desire for others. Zizi's behavior, described above, is a good example. Virgil was not one of her friends, but Cyclops was. Usually, however, females expressed preferences and aversions more subtly. For example, Delphi, a petite adolescent female, found herself pursued by Hector, a middle-aged adult male. She did not run away or refuse to mate with him, but whenever he wasn't watching, she looked around for her friend Homer, an adolescent male. When she succeeded in catching Homer's eye, she narrowed her eyes and flattened her ears against her skull, the friendliest face one baboon can send another. This told Homer she would rather be with him. Females expressed satisfaction with a current consort partner by staying close to him, initiating copulations, and not making advances toward other males. Baboons are very sensitive to such cues, as indicated by an experimental study in which rival hamadryas baboons rarely challenged a male-female pair if the female strongly preferred her current partner. Similarly, in Eburru Cliffs, males were less apt to challenge consorts involving a pair that shared a long-term friendship.

Even though females usually consorted with their friends, they also mated with other males, so it is not surprising that friendships were most vulnerable during periods of sexual activity. In a few cases, the female consorted with another male more often than with her friend, but the friendship survived nevertheless. One female, however, formed a strong sexual bond with a new male. This bond persisted after conception, replacing her previous friendship. My observations suggest that adolescent and young adult females tend to have shorter, less stable friendships than do older females. Some friendships, however, last a very long time. When I returned to Eburru Cliffs six years after my study began, five couples were still together. It is possible that friendships occasionally last for life (baboons probably live twenty to thirty years in the wild), but it will require longer studies, and some very patient scientists to find out.

By increasing both the male's chances of mating in the future and the likelihood that a female's infant will survive, friendship contributes to the reproductive success of both partners. This clarifies the evolutionary basis of friendship-forming tendencies in baboons, but what does friendship mean to a baboon? To answer this question we need to view baboons as sentient beings with feelings and goals not unlike our own in similar circumstances. Consider, for example, the friendship between Thalia and Alexander.

The affair began one evening as Alex and Thalia sat about fifteen feet apart on the sleeping cliffs. It was like watching two novices in a singles bar. Alex stared at Thalia until she turned and almost caught him looking at her. He glanced away immediately, and then she stared at him until his head began to turn toward her. She suddenly became engrossed in grooming her toes. But as soon as Alex looked away, her gaze returned to him. They went on like this for more than fifteen minutes, always with split-second timing. Finally, Alex managed to catch Thalia looking at him. He made the friendly eyes-narrowed, ears-back face and smacked his lips together rhythmically. Thalia froze, and for a second she looked into his eyes. Alex approached, and Thalia, still nervous, groomed him. Soon she calmed down, and I found them still together on the cliffs the next morning. Looking back on this event months later, I realized that it marked the beginning of their friendship. Six years later, when I returned to Eburru Cliffs, they were still friends.

If flirtation forms an integral part of baboon friendship, so does jealousy. Overt displays of jealousy, such as chasing a friend away from a potential rival, occur occasionally, but like humans, baboons often express their emotions in more subtle ways. One evening a colleague and I climbed the cliffs and settled down near Sherlock, who was friends with Cybelle, a middle-aged female still foraging on the ground below the cliffs. I observed Cybelle while my colleague watched Sherlock, and we kept up a running commentary. As long as Cybelle was feeding or interacting with females, Sherlock was relaxed, but each time she approached another male, his body would stiffen, and he would stare intently at the scene below. When Cybelle presented politely to a male who had recently tried to befriend her, Sherlock even made threatening sounds under his breath. Cybelle was not in estrus at the time, indicating that male baboon jealousy extends beyond the sexual arena to include affiliative interactions between a female friend and other males.

Because baboon friendships are embedded in a network of friendly and antagonistic relationships, they inevitably lead to repercussions extending beyond the pair. For example, Virgil once provoked his weaker rival Cyclops into a fight by first attacking Cyclops's friend Phoebe. On another occasion, Sherlock chased Circe, Hector's best friend, just after Hector had chased Antigone, Sherlock's friend.

In another incident, the prime adult male Triton challenged Cyclops's possession of meat. Cyclops grew increasingly tense and seemed about to abandon the prey to the younger male. Then Cyclops's friend Phoebe appeared with her infant Phyllis. Phyllis wandered over to Cyclops. He immediately grabbed her, held her close, and threatened Triton away from the prey. Because any challenge to Cyclops now involved a threat to Phyllis as well, Triton risked being mobbed by Phoebe and her relatives and friends. For this reason, he backed down. Males frequently use the infants of their female friends as buffers in this way. Thus, friendship involves costs as well as benefits because it makes the participants vulnerable to social manipulation or redirected aggression by others.

Finally, as with humans, friendship seems to mean something different to each baboon. Several females in Eburru Cliffs had only one friend. They were devoted companions. Louise and Pandora, for example, groomed their friend Virgil and no other male. Then there was Leda, who, with five friends, spread herself more thinly than any other female. These contrasting patterns of friendship were associated with striking personality differences. Louise and Pandora were unobtrusive females who hung around quietly with Virgil and their close relatives. Leda seemed to be everywhere at once, playing with infants, fighting with juveniles, and making friends with males. Similar differences were apparent among the males. Some devoted a great deal of time and energy to cultivating friendships with females, while others focused more on challenging other males. Although we probably will never fully understand the basis of these individual differences, they contribute immeasurably to the richness and complexity of baboon society.

Male-female friendships may be widespread among primates. They have been reported for many other groups of savanna baboons, and they also occur in rhesus and Japanese Macaques, capuchin monkeys, and perhaps in bonobos (pygmy chimpanzees). These relationships should give us pause when considering popular scenarios for the evolution of male-female relationships in humans. Most of these scenarios assume that, except for mating, males and females had little to do with one another until the development of a sexual division of labor, when, the story goes, females began to rely on males to provide meat in exchange for gathered food. This, it has been argued, set up new selection pressures favoring the development of long-term bonds between individual males and females, female sexual fidelity, and as paternity certainty increased, greater male investment in the offspring of these unions. In other words, once women began to gather and men to hunt, presto—we had the nuclear family.

This scenario may have more to do with cultural biases about women's economic dependence on men and idealized views of the nuclear family than with the actual behavior of our hominid ancestors. The nonhuman primate evidence challenges this story in at least three ways.

First, long-term bonds between the sexes can evolve in the absence of a sexual division of labor of food sharing. In our primate relatives, such relationships rest on exchanges of social, not economic, benefits.

Second, primate research shows that highly differentiated, emotionally intense male-female relationships can occur without sexual exclusivity. Ancestral men and women may have experienced intimate friendships long before they invented marriage and norms of sexual fidelity.

Third, among our closest primate relatives, males clearly provide mothers and infants with social benefits even when they are unlikely to be the fathers of those infants. In return, females provide a variety of benefits to the friendly males, including acceptance into the group and, at least in baboons, increased mating opportunities in the future. This suggests that efforts to reconstruct the evolution of hominid societies may have overemphasized what the female must supposedly do (restrict her mating to just one male) in order to obtain male parental investment.

Maybe it is time to pay more attention to what the male must do (provide benefits to females and young) in order to obtain female cooperation. Perhaps among our ancestors, as in baboons today, sex and friendship went hand in hand. As for marriage—well, that's another story.

What's Love Got to Do With It?

Sex Among Our Closest Relatives Is a Rather Open Affair

MEREDITH F. SMALL

Maiko and Lana are having sex. Maiko is on top, and Lana's arms and legs are wrapped tightly around his waist. Lina, a friend of Lana's, approaches from the right and taps Maiko on the back, nudging him to finish. As he moves away, Lina enfolds Lana in her arms, and they roll over so that Lana is now on top. The two females rub their genitals together, grinning and screaming in pleasure.

This is no orgy staged for an X-rated movie. It doesn't even involve people—or rather, it involves them only as observers. Lana, Maiko, and Lina are bonobos, a rare species of chimplike ape in which frequent couplings and casual sex play characterize every social relationship—between males and females, members of the same sex, closely related animals, and total strangers. Primatologists are beginning to study the bonobos' unrestrained sexual behavior for tantalizing clues to the origins of our own sexuality.

In reconstructing how early man and woman behaved, researchers have generally looked not to bonobos but to common chimpanzees. Only about 5 million years ago human beings and chimps shared a common ancestor, and we still have much behavior in common: namely, a long period of infant dependency, a reliance on learning what to eat and how to obtain food, social bonds that persist over generations, and the need to deal as a group with many everyday conflicts. The assumption has been that chimp behavior today may be similar to the behavior of human ancestors.

Bonobo behavior, however, offers another window on the past because they, too, shared our 5-million-year-old ancestor, diverging from chimps just 2 million years ago. Bonobos have been less studied than chimps for the simple reason that they are difficult to find. They live only on a small patch of land in Zaire, in central Africa. They were first identified, on the basis of skeletal material, in the 1920s, but it wasn't until the 1970s that their behavior in the wild was studied, and then only sporadically.

Bonobos, also known as pygmy chimpanzees, are not really pygmies but welterweights. The largest males are as big as chimps, and the females of the two species are the same size. But bonobos are more delicate in build, and their arms and legs are long and slender.

On the ground, moving from fruit tree to fruit tree, bonobos often stand and walk on two legs—behavior that makes them seem more like humans than chimps. In some ways their sexual behavior seems more human as well, suggesting that in the sexual arena, at least, bonobos are the more appropriate ancestral model. Males and females frequently copulate face-to-face, which is an uncommon position in animals other than humans. Males usually mount females from behind, but females seem to prefer sex face-to-face. "Sometimes the female will let a male start to mount from behind," says Amy Parish, a graduate student at the University of California at Davis who's been watching female bonobo sexual behavior in several zoo colonies around the world. "And then she'll stop, and of course he's really excited, and then she continues face-to-face." Primatologists assume the female preference is dictated by her anatomy: her enlarged clitoris and sexual swellings are oriented far forward. Females presumably prefer face-to-face contact because it feels better.

"Sex is fun. Sex makes them feel good and keeps the group together."

Like humans but unlike chimps and most other animals, bonobos separate sex from reproduction. They seem to treat sex as a pleasurable activity, and they rely on it as a sort of social glue, to make or break all sorts of relationships. "Ancestral humans behaved like this," proposes Frans de Waal, an ethologist at the Yerkes Regional Primate Research Center at Emory University. "Later, when we developed the family system, the use of sex for this sort of purpose became more limited, mainly occurring within families. A lot of the things we see, like pedophilia and homosexuality, may be leftovers that some now consider unacceptable in our particular society."

Depending on your morals, watching bonobo sex play may be like watching humans at their most extreme and perverse. Bonobos seem to have sex more often and in more combinations than the average person in any culture, and most of the time bonobo sex has nothing to do with making babies. Males mount females and females sometimes mount them back; females rub against other females just for fun; males stand rump to rump and press their scrotal areas together. Even juveniles participate by rubbing their genital areas against adults, although ethologists don't think that males actually insert their penises into juvenile females. Very young animals also have sex with each other: little males suck on each other's penises or French-kiss. When two animals initiate sex, others freely join in by poking their fingers and toes into the moving parts.

One thing sex does for bonobos is decrease tensions caused by potential competition, often competition for food. Japanese primatologists observing bonobos in Zaire were the first to notice that when bonobos come across a large fruiting tree or encounter piles of provisioned sugarcane, the sight of food triggers a binge of sex. The atmosphere of this sexual free-for-all is decidedly friendly, and it eventually calms the group down. "What's striking is how rapidly the sex drops off," says Nancy Thompson-Handler of the State University of New York at Stony Brook, who has observed bonobos at a site in Zaire called Lomako. "After ten minutes, sexual behavior decreases by fifty percent." Soon the group turns from sex to feeding.

But it's tension rather than food that causes the sexual excitement. "I'm sure the more food you give them, the more sex you'll get," says De Waal. "But it's not really the food, it's competition that triggers this. You can throw in a cardboard box and you'll get sexual behavior." Sex is just the way bonobos deal with competition over limited resources and with the normal tensions caused by living in a group. Anthropologist Frances White of Duke University, a bonobo observer at Lomako since 1983, puts it simply: "Sex is fun. Sex makes them feel good and therefore keeps the group together."

"Females rule the business. It's a good species for feminists, I think."

Sexual behavior also occurs after aggressive encounters, especially among males. After two males fight, one may reconcile with his opponent by presenting his rump and backing up against the other's testicles. He might grab the penis of the other male and stroke it. It's the male bonobo's way of shaking hands and letting everyone know that the conflict has ended amicably.

Researchers also note that female bonobo sexuality, like the sexuality of female humans, isn't locked into a monthly cycle.

In most other animals, including chimps, the female's interest in sex is tied to her ovulation cycle. Chimp females sport pink swellings on their hind ends for about two weeks, signaling their fertility, and they're only approachable for sex during that time. That's not the case with humans, who show no outward signs that they are ovulating, and can mate at all phases of the cycle. Female bonobos take the reverse tack, but with similar results. Their large swellings are visible for weeks before and after their fertile periods, and there is never any discernibly wrong time to mate. Like humans, they have sex whether or not they are ovulating.

What's fascinating is that female bonobos use this boundless sexuality in all their relationships. "Females rule the business—sex and food," says De Waal. "It's a good species for feminists, I think." For instance, females regularly use sex to cement relationships with other females. A genital-genital rub, better known as GG-rubbing by observers, is the most frequent behavior used by bonobo females to reinforce social ties or relieve tension. GG-rubbing takes a variety of forms. Often one female rolls on her back and extends her arms and legs. The other female mounts her and they rub their swellings right and left for several seconds, massaging their clitorises against each other. GG-rubbing occurs in the presence of food because food causes tension and excitement, but the intimate contact has the effect of making close friends.

Sometimes females would rather GG-rub with each other than copulate with a male. Parish filmed a 15-minute scene at a bonobo colony at the San Diego Wild Animal Park in which a male, Vernon, repeatedly solicited two females, Lisa and Loretta. Again and again he arched his back and displayed his erect penis—the bonobo request for sex. The females moved away from him, tactfully turning him down until they crept behind a tree and GG-rubbed with each other.

Unlike most primate species, in which males usually take on the dangerous task of leaving home, among bonobos females are the ones who leave the group when they reach sexual maturity, around the age of eight, and work their way into unfamiliar groups. To aid in their assimilation into a new community, the female bonobos make good use of their endless sexual favors. While watching a bonobo group at a feeding tree, White saw a young female systematically have sex with each member before feeding. "An adolescent female, presumably a recent transfer female, came up to the tree, mated with all five males, went into the tree, and solicited GG-rubbing from all the females present," says White.

Once inside the new group, a female bonobo must build a sisterhood from scratch. In groups of humans or chimps, unrelated females construct friendships through the rituals of shopping together or grooming. Bonobos do it sexually. Although

Hidden Heat

Standing upright is not a position usually—or easily—associated with sex. Among people, at least, anatomy and gravity prove to be forbidding obstacles. Yet our two-legged stance may be the key to a distinctive aspect of human sexuality: the independence of women's sexual desires from a monthly calendar.

Males in the two species most closely related to us, chimpanzees and bonobos, don't spend a lot of time worrying, "Is she interested or not?" The answer is obvious. When ovulatory hormones reach a monthly peak in female chimps and bonobos, and their eggs are primed for fertilization, their genital area swells up, and both sexes appear to have just one thing on their mind. "These animals really turn on when this happens. Everything else is dropped," says primatologist Frederick Szalay of Hunter College in New York.

Women, however, don't go into heat. And this departure from our relatives' sexual behavior has long puzzled researchers. Clear signals of fertility and the willingness to do something about it bring major evolutionary advantages: ripe eggs lead to healthier pregnancies, which leads to more of your genes in succeeding generations, which is what evolution is all about. In addition, male chimps give females that are waving these red flags of fertility first chance at high-protein food such as meat.

So why would our ancestors give this up? Szalay and graduate student Robert Costello have a simple explanation. Women gave heat up, they say, because our ancestors stood up.

Fossil footprints indicate that somewhere around 3.5 million years ago hominids—non-ape primates—began walking on two legs. "In hominids, something dictated getting up. We don't know what it was," Szalay says. "But once it did, there was a problem with the signaling system." The problem was that it didn't work. Swollen genital areas that were visible when their owners were down on all fours became hidden between the legs. The mating signal was lost.

"Uprightness meant very tough times for females working with the old ovarian cycle," Szalay says. Males wouldn't notice them, and the swellings themselves, which get quite large, must have made it hard for two-legged creatures to walk around.

Those who found a way out of this quandary, Szalay suggests, were females with small swellings but with a little less hair on their rears and a little extra fat. It would have looked a bit like the time-honored mating signal. They got more attention, and produced more offspring. "You don't start a completely new trend in signaling," Szalay says. "You have a little extra fat, a little nakedness to mimic the ancestors. If there was an ever-so-little advantage because, quite simply, you look good, it would be selected for."

And if a little nakedness and a little fat worked well, Szalay speculates, then a lot of both would work even better. "Once you start a trend in sexual signaling, crazy things happen," he notes. "It's almost like: let's escalate, let's add more. That's what happens in horns with sheep. It's a

particular part of the body that brings an advantage." In a few million years human ancestors were more naked than ever, with fleshy rears not found in any other primate. Since these features were permanent, unlike the monthly ups and downs of swellings, sex was free to become a part of daily life.

It's a provocative notion, say Szalay's colleagues, but like any attempt to conjure up the past from the present, there's no real proof of cause and effect. Anthropologist Helen Fisher of the American Museum of Natural History notes that Szalay is merely assuming that fleshy buttocks evolved because they were sex signals. Yet their mass really comes from muscles, which chimps don't have, that are associated with walking. And anthropologist Sarah Blaffer Hrdy of the University of California at Davis points to a more fundamental problem: our ancestors may not have had chimplike swellings that they needed to dispense with. Chimps and bonobos are only two of about 200 primate species, and the vast majority of those species don't have big swellings. Though they are our closest relatives, chimps and bonobos have been evolving during the last 5 million years just as we have, and swollen genitals may be a recent development. The current unswollen human pattern may be the ancestral one.

"Nobody really knows what happened," says Fisher. "Everybody has an idea. You pays your money and you takes your choice."

—Joshua Fischman

pleasure may be the motivation behind a female-female assignation, the function is to form an alliance.

These alliances are serious business, because they determine the pecking order at food sites. Females with powerful friends eat first, and subordinate females may not get any food at all if the resource is small. When times are rough, then, it pays to have close female friends. White describes a scene at Lomako in which an adolescent female, Blanche, benefited from her established friendship with Freda. "I was following Freda and her boyfriend, and they found a tree that they didn't expect to be

there. It was a small tree, heavily in fruit with one of their favorites. Freda went straight up the tree and made a food call to Blanche. Blanche came tearing over—she was quite far away—and went tearing up the tree to join Freda, and they GG-rubbed like crazy."

Alliances also give females leverage over larger, stronger males who otherwise would push them around. Females have discovered there is strength in numbers. Unlike other species of primates, such as chimpanzees or baboons (or, all too often, humans), where tensions run high between males and females, bonobo females are not afraid of males, and the sexes mingle peacefully. "What is consistently different from chimps," says Thompson-Handler, "is the composition of parties. The vast majority are mixed, so there are males and females of all different ages."

Female bonobos cannot be coerced into anything, including sex. Parish recounts an interaction between Lana and a male called Akili at the San Diego Wild Animal Park. "Lana had just been introduced into the group. For a long time she lay on the grass with a huge swelling. Akili would approach her with a big erection and hover over her. It would have been easy for him to do a mount. But he wouldn't. He just kept trying to catch her eye, hovering around her, and she would scoot around the ground, avoiding him. And then he'd try again. She went around full circle." Akili was big enough to force himself on her. Yet he refrained.

In another encounter, a male bonobo was carrying a large clump of branches. He moved up to a female and presented his erect penis by spreading his legs and arching his back. She rolled onto her back and they copulated. In the midst of their joint ecstasy, she reached out and grabbed a branch from the male. When he pulled back, finished and satisfied, she moved away, clutching the branch to her chest. There was no tension between them, and she essentially traded copulation for food. But the key here is that the male allowed her to move away with the branch—it didn't occur to him to threaten her, because their status was virtually equal.

Although the results of sexual liberation are clear among bonobos, no one is sure why sex has been elevated to such a high position in this species and why it is restricted merely to reproduction among chimpanzees. "The puzzle for me," says De Waal, "is that chimps do all this bonding with kissing and embracing, with body contact. Why do bonobos do it in a sexual manner?" He speculates that the use of sex as a standard way to underscore relationships began between adult males and adult females as an extension of the mating process and later spread to all members of the group. But no one is sure exactly how this happened.

It is also unclear whether bonobo sexually became exaggerated only after their split from the human lineage or whether the behavior they exhibit today is the modern version of our common ancestor's sex play. Anthropologist Adrienne Zihlman of the University of California at Santa Cruz, who has used the evidence of fossil bones to argue that our earliest known non-ape ancestors, the australopithecines, had body proportions similar to those of bonobos, says, "The path of evolution is not a straight line from either species, but what I think is important is that the bonobo information gives us more possibilities for looking at human origins."

Some anthropologists, however, are reluctant to include the details of bonobo life, such as wide-ranging sexuality and a strong sisterhood, into scenarios of human evolution. "The researchers have all these commitments to male dominance [as in chimpanzees], and yet bonobos have egalitarian relationships," says De Waal. "They also want to see humans as unique, yet bonobos fit very nicely into many of the scenarios, making humans appear less unique."

Our divergent, non-ape path has led us away from sex and toward a culture that denies the connection between sex and social cohesion. But bonobos, with their versatile sexuality, are here to remind us that our heritage may very well include a primordial urge to make love, not war.

Apes of Wrath

BARBARA SMUTS

Nearly 20 years ago I spent a morning dashing up and down the hills of Gombe National Park in Tanzania, trying to keep up with an energetic young female chimpanzee, the focus of my observations for the day. On her rear end she sported the small, bright pink swelling characteristic of the early stages of estrus, the period when female mammals are fertile and sexually receptive. For some hours our run through the park was conducted in quiet, but then, suddenly, a chorus of male chimpanzee pant hoots shattered the tranquility of the forest. My female rushed forward to join the males. She greeted each of them, bowing and then turning to present her swelling for inspection. The males examined her perfunctorily and resumed grooming one another, showing no further interest.

Some female primates use social bonds to escape male aggression. Can women?

At first I was surprised by their indifference to a potential mate. Then I realized that it would be many days before the female's swelling blossomed into the large, shiny sphere that signals ovulation. In a week or two, I thought, these same males will be vying intensely for a chance to mate with her.

The attack came without warning. One of the males charged toward us, hair on end, looking twice as large as my small female and enraged. As he rushed by he picked her up, hurled her to the ground, and pummeled her. She cringed and screamed. He ran off, rejoining the other males seconds later as if nothing had happened. It was not so easy for the female to return to normal. She whimpered and darted nervous glances at her attacker, as if worried that he might renew his assault.

In the years that followed I witnessed many similar attacks by males against females, among a variety of Old World primates, and eventually I found this sort of aggression against females so puzzling that I began to study it systematically—something that has rarely been done. My long-term research on olive baboons in Kenya showed that, on average, each pregnant or lactating female was attacked by an adult male about once a week and seriously injured about once a year. Estrous females were the target of even more aggression. The obvious question was, Why?

In the late 1970s, while I was in Africa among the baboons, feminists back in the United States were turning their attention to male violence against women. Their concern stimulated a wave of research documenting disturbingly high levels of battering, rape, sexual harassment, and murder. But although scientists investigated this kind of behavior from many perspectives, they mostly ignored the existence of similar behavior in other animals. My observations over the years have convinced me that a deeper understanding of male aggression against females in other species can help us understand its counterpart in our own.

Researchers have observed various male animals—including insects, birds, and mammals—chasing, threatening, and attacking females. Unfortunately, because scientists have rarely studied such aggression in detail, we do not know exactly how common it is. But the males of many of these species are most aggressive toward potential mates, which suggests that they sometimes use violence to gain sexual access.

Jane Goodall provides us with a compelling example of how males use violence to get sex. In her 1986 book, *The Chimpanzees of Gombe*, Goodall describes the chimpanzee dating game. In one of several scenarios, males gather around attractive estrous females and try to lure them away from other males for a one-on-one sexual expedition that may last for days or weeks. But females find some suitors more appealing than others and often resist the advances of less desirable males. Males often rely on aggression to counter female resistance. For example, Goodall describes how Evered, in "persuading" a reluctant Winkle to accompany him into the forest, attacked her six times over the course of five hours, twice severely.

Sometimes, as I saw in Gombe, a male chimpanzee even attacks an estrous female days before he tries to mate with her. Goodall thinks that a male uses such aggression to train a female to fear him so that she will be more likely to surrender to his subsequent sexual advances. Similarly, male hamadryas baboons, who form small harems by kidnapping child brides, maintain a tight rein over their females through threats and intimidation. If, when another male is nearby, a hamadryas female strays even a few feet from her mate, he shoots her a threatening stare and raises his brows. She usually responds by rushing to his side; if not, he bites the back of her neck. The neck bite is ritualized—the male does not actually sink his razor-sharp canines into her flesh—but the threat of injury is clear. By repeating this behavior hundreds of times, the male lays claim to particular females months or even years before mating with them. When a female comes into estrus, she solicits sex only from her harem master, and other males rarely challenge his sexual rights to her.

In some species, females remain in their birth communities their whole lives, joining forces with related females to defend vital food resources against other females.

These chimpanzee and hamadryas males are practicing sexual coercion: male use of force to increase the chances that a female victim will mate with him, or to decrease the chances that she will mate with someone else. But sexual coercion is much more common in some primate species than in others. Orangutans and chimpanzees are the only nonhuman primates whose males in the wild force females to copulate, while males of several other species, such as vervet monkeys and bonobos (pygmy chimpanzees), rarely if ever try to coerce females sexually. Between the two extremes lie many species, like hamadryas baboons, in which males do not force copulation but nonetheless use threats and intimidation to get sex.

These dramatic differences between species provide an opportunity to investigate which factors promote or inhibit sexual coercion. For example, we might expect to find more of it in species in which males are much larger than females—and we do. However, size differences between the sexes are far from the whole story. Chimpanzee and bonobo males both have only a slight size advantage, yet while male chimps frequently resort to force, male bonobos treat the fair sex with more respect. Clearly, then, although size matters, so do other factors. In particular, the social relationships females form with other females and with males appear to be as important.

In some species, females remain in their birth communities their whole lives, joining forces with related females to defend vital food resources against other females. In such "female bonded" species, females also form alliances against aggressive males. Vervet monkeys are one such species, and among these small and exceptionally feisty African monkeys, related females gang up against males. High-ranking females use their dense network of female alliances to rule the troop; although smaller than males, they slap persistent suitors away like annoying flies. Researchers have observed similar alliances in many other female-bonded species, including other Old World monkeys such as macaques, olive baboons, patas and rhesus monkeys, and gray langurs; New World monkeys such as the capuchin; and prosimians such as the ring-tailed lemur.

Females in other species leave their birth communities at adolescence and spend the rest of their lives cut off from their female kin. In most such species, females do not form strong bonds with other females and rarely support one another against males. Both chimpanzees and hamadryas baboons exhibit this

pattern, and, as we saw earlier, in both species females submit to sexual control by males.

Some of the factors that influence female vulnerability to male sexual coercion in different species may also help explain such variation among different groups in the same species.

This contrast between female-bonded species, in which related females gang together to thwart males, and non-female-bonded species, in which they don't, breaks down when we come to the bonobo. Female bonobos, like their close relatives the chimpanzees, leave their kin and live as adults with unrelated females. Recent field studies show that these unrelated females hang out together and engage in frequent homoerotic behavior, in which they embrace face-to-face and rapidly rub their genitals together; sex seems to cement their bonds. Examining these studies in the context of my own research has convinced me that one way females use these bonds is to form alliances against males, and that, as a consequence, male bonobos do not dominate females or attempt to coerce them sexually. How and why female bonobos, but not chimpanzees, came up with this solution to male violence remains a mystery.

Female primates also use relationships with males to help protect themselves against sexual coercion. Among olive baboons, each adult female typically forms long-lasting "friendships" with a few of the many males in her troop. When a male baboon assaults a female, another male often comes to her rescue; in my troop, nine times out of ten the protector was a friend of the female's. In return for his protection, the defender may enjoy her sexual favors the next time she comes into estrus. There is a dark side to this picture, however. Male baboons frequently threaten or attack their female friends—when, for example, one tries to form a friendship with a new male. Other males apparently recognize friendships and rarely intervene. The female, then, becomes less vulnerable to aggression from males in general, but more vulnerable to aggression from her male friends.

As a final example, consider orangutans. Because their food grows so sparsely adult females rarely travel with anyone but their dependent offspring. But orangutan females routinely fall victim to forced copulation. Female orangutans, it seems, pay a high price for their solitude.

Some of the factors that influence female vulnerability to male sexual coercion in different species may also help explain such variation among different groups in the same species. For example, in a group of chimpanzees in the Taï Forest in the Ivory Coast, females form closer bonds with one another than do females at Gombe. Taï females may consequently have more egalitarian relationships with males than their Gombe counterparts do.

Such differences between groups especially characterize humans. Among the South American Yanomamö, for instance, men frequently abduct and rape women from neighboring villages and severely beat their wives for suspected adultery. However, among the Aka people of the Central African Republic, male aggression against women has never been observed. Most human societies, of course, fall between these two extremes.

How are we to account for such variation? The same social factors that help explain how sexual coercion differs among nonhuman primates may deepen our understanding of how it varies across different groups of people. In most traditional human societies, a woman leaves her birth community when she marries and goes to live with her husband and his relatives. Without strong bonds to close female kin, she will probably be in danger of sexual coercion. The presence of close female kin, though, may protect her. For example, in a community in Belize, women live near their female relatives. A man will sometimes beat his wife if he becomes jealous or suspects her of infidelity, but when this happens, onlookers run to tell her female kin. Their arrival on the scene, combined with the presence of other glaring women, usually shames the man enough to stop his aggression.

Even in societies in which women live away from their families, kin may provide protection against abusive husbands, though how much protection varies dramatically from one society to the next. In some societies a woman's kin, including her father and brothers, consistently support her against an abusive husband, while in others they rarely help her. Why?

The key may lie in patterns of male-male relationships. Alliances between males are much more highly developed in humans than in other primates, and men frequently rely on such alliances to compete successfully against other men. They often gain more by supporting their male allies than they do by supporting female kin. In addition, men often use their alliances to defeat rivals and abduct or rape their women, as painfully illustrated by recent events in Bosnia. When women live far from close kin, among men who value their alliances with other men more than their bonds with women, they may be even more vulnerable to sexual coercion than many nonhuman primate females.

Even in societies in which women live away from their families, kin may provide protection against abusive husbands.

Like nonhuman primate females, many women form bonds with unrelated males who may protect them from other males. However, reliance on men exacts a cost—women and other primate females often must submit to control by their protectors. Such control is more elaborate in humans because allied men agree to honor one another's proprietary rights over women. In most of the world's cultures, marriage involves not only the exclusion of other men from sexual access to a man's wife— which protects the woman against rape by other men—but also entails the husband's right to complete control over his wife's sexual life, including the right to punish her for real or suspected adultery, to have sex with her whenever he wants, and even to restrict her contact with other people, especially men.

In modern industrial society, many men—perhaps most—maintain such traditional notions of marriage. At the same time, many of the traditional sources of support for women, including censure of abusive husbands by the woman's kinfolk or other community members, are eroding as more and more people end up without nearby kin or long-term neighbors. The increased vulnerability of women isolated from their birth communities, however, is not just a by-product of modern living. Historically, in highly patriarchal societies like those found in China and northern India, married women lived in households ruled by their husband's mother and male kin, and their ties with their own kin were virtually severed. In these societies, today as in the past, the husband's female kin often view the wife as a competitor for resources. Not only do they fail to support her against male coercive control, but they sometimes actively encourage it. This scenario illustrates an important point: women do not invariably support other women against men, in part because women may perceive their interests as best served through alliances with men, not with other women. When men have most of the power and control most of the resources, this looks like a realistic assessment.

Decreasing women's vulnerability to sexual coercion, then, may require fundamental changes in social alliances. Women gave voice to this essential truth with the slogan SISTERHOOD IS POWERFUL—a reference to the importance of women's ability to cooperate with unrelated women as if they were indeed sisters. However, among humans, the male-dominant social system derives support from political, economic, legal, and ideological institutions that other primates can't even dream of. Freedom from male control—including male sexual coercion— therefore requires women to form alliances with one another (and with like-minded men) on a scale beyond that shown by nonhuman primates and humans in the past. Although knowledge of other primates can provide inspiration for this task, its achievement depends on the uniquely human ability to envision a future different from anything that has gone before.

BARBARA SMUTS is a professor of psychology and anthropology at the University of Michigan. She has been doing fieldwork in animal behavior since the early 1970s, studying baboons, chimps, and dolphins. "In my work I combine research in animal behavior with an abiding interest in feminist perspectives on science," says Smuts. She is the author of *Sex and Friendship in Baboons*.

Mothers and Others

ABSTRACT: From Queen Bees To Elephant Matriarchs, Many Animal Mothers Are Assisted By Others In Rearing Offspring. Anthropologist Sarah Blaffer Hrdy Maintains That Our Human Ancestors, Too, Were "Cooperative Breeders"—A Mode Of Life That Enabled Them To Thrive In Many New Environments. Today, Argues Hrdy, Our Continued Ability To Raise Emotionally Healthy Children May Well Depend On How Well We Understand The Cooperative Aspect Of Our Evolutionary Heritage.

SARAH BLAFFER HRDY

Mother apes—chimpanzees, gorillas, orangutans, humans—dote on their babies. And why not? They give birth to an infant after a long gestation and, in most cases, suckle it for years. With humans, however, the job of providing for a juvenile goes on and on. Unlike all other ape babies, ours mature slowly and reach independence late. A mother in a foraging society may give birth every four years or so, and her first few children remain dependent long after each new baby arrives; among nomadic foragers, grown-ups may provide food to children for eighteen or more years. To come up with the 10–13 million calories that anthropologists such as Hillard Kaplan calculate are needed to rear a young human to independence, a mother needs help.

So how did our prehuman and early human ancestresses living in the Pleistocene Epoch (from 1.6 million until roughly 10,000 years ago) manage to get those calories? And under what conditions would natural selection allow a female ape to produce babies so large and slow to develop that they are beyond her means to rear on her own?

The old answer was that fathers helped out by hunting. And so they do. But hunting is a risky occupation, and fathers may die or defect or take up with other females. And when they do, what then? New evidence from surviving traditional cultures suggests that mothers in the Pleistocene may have had a significant degree of help—from men who thought they just might have been the fathers, from grandmothers and great-aunts, from older children.

These helpers other than the mother, called allomothers by sociobiologists, do not just protect and provision youngsters. In groups such as the Efe and Aka Pygmies of central Africa, allomothers actually hold children and carry them about. In these tight-knit communities of communal foragers—within which men, women, and children still hunt with nets, much as humans are thought to have done tens of thousands of years ago—siblings, aunts, uncles, fathers, and grandmothers hold newborns on the first day of life. When University of New Mexico anthropologist Paula Ivey asked an Efe woman, "Who cares for babies?" the immediate answer was, "We all do!" By three weeks

of age, the babies are in contact with allomothers 40 percent of the time. By eighteen weeks, infants actually spend more time with allomothers than with their gestational mothers. On average, Efe babies have fourteen different caretakers, most of whom are close kin. According to Washington State University anthropologist Barry Hewlett, Aka babies are within arm's reach of their fathers for more than half of every day.

Accustomed to celebrating the antiquity and naturalness of mother-centered models of child care, as well as the nuclear family in which the mother nurtures while the father provides, we Westerners tend to regard the practices of the Efe and the Aka as exotic. But to sociobiologists, whose stock in trade is comparisons across species, all this helping has a familiar ring. It's called cooperative breeding. During the past quarter century, as anthropologists and sociobiologists started to compare notes, one of the spectacular surprises has been how much allomaternal care goes on, not just within various human societies but among animals generally. Evidently, diverse organisms have converged on cooperative breeding for the best of evolutionary reasons.

A broad look at the most recent evidence has convinced me that cooperative breeding was the strategy that permitted our own ancestors to produce costly, slow-maturing infants at shorter intervals, to take advantage of new kinds of resources in habitats other than the mixed savanna-woodland of tropical Africa, and to spread more widely and swiftly than any primate had before. We already know that animal mothers who delegate some of the costs of infant care to others are thereby freed to produce more or larger young or to breed more frequently. Consider the case of silver-backed jackals. Patricia Moehlman, of the World Conservation Union, has shown that for every extra helper bringing back food, jackal parents rear one extra pup per litter. Cooperative breeding also helps various species expand into habitats in which they would normally not be able to rear any young at all. Florida scrub-jays, for example, breed in an exposed landscape where unrelenting predation from hawks and snakes usually precludes the fledging of young; survival in this habitat is possible only because older siblings help guard and feed the young. Such cooperative arrangements permit animals as different as naked mole rats (the social insects of the mammal world) and wolves to move into new habitats and sometimes to spread over vast areas.

When animal mothers delegate some infant-care costs to others, they can produce more or larger young and raise them in less-than-ideal habitats.

What does it take to become a cooperative breeder? Obviously, this lifestyle is an option only for creatures capable of living in groups. It is facilitated when young but fully mature individuals (such as young Florida scrub-jays) do not or cannot immediately leave their natal group to breed on their own and instead remain among kin in their natal location. As with delayed maturation, delayed dispersal of young means that teenagers, "spinster" aunts, real and honorary uncles will be on hand to help their kin rear young. Flexibility is another criterion for cooperative breeders. Helpers must be ready to shift to breeding mode should the opportunity arise. In marmosets and tamarins—the little South American monkeys that are, besides us, the only full-fledged cooperative breeders among primates—a female has to be ready to be a helper this year and a mother the next. She may have one mate or several. In canids such as wolves or wild dogs, usually only the dominant, or alpha, male and female in a pack reproduce, but younger group members hunt with the mother and return to the den to regurgitate predigested meat into the mouths of her pups. In a fascinating instance of physiological flexibility, a subordinate female may actually undergo hormonal transformations similar to those of a real pregnancy: her belly swells, and she begins to manufacture milk and may help nurse the pups of the alpha pair. Vestiges of cooperative breeding crop up as well in domestic dogs, the distant descendants of wolves. After undergoing a pseudopregnancy, my neighbors' Jack Russell terrier chased away the family's cat and adopted and suckled her kittens. To suckle the young of another species is hardly what Darwinians call an adaptive trait (because it does not contribute to the surrogate's own survival). But in the environment in which the dog family evolved, a female's tendency to respond when infants signaled their need—combined with her capacity for pseudopregnancy—would have increased the survival chances for large litters born to the dominant female.

According to the late W.D. Hamilton, evolutionary logic predicts that an animal with poor prospects of reproducing on his or her own should be predisposed to assist kin with better prospects so that at least some of their shared genes will be perpetuated. Among wolves, for example, both male and female helpers in the pack are likely to be genetically related to the alpha litter and to have good reasons for not trying to reproduce on their own: in a number of cooperatively breeding species (wild dogs, wolves, hyenas, dingoes, dwarf mongooses, marmosets), the helpers do try, but the dominant female is likely to

bite their babies to death. The threat of coercion makes postponing ovulation the better part of valor, the least-bad option for females who must wait to breed until their circumstances improve, either through the death of a higher-ranking female or by finding a mate with an unoccupied territory.

One primate strategy is to line up extra fathers. Among common marmosets and several species of tamarins, females mate with several males, all of which help rear her young. As primatologist Charles T. Snowdon points out, in three of the four genera of Callitrichidae (*Callithrix, Saguinus,* and *Leontopithecus*), the more adult males the group has available to help, the more young survive. Among many of these species, females ovulate just after giving birth, perhaps encouraging males to stick around until after babies are born. (In cotton-top tamarins, males also undergo hormonal changes that prepare them to care for infants at the time of birth.) Among cooperative breeders of certain other species, such as wolves and jackals, pups born in the same litter can be sired by different fathers.

Human mothers, by contrast, don't ovulate again right after birth, nor do they produce offspring with more than one genetic father at a time. Ever inventive, though, humans solve the problem of enlisting help from several adult males by other means. In some cultures, mothers rely on a peculiar belief that anthropologists call partible paternity—the notion that a fetus is built up by contributions of semen from all the men with whom women have had sex in the ten months or so prior to giving birth. Among the Canela, a matrilineal tribe in Brazil studied for many years by William Crocker of the Smithsonian Institution, publicly sanctioned intercourse between women and men other than their husbands—sometimes many men—takes place during villagewide ceremonies. What might lead to marital disaster elsewhere works among the Canela because the men believe in partible paternity. Across a broad swath of South America—from Paraguay up into Brazil, westward to Peru, and northward to Venezuela—mothers rely on this convenient folk wisdom to line up multiple honorary fathers to help them provision both themselves and their children. Over hundreds of generations, this belief has helped children thrive in a part of the world where food sources are unpredictable and where husbands are as likely as not to return from the hunt empty-handed.

The Bari people of Venezuela are among those who believe in shared paternity, and according to anthropologist Stephen Beckerman, Bari children with more than one father do especially well. In Beckerman's study of 822 children, 80 percent of those who had both a "primary" father (the man married to their mother) and a "secondary" father survived to age fifteen, compared with 64 percent survival for those with a primary father alone. Not surprisingly, as soon as a Bari woman suspects she is pregnant, she accepts sexual advances from the more successful fishermen or hunters in her group. Belief that fatherhood can be shared draws more men into the web of possible paternity, which effectively translates into more food and more protection.

But for human mothers, extra mates aren't the only source of effective help. Older children, too, play a significant role in family survival. University of Nebraska anthropologists Patricia Draper and Raymond Hames have just shown that among !Kung hunters and gatherers living in the Kalahari Desert, there is a significant correlation between how many children a parent successfully raises and how many older siblings were on hand to help during that person's own childhood.

One primate strategy is to line up extra "fathers." In some species of marmosets, females mate with several males, all of which help her raise her young.

Older matrilineal kin may be the most valuable helpers of all. University of Utah anthropologists Kristen Hawkes and James O'Connell and their UCLA colleague Nicholas Blurton Jones, who have demonstrated the important food-gathering role of older women among Hazda hunter-gatherers in Tanzania, delight in explaining that since human life spans may extend for a few decades after menopause, older women become available to care for—and to provide vital food for—children born to younger kin. Hawkes, O'Connell, and Blurton Jones further believe that dating from the earliest days of Homo erectus, the survival of weaned children during food shortages may have depended on tubers dug up by older kin.

At various times in human history, people have also relied on a range of customs, as well as on coercion, to line up allomaternal assistance—for example, by using slaves or hiring poor women as wet nurses. But all the helpers in the world are of no use if they're not motivated to protect, carry, or provision babies. For both humans and nonhumans, this motivation arises in three main ways: through the manipulation of information about kinship; through appealing signals coming from the babies themselves; and, at the heart of it all, from the endocrinological and neural processes that induce individuals to respond to infants' signals. Indeed, all primates and many other mammals eventually respond to infants in a nurturing way if exposed long enough to their signals. Trouble is, "long enough" can mean very different things in males and females, with their very different response thresholds.

For decades, animal behaviorists have been aware of the phenomenon known as priming. A mouse or rat encountering a strange pup is likely to respond by either ignoring the pup or eating it. But presented with pup after pup, rodents of either sex eventually become sensitized to the baby and start caring for it. Even a male may gather pups into a nest and lick or huddle over them. Although nurturing is not a routine part of a male's repertoire, when sufficiently primed he behaves as a mother would. Hormonal change is an obvious candidate for explaining this transformation. Consider the case of the cooperatively breeding Florida scrub-jays studied by Stephan Schoech, of the University of Memphis. Prolactin, a protein hormone that initiates the secretion of milk in female mammals, is also present in male mammals and in birds of both sexes. Schoech showed that levels of prolactin go up in a male and female jay as they build their nest and incubate eggs and that these levels reach a peak when they feed their young. Moreover, prolactin levels rise in the jays' nonbreeding helpers and are also at their highest when they assist in feeding nestlings.

As it happens, male, as well as immature and nonbreeding female, primates can respond to infants' signals, although quite different levels of exposure and stimulation are required to get them going. Twenty years ago, when elevated prolactin levels were first reported in common marmoset males (by Alan Dixson, for *Callithrix jacchus*), many scientists refused to believe it. Later, when the finding was confirmed, scientists assumed this effect would be found only in fathers. But based on work by Scott Nunes, Jeffrey Fite, Jeffrey French, Charles Snowdon, Lucille Roberts, and many others—work that deals with a variety of species of marmosets and tamarins—we now know that all sorts of hormonal changes are associated with increased nurturing in males. For example, in the tufted-eared marmosets studied by French and colleagues, testosterone levels in males went down as they engaged in caretaking after the birth of an infant. Testosterone levels tended to be lowest in those with the most paternal experience.

Genetic relatedness alone, in fact, is a surprisingly unreliable predictor of love. What matters are cues from infants and how we process these cues emotionally.

The biggest surprise, however, has been that something similar goes on in males of our own species. Anne Storey and colleagues in Canada have reported that prolactin levels in men who were living with pregnant women went up toward the end of the pregnancy. But the most significant finding was a 30 percent drop in testosterone in men right after the birth. (Some endocrinologically literate wags have proposed that this drop in testosterone levels is due to sleep deprivation, but this would probably not explain the parallel testosterone drop in marmoset males housed with parturient females.) Hormonal changes during pregnancy and lactation are, of course, indisputably more pronounced in mothers than in the men consorting with them, and no one is suggesting that male consorts are equivalent to mothers. But both sexes are surprisingly susceptible to infant signals—explaining why fathers, adoptive parents, wet nurses, and day-care workers can become deeply involved with the infants they care for.

Genetic relatedness alone, in fact, is a surprisingly unreliable predictor of love. What matters are cues from infants and how these cues are processed emotionally. The capacity for becoming emotionally hooked—or primed—also explains how a fully engaged father who is in frequent contact with his infant can become more committed to the infant's well-being than a detached mother will.

But we can't forget the real protagonist of this story: the baby. From birth, newborns are powerfully motivated to stay close, to root—even to creep—in quest of nipples, which they instinctively suck on. These are the first innate behaviors that any of us engage in. But maintaining contact is harder for little humans to do than it is for other primates. One problem is that human mothers are not very hairy, so a human mother not only has to position the baby on her breast but also has to keep him there. She must be motivated to pick up her baby even *before* her milk comes in, bringing with it a host of hormonal transformations.

Within minutes of birth, human babies can cry and vocalize just as other primates do, but human newborns can also read facial expressions and make a few of their own. Even with blurry vision, they engage in eye-to-eye contact with the people around them. Newborn babies, when alert, can see about eighteen inches away. When people put their faces within range, babies may reward this attention by looking back or even imitating facial expressions. Orang and chimp babies, too, are strongly attached to and interested in their mothers' faces. But unlike humans, other ape mothers and infants do not get absorbed in gazing deeply into each other's eyes.

To the extent that psychiatrists and pediatricians have thought about this difference between us and the other apes, they tend to attribute it to human mental agility and our ability to use language. Interactions between mother and baby, including vocal play and babbling, have been interpreted as protoconversations: revving up the baby to learn to talk. Yet even babies who lack face-to-face stimulation—babies born blind, say—learn to talk. Furthermore, humans are not the only primates to engage in the continuous rhythmic streams of vocalization known as babbling. Interestingly, marmoset and tamarin babies also babble. It may be that the infants of cooperative breeders are specially equipped to communicate with caretakers. This is not to say that babbling is not an important part of learning to talk, only to question which came first—babbling so as to develop into a talker, or a predisposition to evolve into a talker because among cooperative breeders, babies that babble are better tended and more likely to survive.

If humans evolved as cooperative breeders, the degree of a human mother's commitment to her infant should be linked to how much social support she herself can expect. Mothers in cooperatively breeding primate species can afford to bear and rear such costly offspring as they do only if they have help on hand. Maternal abandonment and abuse are very rarely observed among primates in the wild. In fact, the only primate species in which mothers are anywhere near as likely to abandon infants at birth as mothers in our own species are the other cooperative breeders. A study of cotton-top tamarins at the New England Regional Primate Research Center showed a 12 percent chance of abandonment if mothers had older siblings on hand to help them rear twins, but a 57 percent chance when no help was available. Overburdened mothers abandoned infants within seventy-two hours of birth.

This new way of thinking about our species' history, with its implications for children, has made me concerned about the future. So far, most Western researchers studying infant development have presumed that living in a nuclear family with a fixed division of labor (mom nurturing, dad providing) is the normal human adaptation. Most contemporary research on children's psychosocial development is derived from John Bowlby's theories of attachment and has focused on such variables as how available and responsive the mother is, whether the father is present or absent, and whether the child is in the mother's care or in day care. Sure enough, studies done with this model in mind always show that children with less responsive mothers are at greater risk.

In cooperative breeders, the degree of a mother's commitment to her infant should correlate with how much social support she herself can expect.

It is the baby, first and foremost, who senses how available and how committed its mother is. But I know of no studies that take into account the possibility that humans evolved as cooperative breeders and that a mother's responsiveness also happens to be a good indicator of her social supports. In terms of developmental outcomes, the most relevant factor might not be how securely or insecurely attached to the mother the baby is—the variable that developmental psychologists are trained to measure—but rather how secure the baby is in relation to all the people caring for him or her. Measuring attachment this way might help explain why even children whose relations with their mother suggest they are at extreme risk manage to do fine because of the interventions of a committed father, an older sibling, or a there-when-you-need-her grandmother.

The most comprehensive study ever done on how nonmaternal care affects kids is compatible with both the hypothesis that humans evolved as cooperative breeders and the conventional hypothesis that human babies are adapted to be reared exclusively by mothers. Undertaken by the National Institute of Child Health and Human Development (NICHD) in 1991, the seven-year study included 1,364 children and their families (from diverse ethnic and economic backgrounds) and was conducted in ten different U.S. locations. This extraordinarily ambitious study was launched because statistics showed that 62 percent of U.S. mothers with children under age six were working outside the home and that the majority of them (willingly or unwillingly) were back at work within three to five months of giving birth. Because this was an entirely new social phenomenon, no one really knew what the NICHD's research would reveal.

The study's main finding was that both maternal and hired caretakers' sensitivity to infant needs was a better predictor of a child's subsequent development and behavior (such traits as social "compliance," respect for others, and self-control were measured) than was actual time spent apart from the mother. In other words, the critical variable was not the continuous presence of the mother herself but rather how secure infants felt when cared for by someone else. People who had been convinced that babies need full-time care from mothers to develop normally were stunned by these results, while advocates of day care felt vindicated. But do these and other, similar findings mean that day care is not something we need to worry about anymore?

Not at all. We should keep worrying. The NICHD study showed only that day care was better than mother care if the mother was neglectful or abusive. But excluding such worst-case scenarios, the study showed no detectable ill effects from day care only when infants had a secure relationship with parents to begin with (which I take to mean that babies felt wanted) and only when the day care was of high quality. And in this study's context, "high quality" meant that the facility had a high ratio of caretakers to babies, that it had the same caretakers all the time, and that the caretakers were sensitive to infants' needs—in other words, that the day care staff acted like committed kin.

Bluntly put, this kind of day care is almost impossible to find. Where it exists at all, it's expensive. Waiting lists are long, even for cheap or inadequate care. The average rate of staff turnover in day care centers is 30 percent per year, primarily because these workers are paid barely the minimum wage (usually less, in fact, than parking-lot attendants). Furthermore, day care tends to be age-graded, so even at centers where staff members stay put, kids move annually to new teachers. This kind of day care is unlikely to foster trusting relationships.

What conclusion can we draw from all this? Instead of arguing over "mother care" versus "other care," we need to make day care better. And this is where I think today's evolution-minded researchers have something to say. Impressed by just how variable child-rearing conditions can be in human societies, several anthropologists and psychologists (including Michael Lamb, Patricia Draper, Henry Harpending, and James Chisholm) have suggested that babies are up to more than just maintaining the relationship with their mothers. These researchers propose that babies actually monitor mothers to gain information about the world they have been born into. Babies ask, in effect, Is this world filled with people who are going to provide for me and help me survive? Can I count on them to care about me? If the answer to those questions is yes, they begin to sense that developing a conscience and a capacity for compassion would be a great idea. If the answer is no, they may then be asking, Can I not afford to count on others? Would I be better off just grabbing what I need, however I can? In this case, empathy, or thinking about others' needs, would be more of a hindrance than a help.

For a developing baby and child, the most practical way to behave might vary drastically, depending on whether the mother has kin who help, whether the father is around, whether foster parents are well-meaning or exploitative. These factors, however unconsciously perceived by the child, affect important developmental decisions. Being extremely self-centered or selfish, being oblivious to others or lacking in conscience—traits that psychologists and child-development theorists may view as pathological—are probably quite adaptive traits for an individual who is short on support from other group members.

If I am right that humans evolved as cooperative breeders, Pleistocene babies whose mothers lacked social support and were less than fully committed to infant care would have been unlikely to survive. But once people started to settle down—10,000 or 20,000 or perhaps 30,000 years ago—the picture changed. Ironically, survival chances for neglected children increased. As people lingered longer in one place, eliminated predators, built walled houses, stored food—not to mention inventing things such as rubber nipples and pasteurized milk—infant survival became decoupled from continuous contact with a caregiver.

Since the end of the Pleistocene, whether in preindustrial or industrialized environments, some children have been surviving levels of social neglect that previously would have meant certain death. Some children get very little attention, even in the most benign of contemporary homes. In the industrialized world, children routinely survive caretaking practices that an Efe or a !Kung mother would find appallingly negligent. In traditional societies, no decent mother leaves her baby alone at any time, and traditional mothers are shocked to learn that Western mothers leave infants unattended in a crib all night.

In effect, babies ask: Is this world filled with people who are going to provide for me and help me survive? Can I count on them to care about me?

Without passing judgment, one may point out that only in the recent history of humankind could infants deprived of supportive human contact survive to reproduce themselves. Certainly there are a lot of humanitarian reasons to worry about this situation: one wants each baby, each child, to be lovingly cared for. From my evolutionary perspective, though, even more is at stake.

Even if we manage to survive what most people are worrying about—global warming, emergent diseases, rogue viruses, meteorites crashing into earth—will we still be human thousands of years down the line? By that I mean human in the way we currently define ourselves. The reason our species has managed to survive and proliferate to the extent that 6 billion people currently occupy the planet has to do with how readily we can learn to cooperate when we want to. And our capacity for empathy is one of the things that made us good at doing that.

At a rudimentary level, of course, all sorts of creatures are good at reading intentions and movements and anticipating what other animals are going to do. Predators from gopher snakes to lions have to be able to anticipate where their quarry will dart. Chimps and gorillas can figure out what another individual is likely to know or not know. But compared with that of humans, this capacity to entertain the psychological perspective of other individuals is crude.

During early childhood, through relationships with mothers and other caretakers, individuals learn to look at the world from someone else's perspective.

The capacity for empathy is uniquely well developed in our species, so much so that many people (including me) believe that along with language and symbolic thought, it is what makes us human. We are capable of compassion, of understanding other people's "fears and motives, their longings and griefs and vanities," as novelist Edmund White puts it. We spend time and energy worrying about people we have never even met, about babies left in dumpsters, about the existence of more than 12 million AIDS orphans in Africa.

Psychologists know that there is a heritable component to emotional capacity and that this affects the development of compassion among individuals. By fourteen months of age, identical twins (who share all genes) are more alike in how they react to an experimenter who pretends to painfully pinch her finger on a clipboard than are fraternal twins (who share only half their genes). But empathy also has a learned component, which has more to do with analytical skills. During the first years of life, within the context of early relationships with mothers and other committed caretakers, each individual learns to look at the world from someone else's perspective.

And this is why I get so worried. Just because humans have evolved to be smart enough to chronicle our species' histories, to speculate about its origins, and to figure out that we have about 30,000 genes in our genome is no reason to assume that evolution has come to a standstill. As gene frequencies change, natural selection acts on the outcome, the expression of those genes. No one doubts, for instance, that fish benefit from being able to see. Yet species reared in total darkness—as are the small, cave-dwelling characin of Mexico—fail to develop their visual capacity. Through evolutionary time, traits that are unexpressed are eventually lost. If populations of these fish are isolated in caves long enough, youngsters descended from those original populations will no longer be able to develop eyesight at all, even if reared in sunlight.

If human compassion develops only under particular rearing conditions, and if an increasing proportion of the species survives to breeding age without developing compassion, it won't make any difference how useful this trait was among our ancestors. It will become like sight in cave-dwelling fish.

No doubt our descendants thousands of years from now (should our species survive) will still be bipedal, symbol-generating apes. Most likely they will be adept at using sophisticated technologies. But will they still be human in the way we, shaped by a long heritage of cooperative breeding, currently define ourselves?

This article was adapted from "Cooperation, Empathy, and the Needs of Human Infants," a Tanner Lecture delivered at the University of Utah. It is used with the permission of the Tanner Lectures on Human Values, a Corporation, University of Utah, Salt Lake City.

This article was reprinted from *Natural History*, May 2001, pp. 50-62, adapted from "Cooperation, Empathy, and the Needs of Human Infants," a Tanner Lecture. Copyright © 2001 Tanner Lectures on Human Values, Inc., University of Utah, Salt Lake City.

Had King Henry VIII's Wives Only Known …

Want a boy at all costs? The secret may lie in your glucose levels

MARY DUENWALD

Some couples worry—a lot—about whether they will have a boy or a girl. So they try any number of strategies to influence nature's choice. Folklore says eating more red meat increases the chance of having a boy. So does having sex standing up or during the quarter moon or on odd-numbered dates. A diet of fish and vegetables, on the other hand, is said to produce girls—as does having sex when the moon is full or the date is even. Some popular books say having sex during ovulation is likely to produce a girl, others that it more often leads to a boy.

These techniques may seem laughably unscientific, but the idea of influencing a child's sex is not. Hundreds of scientific studies have shown that insects, reptiles, birds, and mammals unconsciously influence the sex of their offspring, producing more males at times and more females at other times.

In 1973 two Harvard scientists, biologist Robert Trivers and mathematician Dan Willard, came up with an evolutionary theory to explain this behavior. If a pregnant woman is strong and likely to bear a healthy child, they noted, she's better off having a boy: Healthy males tend to have many more offspring than weaker males. If a mother is weak and apt to bear a weak child, it is to her evolutionary advantage to have a girl: Even the least robust females tend to have some offspring, whereas the weakest males may never mate. Natural selection should therefore encourage mothers in poor condition to bear daughters and those in prime shape to have sons.

In the 32 years since the Trivers-Willard hypothesis was published, it has spawned more than 1,000 reports of evidence for it and (less often) against it, in animals and people. A review of 10,000 human births in Italy, for instance, found that mothers who weighed the least before becoming pregnant had 3 percent more daughters than heavier women did. Among women living in a small Ethiopian community, those with the most fat and muscle mass in their upper arms were more than twice as likely to have boy babies as those women with the thinnest arms. Single American mothers, who tend to have fewer resources than those who are married, have boys less than half the time, while married women have them 51.5 percent of the time. Other research has shown that women who smoke cigarettes or ingest a lot of PCB-contaminated fish bear more girls.

A few years ago, John Lazarus, a biologist at the University of Newcastle upon Tyne in England, reviewed 54 studies of sex ratios in humans. He found that 26 supported the Trivers-Willard hypothesis, one found evidence against it, and the rest found no effect. The evidence in animals follows the same equivocal yet supportive trend. In a recent review of 422 mammal studies, Elissa Cameron, a mammal ecologist at the University of Nevada at Reno, found that 34 percent of the studies supported the idea that a mother's condition can affect the sex of her offspring. Only 8.5 percent found the opposite. Most of the studies found no proof either way.

Cameron had earlier spent four years as a graduate student at Massey University in New Zealand, observing wild horses in the Kaimanawa Mountains. She correlated sex ratios of foals with their mothers' condition at three different points: at conception, halfway through their 11-month pregnancies, and at the foal's birth. "The only consistent result I got was with the measure at conception," Cameron says. "Condition at conception was strongly linked to the birth of a male or female foal."

With this in mind, she took a closer look at the other studies she had reviewed. Of those that correlated sex ratios with the mothers' condition at the time of conception, she found that three-fourths supported Trivers and Willard.

Humans show a similar pattern. Scientists have observed, for instance, that couples who have sex a few days before or after the woman ovulates tend to have boys, while those who have sex at or very near the point of ovulation tend to have girls. Some of the evidence for this comes from studies that looked at frequency of intercourse: Couples who have a lot of sex are more likely to conceive early in the woman's cycle. One study looked at births in Australia from 1908 to 1967 among couples who conceived during the first month of their marriages, a time of frequent sex. These couples had 3 percent more boys than average. Similarly, in countries at war, sex ratios tilt toward boys, presumably because of frequent sex during home leaves and after demobilization.

How does this work? One theory holds that shifting hormone levels affect the viscosity of fluids in a woman's reproductive tract, giving either the X-bearing sperm or the smaller Y-bearing ones an advantage in the race for the egg. When luteinizing hormone is released at ovulation, for example, it may somehow make the chase easier for the X sperm. Animal studies suggest the mother's diet makes a difference. Dairy cows fed nutritious diets bear more bull calves than cows with poorer nutrition. Rat and mouse mothers on high-fat diets have more male offspring than those on high-carbohydrate diets.

After reviewing many such studies, Cameron suspects that high-fat diets make a difference by raising blood glucose levels. Glucose, she says, aids the survival of male embryos conceived in laboratory cultures. When glucose is added to the culture medium for cows and sheep embryos, a greater number of males survive. The same seems to be true for human embryos in vitro: Glucose enhances the growth and development of males but not of females.

Cameron also points to two studies of mice in diabetes research, where blood glucose levels are high. "In both cases, the offspring of the mice had a heavily male-biased sex ratio," she says. Certain findings in humans could also be interpreted to support the importance of glucose in sex determination, Cameron says. Smoking lowers glucose levels, for example, and living in warmer climates raises them.

Cameron plans to do her own mouse study to see if she can skew the sex ratio of offspring by manipulating blood sugar levels. It's not clear that glucose levels in the blood have a direct effect on glucose in the uterus. But if they do, and if the glucose hypothesis proves correct, scientists might one day exploit it. Manipulating human births would be ethically problematic, but livestock and lab animals would be natural targets. "In dairy cows, for example, you would like to get more female calves," says Cheryl Rosenfeld, a veterinarian and biomedical researcher at the University of Missouri at Columbia. "In beef cattle, on the other hand, breeders would like to increase the ratio of male calves." In laboratories, medical researchers would like to manipulate the sex ratios of rats and mice to make it easier to conduct studies that focus on a specific sex.

When it comes to people, could diet somehow work in tandem with the timing of intercourse to influence gender? Perhaps, Cameron and others say. Both are tied to hormone levels and thus affect conditions inside the uterus. As for all those diet strategies (eat meat for a boy, vegetables for a girl), Cameron says she wouldn't be entirely surprised if they affected conditions in the uterus enough to make a difference: "We can't know if they are hogwash until we understand how this really works."

UNIT 4
The Fossil Evidence

Unit Selections

Key Points to Consider

- Why is the naming of species as if they were discontinuously separate categories a "convenient fiction? "

- When and where did our ancestors split off from the apes?

- What did the last common ancestor hominids and chimpanzees look like?

- Under what circumstances did bipedalism evolve? What are its advantages?

- What is the "man the hunter " hypothesis, and how might the"scavenging theory " better suit the early hominid data?

- How would you draw the early hominid family tree?

- How did the Piltdown Hoax affect the interpretation of hominid fossil evidence for forty years?

- Was "Peking Man " a cannibal, a hunter, or the hunted?

Student Website

www.mhcls.com/online

Internet References

Further information regarding these websites may be found in this book's preface or online.

The African Emergence and Early Asian Dispersals of the Genus *Homo*
   http://www.uiowa.edu/~bioanth/homo.html

Anthropology, Archaeology, and American Indian Sites on the Internet
   http://dizzy.library.arizona.edu/library/teams/sst/anthro/

Long Foreground: Human Prehistory
   http://www.wsu.edu/gened/learn-modules/top_longfor/lfopen-index.html

A primary focal point of this book, as well as of the whole of biological anthropology, is the search for and interpretation of fossil evidence for hominid (meaning human or humanlike) evolution. Paleoanthropologists are those who carry out this task by conducting the painstaking excavations and detailed analyses that serve as a basis for understanding our past. Every fragment found is cherished like a ray of light that may help to illuminate the path taken by our ancestors in the process of becoming "us." At least, that is what we would like to believe. In reality, each discovery leads to further mystery, and for every fossil-hunting paleoanthropologist who thinks his or her find supports a particular theory, there are many others anxious to express their disagreement. (See "African Trailblazers" and "Hunting the First Hominid".)

How wonderful it would be, we sometimes think in moments of frustration over inconclusive data, if the fossils would just speak for themselves, and every primordial piece of humanity were to carry with it a self-evident explanation for its place in the evolutionary story. Paleoanthropology would then be more a quantitative problem of amassing enough material to reconstruct our ancestral development than a qualitative problem of inter-preting what it all means. It would certainly be a simpler process, but would it be as interesting?

Most scientists tolerate, welcome, or even (dare it be said?) thrive on controversy, recognizing that diversity of opinion refreshes the mind, rouses students, and captures the imagination of the general public. After all, where would paleoanthropology be without the gadflies, the near-mythic heroes, and, lest we forget, the research funds they generate? Consider, for example, the issue of the differing roles played by males and females in the transition to humanity and all that it implies with regard to bipedalism, tool making, and the origin of the family. Did bipedalism really evolve in the grasslands? Did bipedalism develop as a means of pursuing game animals? Should the primary theme of human evolution be summed up as "man the hunter" or "woman the gatherer"?

Not all the research and theoretical speculation taking place in the field of paleoanthropology is so controversial. Most scientists, in fact, go about their work quietly and methodically, generating hypotheses that are much less explosive and yet have the cumulative effect of enriching our understanding of the details of human evolution. In "The Salamander's Tale," Richard Dawkins reminds us that we should not take the names we give to "species" too seriously. In "Digital Ancestors Walk Again" and in "The Scavenging of 'Peking Man'," we learn how new methods of analysis are enhancing our understanding of the fossil evidence. In "Scavenger Hunt," Pat Shipman tells how modern technology, in the form of the scanning electron microscope, combined with meticulous detailed analysis of cut marks on fossil animal bones, can help us better understand the locomotor and food-getting adaptations of our early hominid ancestors. In one stroke, Shipman is able to challenge the traditional "man the hunter" theme that has pervaded most early hominid research and writing and to simultaneously set forth an alternative hypothesis that will, in turn, inspire further research.

As we mull over the controversies outlined in this unit, we should not take them as reflecting an inherent weakness of the field of paleoanthropology, but rather as symbolic of its strength: the ability and willingness to scrutinize, question, and reflect (seemingly endlessly) on every bit of evidence.

Contrary to the way that the way that the proponents of creationism or "intelligent design theory" would have it, an admission of doubt is not an expression of ignorance but simply a frank recognition of the imperfect state of our knowledge. If we are to increase our understanding of ourselves, we must maintain an atmosphere of free inquiry without preconceived notions and an unquestioning commitment to a particular point of view.

To paraphrase anthropologist Ashley Montagu, *whereas creationism seeks certainty without proof, science seeks proof without certainty.*

The Salamander's Tale

RICHARD DAWKINS

Names are a menace in evolutionary history. It is no secret that palaeontology is a controversial subject in which there are even some personal enmities. At least eight books called *Bones of Contention* are in print. And if you look at what two palaeontologists are quarrelling about, as often as not it turns out to be a name. Is this fossil *Homo erectus,* or is it an archaic *Homo sapiens*? Is this one an early *Homo habilis* or a late *Australopithecus*? People evidently feel strongly about such questions they often turn out to be splitting hairs. Indeed, they resemble theological questions, which I suppose gives a clue to why they arouse such passionate disagreements. The Obsession with discrete names is an example of what I call the tyranny of the discontinuous mind. The Salamander's Tale strikes a blow against the discontinuous mind.

The Central Valley runs much of the length of California, bound the Coastal Range to the west and by the Sierra Nevada to the east. These long mountain ranges link up at the north and the south ends of the valley, which is therefore surrounded by high ground. Throughout this high ground lives a genus of salamanders called *Ensatina*. The Central Valley itself, about 40 miles wide, is not friendly to salamanders, and they are not found there. They can move all round the valley but normally not across it, in an elongated ring of more or less continuous population. In practice any one salamander's short legs in its short lifetime don't carry it far from its birthplace. But genes, persisting through a longer timescale are another matter. Individual salamanders can interbreed with neighbours whose parents may have interbred with neighbours further round the ring, and so on. There is therefore potentially gene flow all around the ring. Potentially. What happens in practice has been elegantly worked out by the research of my old colleagues at the University of California at Berkeley, initiated by Robert Stebbins and continued by David Wake.

In a study area called Camp Wolahi, in the mountains to the south of the valley, there are two clearly distinct species of *Ensatina* which do not interbreed. One is conspicuously marked with yellow and black blotches. The other is a uniform light brown with no blotches. Camp Wolahi is in a zone of overlap, but wider sampling shows that the blotched species is typical of the eastern side of the Central Valley which, here in Southern California, is known as the San Joaquin Valley. The light brown species, on the contrary, is typically found on the western side of the San Joaquin.

Non-interbreeding is the recognised criterion for whether two populations deserve distinct species names. It therefore should be straightforward to use the name *Ensatina eschscholtzii* for the plain western species and *Ensatina klauberi* for the blotched eastern species—straightforward but for one remarkable circumstance, which is the nub of the tale.

If you go up to the mountains that bound the north end of the Central Valley, which up there is called the Sacramento Valley, you'll find only one species of *Ensatina*. Its appearance is intermediate between the blotched and the plain species: mostly brown, with rather indistinct blotches. It is not a hybrid between the two: that is the wrong way to look at it. To discover the right way, make two expeditions south, sampling the salamander populations as they fork to west and east on either side of Central Valley. On the east side, they become progressively more pitched until they reach the extreme of *klauberi* in the far south. On the west side, the salamanders become progressively more like the plain *escholtzii* that we met in the zone of overlap at Camp Wolahi.

This is why it is hard to treat *Ensatina eschscholtzii* and *Ensatina klauberi* with confidence as separate species. They constitute a 'ring species'. You'll recognise them as separate species if you only sample in the south. Move north, however, and they gradually turn into each other. Zoogists normally follow Stebbins's lead and place them all in the same species, *Ensatina eschscholtzii*, but give them a range of subspecies names. Starting in the far south with *Ensatina eschscholtzii eschscholtzii*, the plain brown form, we move up the west side of the valley through *Ensatina eschscholtzii xanthoptica* and *Ensatina eschscholtzii oregonensis* which as its name suggests, is also found further north in Oregon and Washington. At the north end of California's Central Valley is *Ensatina eschscholtzii picta*, the semi-blotched form mentioned before. Moving on round the ring and down the east side of the valley, we pass through *Ensatina eschscholtzii platensis* which is a bit more blotched than *picta,* then *Ensatina eschscholtzii croceater* until we reach *Ensatina eschsd klauberi* (which is the very blotched one that we previously called *Ensatina klauberi* when we were considering it to be a separate species).

Stebbins believes that the ancestors of *Ensatina* arrived at the north end of the Central Valley and evolved gradually down the two sides the valley, diverging as they went. An alternative possibility is that started in the south as, say, *Ensatina eschscholtzii eschscholtzii*, then evolved their way up the west side of the valley, round the top and down the other side, ending up as *Ensatina eschscholtzii klauberi* at the other end of the ring. Whatever the history, what happens today is that there is hybridization all round the ring, except where the two ends of the line meet, in the far south of California.

As a complication, it seems that the Central Valley is not a total barrier to gene flow. Occasionally, salamanders seem to have made it across for there are populations of, for example, *xanthoptica*, one of the western subspecies, on the eastern side of the valley, where they hybridise with the eastern subspecies, *platensis*. Yet another complication is that there is a small break near the south end of the ring, where there seem to be no salamanders at all. Presumably they used to be there, but have died out. Or maybe they are still there but have not been found: I am told that the mountains in this area are rugged and hard to search. The ring is complicated, but a ring of continuous gene flow is, nevertheless, the predominant pattern in this genus, as it is with the better-known case of herring gulls and lesser black-backed gulls around the Arctic Circle.

In Britain the herring gull and the lesser black-backed gull are clearly distinct species. Anybody can tell the difference, most easily by the colour of the wing backs. Herring gulls have silver-grey wing backs, lesser black-backs, dark grey, almost black. More to the point, the birds themselves can tell the difference too, for they don't hybridise although they often meet and sometimes even breed alongside one another in mixed colonies. Zoologists therefore feel fully justified in giving them different names, *Larus argentatus* and *Larus fuscus*.

But now here's the interesting observation, and the point of resemblance to the salamanders. If you follow the population of herring gulls westward to North America, then on around the world across Siberia and back to Europe again, you notice a curious fact. The 'herring gulls', as you move round the pole, gradually become less and less like herring gulls and more and more like lesser black-backed gulls until it turns out that our Western European lesser black-backed gulls actually are the other end of a ring-shaped continuum which started with herring gulls At every stage around the ring, the birds are sufficiently similar to their immediate neighbours in the ring to interbreed with them. Until, that is the ends of the continuum are reached, and the ring bites itself in the tail. The herring gull and the lesser black-backed gull in Europe never interbreed, although they are linked by a continuous series of interbreeding colleagues all the way round the other side of the world.

Ring species like the salamanders and the gulls are only showing us in the spatial dimension something that must always happen in the time dimension. Suppose we humans, and the chimpanzees, were a ring species. It could have happened: a ring perhaps moving up one side of the Rift Valley, and down the other side, with two completely separate species coexisting at the southern end of the ring, but an unbroken continuum of interbreeding all the way up and back round the other side. If this were true, what would it do to our attitudes to other species? To apparent discontinuities generally?

Many of our legal and ethical principles depend on the separation between *Homo sapiens* and all other species. Of the people who regard abortion as a sin, including the minority who go to the lengths of assassinating doctors and blowing up abortion clinics, many are unthinking meat-eaters, and have no worries about chimpanzees being imprisoned in zoos and sacrificed in laboratories. Would they think again, if we could lay out a living continuum of intermediates between ourselves and chimpanzees, linked in an unbroken chain of interbreeders like the Californian salamanders? Surely they would. Yet it is the merest accident that the intermediates all happen to be dead. It is only because of this accident that we can comfortably and easily imagine a huge gulf between our species—or between any two species, for that matter.

I have previously recounted the case of the puzzled lawyer who questioned me after a public lecture. He brought the full weight of his legal acumen to bear on the following nice point. If species A evolves into species B, he reasoned closely, there must come a point when a child belongs to the new species B but his parents still belong to the old species A. Members of different species cannot, by definition, interbreed with one another, yet surely a child would not be so different from its parents as to be incapable of interbreeding with their kind. Doesn't this, he wound up, wagging his metaphorical finger in the special way that lawyers, at least in courtroom dramas, have perfected as their own, undermine the whole idea of evolution?

That is like saying, 'When you heat a kettle of cold water, there is no particular moment when the water ceases to be cold and becomes hot, therefore it is impossible to make a cup of tea.' Since I always try to turn questions in a constructive direction, I told my lawyer about the herring gulls, and I think he was interested. He had insisted on placing individuals firmly in this species or that. He didn't allow for the possibility that an individual might lie half way between two species, or a tenth of the way from species A to species B. Exactly the same limitation of thought hamstrings the endless debates about exactly when in the development of an embryo it becomes human (and when, by implication, abortion should be regarded as tantamount to murder). It is no use saying to these people that, depending upon the human characteristic that interests you, a focus can be 'half human' or 'a hundredth human'. 'Human', to the qualitative, absolutist mind, is like 'diamond'. There are no halfway houses. Absolutist minds can be a menace. They cause real misery, human misery. This is what I call the tyranny of the discontinuous mind, and it leads me to develop the moral of the Salamander's Tale.

For certain purposes names, and discontinuous categories, are exactly what we need. Indeed, lawyers need them all the time. Children are not allowed to drive; adults are. The law needs to impose a threshold, for example the seventeenth birthday. Revealingly, insurance companies take a very different view of the proper threshold age.

Some discontinuities are real, by any standards. You are a person and I am another person and our names are discontinuous labels that correctly signal our separateness. Carbon monoxide really is distinct from carbon dioxide. There is no overlap. A molecule consists of a carbon and one oxygen, or a carbon and two oxygens. None has a carbon and 1.5 oxygens. One gas is deadly poisonous, the other is needed by plants to make the organic substances that we all depend upon. Gold really is dis-

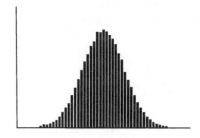

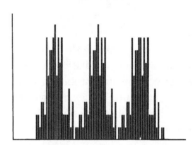

tinct from silver. Diamond crystals really are different from graphite crystals. Both are made of carbon, but the carbon atoms naturally arrange themselves in two quite distinct ways. There are no intermediates.

But discontinuities are often far from so clear. My newspaper carried the following item during a recent flu epidemic. Or was it an epidemic! That question was the burden of the article.

> Official statistics show there are 144 people in every 100,000 suffering from flu, said a spokeswoman for the Department of Health. As the usual gauge of an epidemic is 400 in every 100,000, it is not being officially treated by the Government. But the spokeswoman added: 'Professor Donaldson is happy to stick by his version that this is an epidemic. He believes it is many more than 144 per 100,000. It is very confusing and it depends on which definition you choose. Professor Donaldson has looked at his graph and said it is a serious epidemic.'

What we know is that some particular number of people are suffering from flu. Doesn't that, in itself, tell us what we want to know? Yet for the spokeswomen, the important question is whether this counts as an 'epidemic'. Has the proportion of sufferers crossed the rubicon of 400 per 100,000? This is the great decision which Professor Donaldson had to make, as he pored over his graph. You'd think he might have been better employed trying to do something about it, whether or not it counted officially as an epidemic.

As it happens, in the case of epidemics, for once there really is a natural rubicon: a critical mass of infections above which the virus, or bacterium, suddenly 'takes off' and dramatically increases its rate of spreading. This is why public health officials try so hard to vaccinate more than a threshold proportion of the population against, say whooping cough. The purpose is not just to protect the individuals vaccinated. It is also to deprive the pathogens of the opportunity to reach their own critical mass for 'take off'. In the case of our flu epidemic, what should really worry the spokeswomen for the Ministry of Health is whether the flu virus has yet crossed it rubicon for take-off, and leapt abrupt into high gear in its spread through the population. This should be decided by some means other than reference to magic numbers like 400 per 100,000. Concern with magic numbers is a mark of the discontinuous mine, or qualitative mind. The funny thing is that, in the case, the discontinous mind overlooks a genuine discontinuity, the take-off point for an epidemic. Usually there isn't a genuine discontinuity to overlook.

Many Western countries at present are suffering what is described as an epidemic of obesity. I seem to see evidence of this all around me, but I am not impressed by the preferred way of turning it into numbers. A percentage of population is described as 'clinically obese'. Once again, the discontinuous mind insists on separating people out into the obese on one side of a line, the non-obese on the other. That is not the way real life works. Obesity is continuously distributed. You can measure how obese each individual is, and you can compute group statistics from such measurements. Counts of numbers of people who lie above some arbitrarily defined threshold of obesity are not illuminat-

ing, if only because they immediately prompt a demand for the threshold to be specified and maybe redefined.

The same discontinuous mind also lurks behind all those official figures detailing the numbers of people 'below the poverty line'. You can meaningfully express a family's poverty by telling us their income, preferably expressed in real terms of what they can buy. Or you can say 'X is as poor as a church mouse' or 'y is as rich as Croesus' and everybody will know what you mean. But spuriously precise counts or percentages of people said to fall above or below some arbitrarily defined poverty line are pernicious. They are pernicious because the precision implied by the percentage is instantly belied by the meaningless artificiality of the 'line'. Lines are impositions of the discontinuous mind. Even more politically sensitive is the label 'black', as opposed to 'white', in the context of modern society—especially American society. This is the central issue in the Grasshopper's Tale, and I'll leave it for now, except to say that I believe race is yet another of the many cases where we don't need discontinuous categories, and where we should do without them unless an extremely strong case in their favour is made.

Here's another example. Universities in Britain award degrees that are classified into three distinct classes, First, Second and Third Class. Universities in other countries do something equivalent, if under different names, like A, B, C, etc. Now, my point is this. Students do not really separate neatly into good, middling and poor. There are not discrete and distinct classes of ability or diligence. Examiners go to some trouble to assess students on a finely continuous numerical scale, awarding marks or points that are designed to be added to other such marks, or otherwise manipulated in mathematically continuous ways. The score on such a continuous numerical scale conveys far more information than classification into one of three categories. Nevertheless, only the discontinuous categories are published.

In a very large sample of students, the distribution of ability and prowess would normally be a bell curve with few doing very well, few doing very badly and many in between. It might not actually be a symmetrical bell like the picture, but it would certainly be smoothly continuous and it would become smoother as more and more students are added in.

A few examiners (especially, I hope I'll be forgiven for adding, in non-scientific subjects) seem actually to believe that there really is a discrete entity called the First-Class Mind, or the 'alpha' mind, and a student either definitely has it or definitely hasn't. The task of the examiner is to sort out the Firsts from the Seconds and Seconds from the Thirds, just as one might sort sheep from goats. The likelihood that in reality there is a smooth continuum, sliding from pure sheepiness through all intermediates to pure goatiness, is a difficult one for some kinds of mind to grasp.

If, against all my expectations, it should turn out that the more students you add in, the more the distribution of exam marks approximates to a discontinuous distribution with three peaks, it would be a fascinating result. The awarding of First, Second and Third Class degrees might then actually be justifiable.

But there is certainly no evidence for this, and it would be very surprising given everything we know about human variation. As things are, it is clearly unfair: there is far more difference between the top of one class and the bottom of the same class, than there is between the bottom of one class and the top of the next class. It would be fairer to publish the actual marks obtained, or a rank order based upon those marks. But the discontinuous or qualitative mind insists on forcing people into one or other discrete category.

Returning to our topic of evolution, what about sheep and goats themselves? Are there sharp discontinuities between species, or do they merge into each other like first-class and second-class exam performances? If we look only at surviving animals, the answer is normally yes, are sharp discontinuities. Exceptions like the gulls and the Californian salamanders are rare, but revealing because they translate into the spatial domain the continuity which is normally found only in the temporal domain. People and chimpanzees are certainly linked via a continuous chain of intermediates and a shared ancestor, but the intermediates are extinct: what remains is a discontinuous distribution. The same is true of people and monkeys, and of people and kangaroos, except that the extinct intermediates lived longer ago. Because the intermediates are nearly always extinct, we can usually get away with assuming that there is a sharp discontinuity between every species and every other. But in this book we are concerned with evolutionary history, with the dead as well as the living. When we are talking about all the animals that have ever lived, not just those that are living now, evolution tells us there are lines of gradual continuity linking literally every species to every other. When we are talking history, even apparently discontinuous modern species like sheep and dogs are linked, via their common ancestor, in unbroken lines of smooth continuity.

Ernst Mayr, distinguished elder statesman of twentieth-century evolution, has blamed the delusion of discontinuity—under its philosophical name of Essentialism—as the main reason why evolutionary understanding came so late in human history. Plato, whose philosophy can be seen as the inspiration for Essentialism, believed that actual things are imperfect versions of an ideal archetype of their kind. Hanging somewhere in ideal space is an essential, perfect rabbit, which bears the same relation to a real rabbit as a mathematician's perfect circle bears to a circle drawn in the dust. To this day many people are deeply imbued with the idea that sheep are sheep and goats are goats, and no species can ever give rise to another because to do so they'd have to change their 'essence'.

There is no such thing as essence.

No evolutionist thinks that modern species change into other modern species. Cats don't turn into dogs or vice versa. Rather, cats and dogs have evolved from a common ancestor, who lived tens of millions of years ago. If only all the intermediates were still alive, attempting to separate cats from dogs would be a doomed enterprise, as it is with the salamanders and the gulls. Far from being a question of ideal essences, separating cats from dogs turns out to be possible only because of the lucky (from the point of view of the essentialist) fact that the intermediates happen to be dead. Plato might find it ironic to learn that it is actually an imperfection—the sporadic ill-fortune of death—that makes the ration of anyone species from another possible. This of course applies to the separation of human beings from our nearest relatives—and, indeed, from our more distant relatives too. In a world of perfect and complete information, fossil information as well as recent, discrete names for animals would become impossible. Instead of discrete names we would need sliding scales, just as the words hot, warm, cool and cold are better replaced by a sliding scale such as Celsius or Fahrenheit.

Evolution is now universally accepted as a fact by thinking people, so one might have hoped that essentialist intuitions in biology would have been finally overcome. Alas, this hasn't happened. Essentialism refuses to lie down. In practice, it is usually not a problem. Everyone agrees that *Homo sapiens* is a different species (and most would say a different genus) from *Pan troglodytes*, the chimpanzee. But everyone also agrees that if you follow human ancestry backward to the shared ancestor and then forward to chimpanzees, the intermediates all along the way will form a gradual continuum in which every generation would have been capable mating with its parent or child of the opposite sex.

By the interbreeding criterion every individual is a member of the same species as its parents. This is an unsurprising, not to say platitudinously obvious conclusion, until you realise that it raises an intolerable paradox in the essentialist mind. Most of our ancestors throughout evolutionary history have belonged to different species from us by any criterion and we certainly couldn't have interbred with them. In the Devonian Period our direct ancestors were fish. Yet, although we couldn't interbreed with them, we are linked by an unbroken chain of ancestral generations, every one of which could have interbred with their immediate predecessors and immediate successors in the chain.

In the light of this, see how empty are most of those passionate arguments about the naming of particular hominid fossils. *Homo ergaster* is widely recognised as the predecessor species that gave rise to *Homo sapiens*, so I'll play along with that for what follows. To call *Homo ergaster* a separate species from *Homo sapiens* could have a precise meaning in principle, even if it is impossible to test in practice. It means that if we could go back in our time machine and meet our *Homo ergaster* ancestors, we could not interbreed with them.* But suppose that, instead of zooming directly to the time of *Homo ergaster*, or indeed any other extinct species in our ancestral lineage, we stopped our time machine every thousand years along the way and picked up a young and fertile passenger. We transport this passenger back to the next thousand-year stop and release her (or him: let's take a female and a male at alternate stops). Provided our one-stop time traveller could accommodate to local social and linguistic customs (quite a tall order) there would be no biological barrier to her interbreeding with a member of the opposite sex from 1,000 years earlier. Now we pick up a new passenger, say a male this time, and transport him back another 1,000 years. Once again, he too would be biologically capable of fertilising a female from 1,000 years before his native time. The daisy chain would continue on back to when our ancestors were swimming in the sea. It could go back without a break, to the fishes, and it would still be true that each and every passenger transported, 1,000 years before its own time would be able to interbreed with its predecessors. Yet at some point, which might be a million years back but might be longer or shorter, there would come a time when we moderns could not interbreed with an ancestor, even though our latest one-stop passenger could. At this point we could say that we have travelled back to a different species.

The barrier would not come suddenly. There would never be a generation in which it made sense to say of an individual that he is *Homo sapiens* but his parents are *Homo ergaster*. You can think of it as a paradox if you like, but there is no reason to think that any child was ever a member of a different species from its parents, even though the daisy chain of parents and children stretches back from humans to fish and beyond. Actually it isn't paradoxical to anybody but a dyed-in-the-wool essentialist. It is

no more paradoxical than the statement that there is never a moment when a growing child ceases to be short and becomes tall. Or a kettle ceases to be cold and becomes hot. The legal mind may find it necessary to impose a barrier between childhood and majority—the stroke of midnight on the eighteenth birthday, or whenever it is. But anyone can see that it is a (necessary for some purposes) fiction. If only more people could see that the same applies to when, say, a developing embryo becomes 'human'.

Creationists love 'gaps' in the fossil record. Little do they know, biologists have good reason to love them too. Without gaps in the fossil record, our whole system for naming species would break down. Fossil could not be given names, they'd have to be given numbers, or positions on a graph. Or instead of arguing heatedly over whether a fossil is 'really', say, an early *Homo ergaster* or a late *Homo habilis*, we might call it *habigaster*. There's a lot to be said for this. Nevertheless, perhaps because our brains evolved in a world where most things do fall into discrete categories, and in particular where most of the intermediates between living species are dead, we often feel more comfortable if we can use separate names for things when we talk about them. I am no exception and neither are you, so I shall not bend over backwards to avoid using discontinuous names for species in this book. But the Salamander's Tale explains why this is a human imposition rather than something deeply built into the natural world. Let us use names as if they really reflected a discontinuous reality, but by all means let's privately remember that, at least in the world of evolution, it is no more than a convenient fiction, a pandering to our own limitations.

---

*I am not asserting that as a fact. I don't know if it is a fact, although I suspect that it is. It is an implication of our plausibly agreeing to give *Homo ergaster* a different species name.

African Trailblazers

ANN GIBBONS

*Most scientific problems are far better understood
by studying their history than their logic.*

Ernst Mayr, evolutionary biologist

*I was told as a young student not to waste my time
searching for Early Man in Africa, since "everyone
knew he had started in Asia."*

Louis Leakey, 1966

It was an October morning in 2003. Meave Leakey was driving from Nairobi north along the eastern wall of the Rift Valley in central Kenya, expertly weaving around potholes in the tarmac and dodging oncoming buses that played chicken with smaller vehicles to scare them out of their way. Trucks belched black smoke that stung her eyes, cyclists hitched rides up hills holding on to the backs of buses, and jampacked public shuttles called *matatus* spent almost as much time passing each other as staying on their side of the two-lane road. As Meave negotiated this nerve-racking traffic on the Uplands Road between Nairobi and Nakuru, she calmly recounted the story of how the search for human ancestors began in eastern Africa. "Until the middle of 1959, only a few people seriously believed eastern Africa was a sensible place to look for the earliest human ancestors," she said.

This history is personal for her, because it is the saga of her husband's parents, Louis and Mary Leakey. This formidable pair was among the first to stake their careers on Africa as the birthplace of mankind. For three decades, their work in eastern Africa was an almost solitary pursuit. Even those researchers who found fossils of early ape-men in South Africa during that time had trouble convincing their European colleagues that these primitive fossils were ancestors of humans. Then, in 1959, Mary found a fossil in Olduvai Gorge, Tanzania, that would finally give the Leakeys the hard evidence they had long sought that early humans did indeed evolve in eastern Africa. Louis named the cranium, or partial skull, *Zinjanthropus boisei*—*Zinj* from an Arabic word for eastern Africa, *anthropus* from the Greek word for man, and *boisei* from Charles Boise, a London businessman who was their benefactor. Translated, the name is an assertion: "Man from Eastern Africa." And once the Man from Eastern Africa made his appearance, the push was on to find more extinct men and women. Soon, teams of French and American researchers headed to eastern Africa, like forty-niners to California during the gold rush.

The fossils they found in the Great Rift Valley in the 1960s and 1970s soon made it known as the cradle of humanity.

But Louis's search for the missing link in eastern Africa had started more than thirty years earlier, right in the gullies and rock shelters alongside the Uplands Road where Meave was driving nearly eighty years later, high above the Great Rift Valley. In 1926, the first year that Louis worked in the area, the Uplands Road did not exist, and the trip from Nairobi in Louis's Model T Ford took a half day over muddy tracks. The air was so clear that Louis could see miles across the Great Rift Valley from his camp, down a slope covered with acacia trees and scrub brush to Lake Elmenteita, a shallow alkaline lake rimmed with the pink froth of flamingos. Bush babies, leopards, aardvarks, and ibises lived in the acacia woodlands near the shore, and a herd of hippos wallowed in the lake. Beyond the lake, the jagged calderas of several extinct volcanoes lined up to form the silhouette of a human figure that the local Masai tribesmen called Elngiragata Olmorani, for Sleeping Warrior. A few British settlers were staking out the Masai's traditional grazing grounds for homesteads for cattle ranches, but otherwise the area was still remote and primeval.

Today, Elmenteita is only an hour's drive beyond the shanty sprawl surrounding Nairobi, and much of the land around the lake is fenced in by private owners. The hippos are gone and the bush babies and a few remaining leopards have retreated to a wildlife sanctuary. But the view of the rift valley far below is still stunning, and Meave named the volcanoes visible in the distance as she searched for a familiar turnoff. Spotting it, she jostled down a dirt road, past a quarry where workers mined a crumbly white rock called diatomite, and pulled into a grassy driveway. The sign said: KARIANDUSI MUSEUM, NATIONAL MUSEUMS OF KENYA. It did not look like much: a guard's hut and a whitewashed, single-room museum with some casts of skulls and an exhibit on the formation of the Great Rift Valley. It was clearly off the tourists' safari circuit.

A curator eventually appeared, delighted to find someone who wanted to tour the site on a Monday in October. He was even more surprised to find out that the tall woman with straight, silver-gray hair and hazel eyes spoke Swahili and was a member of the Leakey family—a name that is well-known in Kenya. At sixty-one, Meave had been here many times before and knew the history of Kariandusi by heart. She is long-legged and fit after a lifetime of hard work scrambling over rugged terrain for fossils, and she did not need a guide to lead her into the gulch. She let the curator show her the way to a series of steplike pits anyway, partly because she was curious to learn what he knew. He took her to the first pit, which was covered with a corrugated metal roof. Meave leaned over the rail and pointed to the dirt floor encrusted with hundreds of stone tools, most made of glassy black obsidian, the rock that comes from volcanic lava. There were tear-shaped hand axes, two-sided flakes, and even triplets of round stones that look like black billiard balls. "This is where it all began," said Meave. She was referring to Louis's search for early man in eastern Africa. These shiny black tools at Kariandusi were among the first hard evidence of a sophisticated ancient Stone Age culture in eastern Africa. They were made by people who left them on the shores of the lake almost 500,000 years ago, perhaps when they came to hunt wild animals that were quenching their thirst at dawn or dusk. She climbed down wooden stairs into a deep gully where an ibis was roosting in a tree. Meave remembered a photo from *National Geographic* that showed Louis bending over a cliff there, pointing to stone tools embedded in the wall. This was precisely the spot where Louis and his team found their first ancient hand axes in eastern Africa in 1929.

∎

In 1871, less than sixty years earlier, Charles Darwin had proposed in *The Descent of Man and Selection in Relation to Sex*, that the earliest ancestors of humans probably lived on the African continent. But that prediction was based on absolutely no evidence from fossils. In fact, at the time only one fossil of another type of human being was known, and that was of a Neandertal that had lived in the Neander valley of Germany sometime in the past 70,000 years. Darwin chose Africa because humans' closest cousins in the animal kingdom—chimpanzees and gorillas—lived in Africa; therefore, he wrote, "it is more probable that our early progenitors lived on the African continent than elsewhere." But Darwin admitted that it was "useless to speculate on this subject," since an extinct European ape nearly as large as humans could also have given rise to humans.

That didn't stop Darwin's colleagues from conjecture. His friend and champion Thomas Henry Huxley (also known as Darwin's "bulldog") agreed that humans should be put in the same family as chimpanzees and gorillas, and enthusiastically promoted that view in debates and in his 1863 book *Evidence as to Man's Place in Nature*. (Darwin himself avoided dealing directly with the issue until 1871, when he published *The Descent of Man*.) But a contemporary and admirer of Darwin's, the prominent German biologist Ernst Haeckel, believed that the Asian apes (orangutans and gibbons) were closer relatives of humans than the African apes were. Haeckel proposed this link in his sketches of the human family tree in 1868, drawing a direct line between Asian apes and a new species of fossil human that he proposed and explicitly called the missing link. In his writings and lectures, Haeckel fleshed out this missing link as a hairy, primitive creature half ape, half man, named *Pithecanthropus alalus*. (Literally, "ape-man without speech," from the Greek *pithec*, "ape," *anthropus*, "man," and *alalus*, "without speech.") It walked semierect, had protruding teeth, and was speechless. But there wasn't a bit of hard evidence to support

this vision of an ape-man. Haeckel's missing link was purely theoretical.

One person who heard of Haeckel's ideas on human evolution was a young Dutch medical student, Eugène Dubois, who became the first of a long line of young men obsessed with finding this missing link—and winning honor and fortune. In 1887, when Dubois couldn't get the Dutch government to finance an expedition to the tropics to search for fossils, he quit his job as an anatomist at the University of Amsterdam and joined the Royal Dutch East Indies Army as a military doctor so he could be posted to the Dutch East Indies, now the Indonesian archipelago. Ancient fossils of mammals that had been alive during the earliest stages of the Age of Man (the Pleistocene epoch) had been found there. He thought it most likely that fossils of extinct ancestors of similar age would be preserved there as well. According to his biographer, the anthropologist Pat Shipman, he also reasoned that if apes lived in the tropics today, extinct apes and early ape-men would also have been more likely to live in the tropics.

It was an incredible long shot, but he sailed for the Dutch East Indies at the age of twenty-nine with his young wife and their baby. Dubois was the first of many fossil hunters to risk his life in search of an elusive missing link. He battled malarial fever without modern medicines; his team hemorrhaged workers, who ran away, became ill, or stole fossils to sell as "dragon" bones to traders from China; and they faced bad roads through the overgrown jungles of Java, mosquitoes, hellish heat, and torrential rains. Amazingly, Dubois and his family survived. More incredibly, he found what he was looking for. In August 1891, his crew discovered the molar of a hominid eroding out of the banks of the Solo River near the village of Trinil on the island of Java. Two months later, his crew found a skullcap that was larger than that of a chimpanzee's but smaller than that of a human's. Later, they found a thighbone. Dubois recognized the skullcap as belonging to a species that must have had a brain intermediate in size and development between humans and apes. But the thighbone belonged to a creature that walked upright—even before its brain had expanded.

He pronounced it *Pithecanthropus erectus* (or "erect apeman"). It was an amazing feat. He had never searched for fossils but had nonetheless traveled halfway around the world to an island archipelago where he'd reasoned that such fossils should be found. Today, Dubois's Java man is still recognized as a major discovery—the first fossil found of an early hominid and the first specimen of *Homo erectus* (as it was later renamed), a key human ancestor that arose about 1.8 million years ago, probably in Africa, before migrating to Asia, where it persisted until sometime in the past 250,000 years. This species of human and its descendants may even have lived until as recently as 13,000 years ago in the form of the so-called Hobbit, the dwarf species of human whose remains were found in 2004 on the Indonesian island of Flores.

Convincing his colleagues that he had found the missing link would prove more difficult than finding the fossils themselves. When Dubois announced his discovery of Java man in 1893, he expected honor and scientific recognition. Instead, his monograph on this "man-ape" was met with skepticism and snide comments, some dismissing the fossil as a giant gibbon or an individual whose features had been distorted by disease or a wound. Word reached him in Java in 1894 that his European colleagues questioned many aspects of his monograph on the fossil—from his claim that all the fossils came from the same individual to the way his crew had mapped the fossil site.

Dubois traveled to Europe in 1895 to defend his discovery, winning a few converts as he lectured and displayed the bones themselves. Haeckel, who had inspired him, was one who embraced Java man as a human ancestor. But the theory of evolution was still new and was not universally accepted among scholars. Although Dubois was well educated and a meticulous scientist, perhaps the real problem was that an ancestor that looked so much like an ape was more than the scientific establishment of the late nineteenth century could accept. His biographer Shipman concluded, "In truth, the problem lay more in the prevailing beliefs among his colleagues than in Dubois' shortcomings."

As he battled his colleagues well into the twentieth century, Dubois's own shortcomings also became apparent—he grew secretive and territorial about his fossils, particularly after he gave a cast to a German anatomist who then toured the world with it, giving lectures and publishing a detailed description about Java man before Dubois had finished his own analysis of the skull he had found. After that, he withdrew from his colleagues and even rigged a mirror above his door at home so he could see who was there when his maid answered, turning away prominent scientists who'd traveled from as far as America to see the fossils that he stored in his basement.

History would prove Dubois right about Java man, but he died an angry man, unrecognized and estranged from his wife and friends—all alienated by his increasing irascibility. He was, perhaps, the first fossil hunter to become a victim of his own success in finding a human ancestor, as if the fossil came with a mummy's curse.

It was a bitter omen of the kind of controversy that would swirl around almost every new fossil vying to be a human ancestor. Even experienced researchers often react with more emotion to the discovery of human ancestors than they do to fossils of any other animal, including dinosaurs. New fossils almost always shatter preconceived notions of what our ancestors should look like, revealing our origins as ordinary apes rather than as exalted beings marked from the beginning with a big brain or some other sign of special destiny. Darwin recognized this reflexive denial of our savage past in *The Descent of Man* when he warned, "We must, however, acknowledge, as it seems to me, that man with all his noble qualities, with sympathy which feels for the most debased, with benevolence which extends not only to other men but to the humblest living creature, with his god-like intellect which has penetrated into the movements and constitution of the solar system—with all these exalted powers—Man still bears in his bodily frame the indelible stamp of his lowly origin."

■

Preconceptions of what the missing link should look like, in fact, colored the scientific community's remarkable acceptance of the next fossil proposed as an early hominid, almost twenty years later. This was the Piltdown hoax, a sorry tale that began when a young English soldier and amateur antiquarian, Charles Dawson, "discovered" fragments of a skull, a lower jaw, and some teeth in a gravel pit at Piltdown Sussex, England, over the years between 1908 and 1915. Dawson's fossils were convincing to several leading scientists of the day, including paleontologist Sir Arthur Smith Woodward, who was keeper of geology at the British Museum, and Sir Arthur Keith, the most eminent British anatomist of the day. Keith had spent a year studying the skulls of two hundred primates and come to the conclusion that the essence of being human was to have a big brain relative to the size of the body.

With Piltdown man, Keith got exactly what he was looking for. When put together, the big brain and primitive apelike jaw of Piltdown precisely fit Keith's image of a "missing link" between humans and apes. Keith embraced it as a direct human ancestor, and defended it for decades. There were some skeptics, chiefly Americans who argued that the jaw did not belong with the skull.

It turns out that Piltdown man *had* been made to order. Scientists from Oxford University and the British Museum formally declared it a fake in 1953 after chemical tests showed that the jawbone belonged to an orangutan and that the skullcap was human and older than the jawbone. The forgers had filed the molars to look more human, and they had stained the bones and stone tools to match before planting the "fossils" in the gravel pit. They were obviously familiar with Keith's notion of a human ancestor. Dawson is the front-runner in a line of suspects, but he died in 1916, perhaps before he planned to admit to the hoax.

Unfortunately, Piltdown man continued to cause damage long after Dawson died. For forty years, in fact, fallout from Piltdown would delay the recognition of the authentic fossils, including Dubois's Java man. Java man and fossils found later in South Africa did not fit the image as well as Piltdown for the one missing link between modern humans and apes, so they were relegated to side branches on the family tree—shown as separate species, types or "races" of humanlike creatures that had gone extinct. Worse, some were even thought to have given rise to humans still living in the Stone Age, but not Europeans. Confusion persisted for many years. One popular book, called *Meet Your Ancestors*, by Roy Chapman Andrews, a retired director of the American Museum of Natural History, vividly illustrates a popular view of the human family that still persisted in 1945, when the book was published. On the book's endpapers, a diagram labeled "The Family Tree of Man" shows Java man giving rise to living Australian aborigines. Another Chinese fossil that was considered a close cousin of Java man begets living Mongoloids. Both of those lineages start with a ground ape. But a separate lineage that originated with a forest ape leads to the European cave painters known as the Cro-Magnon, whom the author describes as "wise men" and the ancestors of European *Homo sapiens*. Andrews allows for some inbreeding but describes aborigines as "not much advanced beyond the stage of Neanderthal Man." The message was clear: Europeans were a superior race with their own largely separate ancestry from forest apes, rather than from lowly ground apes.

■

The most significant fossil eclipsed by the long shadow cast by Piltdown was the first early hominid skull from Africa. In 1924, an Australian anatomist named Raymond Dart had taken the post of chair of anatomy at the University of Witswatersrand in Johannesburg, South Africa, after completing his medical studies in London. When he got to Johannesburg, he was dismayed to find no reference collection of skeletons to study. So he announced a competition to see which students could bring in the most interesting bones. He described the outcome years later in his autobiography, *Adventures with the Missing Link:* One student, the only woman in the class, told him she had seen the skull of a fossil baboon on the mantel at a friend's house. Dart told her it was probably not a baboon, since he knew of only two fossil primates that had been found at that time in Sub-Saharan Africa. The student borrowed the skull, which was indeed a fossil baboon. It belonged to her friend, a director of the Northern Lime Company at Taung, where fossils turned up occasionally in the limestone rock. Dart asked the manager to send him any other fossils they might find.

Later that year, in the summer of 1924, two wooden crates were left on Dart's stoop, just as he was struggling to put on a stiff-winged collar for a friend's wedding at his house, in which he was to serve as best man. Dart tore off the collar and pried open the crates with a crowbar, against the admonishments of his wife, who warned that the wedding guests would soon be arriving. In the first crate, there was little of interest. But in the second crate, right on top was the fossilized cast of a brain in limestone—an endocast, which is the imprint of the brain on the inside of the skull, much like a death mask is the plaster impression of a face. This one was complete with the pattern of blood vessels. Having made an intensive study of endocranial casts during his medical studies in London, Dart was thrilled. Ransacking the box for the face that matched the brain cast, he quickly found a large piece of rock and fossilized bone that fit perfectly. He could see that the brain came from a creature with a forebrain that was larger than a chimpanzee's, although it was not big enough to belong to a primitive human. He was standing in the shade admiring it when the groom found him and broke his reverie, telling him he had to finish dressing because the guests were arriving. But, Dart would write in his autobiography, he could scarcely wait until the wedding was over so he could reexamine his "treasures," which he had stored in his wardrobe.

It took him months to clean the fossil, like a sculptor chiseling away stone from a form he sees in marble. Eventually, on December 23, he split the rock with, a hammer and his wife's steel knitting needles, and the face emerged. What Dart saw was a baby's face, with a full set of milk teeth and its first permanent molars just in the process of erupting. "I doubt that there could have been any parent prouder of his offspring than I was of my Taungs baby on that Christmas of 1924," Dart wrote. The baby was neither ape nor human. It had small canines like a four-year-old human, and the position of the head right on top of the spine indicated that it had walked upright. A neuroanatomist

who had not set out to find the missing link, Dart soon realized that he had found a probable early member of the human family. By early 1925, Dart was ready to baptize his baby: he published a paper in the prestigious journal *Nature*, in which he named it *Australopithecus africanus*, or "southern ape from Africa." It soon became known as the Taung baby. A few immediately recognized its importance as a hominid that was even more primitive than Dobois's Java man.

Others, perhaps blinded by the anatomy of Piltdown, dismissed the Taung baby as a young ape, rather than a hominid. It had exactly the opposite gestalt of Piltdown: where Piltdown had a large brain and primitive jaw. Taung had a relatively small brain (although the forebrain was noticeably large) and modern teeth. Both could not have been direct ancestors of modern humans. Keith suggested that Dart had mistakenly identified the Taung baby as a hominid because its young age made it look more modern. In time, he wrote, its face and teeth undoubtedly would have looked more apelike as it developed into an adult ape.

Like Dubois, Dart also traveled to England to rescue the reputation of his fossil hominid. He had anticipated skepticism, but when he got there in 1931, he was not prepared for his delicate skull to be upstaged by a new fossil. This one was from China— Peking man, a skull discovered in 1929 by a young Canadian physician, Davidson Black. The skull resembled Dubois's Java man, but Black gave it a different name, *Sinanthropus pekinensis*, that would persist for years before paleoanthropologists recognized that it and Java man were the same species—the renamed *Homo erectus*. Black was much more effective than Dart or Dubois in spreading the news—sending tactful advance warning to leading scholars he had trained with in England, then touring Europe and the United States immediately with casts and a slide show, which helped build the case that Peking man was a serious contender for the missing link. He claimed that his fossils were intermediate between Dubois's Java man and Neandertals, putting Peking man a step closer to being human.

By the time Dart appeared in England, his presentation on the Taung baby was anticlimactic, and the scientists he visited were much-more interested in talking about the Chinese fossils they had seen and an Asian origin for humans. The Taung baby was a nice fossil, but it was an also-ran, not on the one line of descent to humans. Keith's criticism of the Taung fossil even extended to Dart—he later wrote that he was "rather frightened" of Dart because of "his flightiness, his scorn for accepted opinion, the unorthodoxy of his outlook." Personal politics again colored the reception of a bona fide fossil hominid.

When Dart left London, he was deeply discouraged. He did not fight back. Instead, he put aside his monograph on the fossil and even gave up work on other fossils for many years. He later would explain that he had no burning zeal for fossils and had not come to Africa to search for the missing link. He had other professional interests in neurology and was by then eager to get out of the field of human origins.

■

Dart's experience should have discouraged any young researcher who hoped to find a missing link in Africa. But Louis Leakey, who was a student at St. John's College in Cambridge,

England, liked to challenge conventional wisdom and had unshakable confidence in his own instincts. He had read Darwin's *Descent of Man*, knew about the Taung baby, and had heard about the discovery of a fossil skeleton in 1913 at Oldoway in the British colony Tanganyika—now Olduvai in the nation of Tanzania. These African fossils reinforced his own feelings that Africa was the birthplace of humankind. "I became excited with the idea that everyone was looking in the wrong place," he wrote in his first autobiography, *White African*. And he began to make plans to lead his own small expedition to eastern Africa in the summer of 1926, where he intended to excavate a cave that he and his sisters had discovered when he was a young child growing up in Kabete, in the lush mountains north of Nairopi, where he was born.

Louis was the son of British parents who ran an Anglican mission in Kabete, and he had acquired a missionary's zeal and single-mindedness of purpose in the face of obstacles that would have daunted most men. As a boy, playing with his friends from the Kikuyu tribe, he had found obsidian stone tools in the countryside around Kabete. He had taken some of those tools to a zoologist at Nairobi's Coryndon Museum who was a friend of the family's and who had taught Louis how to classify birds and label museum specimens. The zoologist confirmed that the tools Louis had found were indeed arrowheads. Louis decided then that he wanted to be both a missionary, like his father, and a scientist, like the zoologist. He would one day find out about the Stone Age men who had made the tools he found. At the age of thirteen, he had already set his course.

But when he was sent to boarding school in England at age sixteen, he faced challenges that threatened to derail his plans. Louis later wrote that when he went to school, he was shy and unsophisticated and fit badly into the life of the place. His friends in Kenya had been Kikuyu boys who considered him a blood brother, because he had taken part in their secret initiation ceremonies. Though he had studied Latin and mathematics, and had spoken French at the dinner table in Kenya, Louis related, "In language and mental outlook I was more Kikuyu than English." He did not know Greek; worse, he had never played cricket and did not know how to swim, which brought him ridicule from much younger boys and hazing from older boys. In Kenya, he had been used to a great deal of freedom, building, near his parents' home, his own three-room house, where he lived as a teenager. He earned money by trapping animals and helped his father teach school in Kabete. In England, he chafed at the rules for bedtime and passes for visiting the nearby town, which made him feel like "a child of 10 when I felt like a man of 20." He even had to get permission to stay up late to catch up on his studies. Despite this effort, his headmaster discouraged his ambition to win a scholarship to attend Cambridge University, his father's alma mater.

Perhaps the experience of being different from the other boys contributed to the making of a maverick, because by the time he did get a small scholarship, to St. John's College at Cambridge, he already was charting his own course—convincing administrators to accept Kikuyu as one of two modern-language requirements. He also managed to take time off to search for dinosaur fossils in Tanzania while recovering from a head injury suffered while playing rugby. He got in trouble with the authorities for playing African signal drums on the roof of his rooms in St. John's College, but he earned top marks in anthropology and archaeology and got a small grant to spend Christmas vacation studying collections of ancient artifacts in museums in Europe. While there, he met the prominent German paleontologist Hans Reck, who had found the skeleton in Olduvai. Though the skeleton would eventually prove to be the remains of a modern human that had intruded into older sediments, Louis half seriously made plans that one day they would visit, Olduvai together.

As he studied the artifacts from ancient cultures in Europe, Louis became even more certain that there must be evidence of humans in Africa at least as old as those in Europe. At the time, archaeologists thought that the oldest human culture was one found in Europe—a stone-hand-axe tradition they called the Chellean that had been made by the earliest ape-men. Louis thought he could find similarly ancient stone tools and, perhaps, their toolmakers in Africa: "I was born in Eastern Africa and I've already found traces of early man there. Furthermore, I'm convinced that Africa, not Asia, is the cradle of mankind." One of his professors tried to dissuade him, uttering the now legendary admonishment that he should not waste his time looking for early man in Africa. He should go instead to Asia, home of Java man and Peking man.

Louis ignored the advice, since he was intent on finding out if Darwin was right about Africa being the birthplace of humankind. In 1926, he was awarded a research fellowship by St. John's to return to Kenya for a year to lead the First Eastern African Archaeological Expedition (consisting of himself and a fellow Cambridge graduate). After finding little of importance in caves he had discovered as a child in Kabete, he set up camp in an abandoned pigsty on a farm on the slope above Lake Elmenteita where he had heard about a prehistoric burial mound. Eventually he moved to an abandoned farmhouse where he was joined by his first wife, Frida, and a half-dozen other researchers whose excavations uncovered skeletons of prehistoric humans and tools.

It took them three years, working in rock shelters and on exposed hillsides along the ridges above Lake Elmenteita and the Nakuru-Naivasha basin before they found the evidence Leakey was seeking of a truly ancient culture. Finally, in May 1929, just three weeks before the end of the 1928–29 field season, two members of Louis's team surveyed the gullies at Kariandusi and discovered a cache of tear-shaped hand axes. Louis was delighted: these tools were exactly what he had sought—relics of a culture that appeared to be almost as ancient as the Chellean culture of Europe. He knew they were ancient because they were found along the shorelines of a giant prehistoric lake. He thought the tools dated to the first ice age, which at the time was thought to be 40,000 to 50,000 years ago. Never one for understatement, Leakey wrote in *White African*: "The discovery was, therefore, of the very greatest importance."

Today, the hand axes at Kariandusi are thought to be 500,000 years old, rather than 50,000. But his estimate fit with the best geologic dates known at the time, which were estimated using relative dating, which determines the age of a layer of rock based on the fossils it contains and the rocks and fossils found above and below it. For many years, there was no clear idea of the absolute age of each rock layer or of the earth itself. Therefore, in 1926, the age of the earth was thought to be only 65 million years (now it is dated to be 4.5 billion years), and the age of mankind (the Pleistocene) was estimated at 500,000 years (today the Pleistocene is dated precisely to just under 2 million years). The tools were indeed old and one of the first concentrations found in eastern Africa of the Acheulean tradition, a tool kit that was the Swiss Army knife for *Homo erectus*, the species of Java man and Peking man. Early modern humans, now known as archaic *Homo sapiens*, also used the same type of tool kit, until someone finally invented a more sophisticated stone technology, which caught on about 250,000 years ago.

■

Louis's triumph was short-lived. His stone tools should have been the beginning of a brilliant academic career. They were not. Louis did go to Olduvai with Reck in 1931, and he also worked at two sites in western Kenya, Kanjera and Kanam, where he found a jawbone in 1932. His troubles began when he boldly claimed that the jawbone, which he named *Homo kanamensis*, was "not only the oldest human fragment from Africa, but the most ancient fragment of true *Homo* yet to be discovered anywhere in the world." At first, his colleagues congratulated him, assuming Louis's date of 500,000 years was correct for the jawbone. Louis, not yet thirty, got his first taste of fame.

He returned to Kanam with the eminent British geologist Percy Boswell to prove the ancient age of his jawbone. Instead, he found to his horror that the iron pegs he had left in concrete to mark the spot had been removed, perhaps by the local Luo people for fishing spears. And the photos that Louis had taken of the site had not turned out, so it was impossible to confirm the precise spot where the jawbone was found and thereby date the sediments to nail down the jawbone's age.

When Boswell returned to England and reported to the Royal Society that Louis did not know exactly where he had found his fossil, Louis's reputation suffered. Later, when the fossils were shown to be much younger and Boswell published his doubts in a letter to the journal *Nature*, Louis was labeled as an enthusiast who worked too fast to be trusted as a careful scientist. It also hurt him that he had scandalized Cambridge colleagues by leaving his wife, Frida, in 1934, when she was pregnant with their second child. He had fallen in love with Mary Nicol, a twenty-year-old amateur archaeologist who was illustrating a book Louis was writing on the evolution of man and his culture. The two were living together in a village near Cambridge.

Louis married Mary just as his professional prospects were dimming. His expectations of a fellowship at Cambridge faded, and his ability to obtain grants to search for fossils in Africa deteriorated. The combination of Percy's report on his unreliable science at Kanam and his illicit affair with Mary led to his total ostracism from British academic circles. Louis pressed on anyway, taking Mary to Olduvai in Tanzania in 1935, where they would work off and on for the rest of their lives. There were only tiny grants, and Louis earned just enough money to support his growing family by writing books about his life, human prehistory, and the Kikuyu people. But he and Mary continued to

explore and excavate archaeological and anthropological sites in Kenya. They found stone tools, teeth, and even the jaw of an extinct ape that had lived millions of years ago on the shores of Lake Victoria in Kenya.

It wasn't until more than a decade later, in 1948, that they made a discovery of international importance—the first skull of a fossil ape ever found. It was the partial skull, face and jaw of an 18-million-year-old extinct ape called *Proconsul* from Rusinga Island in Lake Victoria, Kenya. Mary, who had found it, flew with it on her lap to England so that the prominent Oxford University anthropologist Wilfrid Le Gros Clark could examine it. News of the discovery traveled ahead of her to London where newspaper headlines proclaimed, THE LEAKEYS FIND IMPORTANT FOSSIL-MAN ANCESTOR. The headlines were wrong. Although Louis thought *Proconsul* a possible ancestor for later apes and man, he knew it was far too primitive to be a human ancestor. Today it is seen as ancestral to both the lesser and the great apes that came later.

Le Gros Clark was an important friend of the Leakeys. He played a critical role in helping to convince a growing number of prehistorians that Africa was the birthplace of humankind. A year earlier, at the First Pan-African Congress of Prehistory in Nairobi, organized by Louis, Le Gros Clark had openly stated that the Taung skull and other new fossils of its species were indeed near human and belonged in the human family tree. He had traveled to South Africa before the congress and had met with Dart and Robert Broom. Broom, a Scots-born paleontologist who had been trained as a medical doctor, was curator of vertebrate fossils at the Transvaal Museum in South Africa, where he would carry on the paleontological work after Dart became discouraged in the 1930s and 1940s. He and his assistant, John T. Robinson, also found a diverse group of hominids, representing at least three different species, in caves in South Africa. Those fossils included adult specimens of the same species as the Taung baby.

When Le Gros Clark finally saw these fossils from South Africa, he recognized them as hominids. Le Gros, as he was called, gave them his official blessing, and his lead was quickly followed all over the world. The forged Piltdown fossils had recently been given a long-overdue deathblow in 1953 when chemical tests were finally done at the British Museum; the fossils were declared fakes. As Piltdown exited, the stage was cleared for the South African fossils and the entry of the next major player in the prehistory of Africa, the now-famous skull of *Zinjanthropus boisei* that changed Mary and Louis Leakey's lives for good.

■

Mary Leakey discovered Zinj, as it was called, on the morning of July 17, 1959. Louis was in camp with a fever, and she set out by herself on a walk with her two Dalmatians to explore the fossil beds at Olduvai Gorge, where Louis had worked off and on since 1931. A scrap of bone poking out of the dirt caught her eye. She brushed away a little of the sediment and saw two teeth, black-brown and set in the curve of a jaw. She knew right away it was a hominid skull, and rushed back to camp to break the news to Louis. The Leakey biographer Virginia Morell described how Mary told Louis over and over: "I've got him! I've got him! I've got him!"

Louis, who was groggy with fever, was puzzled: "Got what?"

"Him, the man! Our man. The one we've been looking for," she replied.

Louis joked later that he quickly recovered from his fever as soon as he saw the fossil. He was disappointed to note that the skull had teeth just like those of the south African australopithecines, which he had considered an evolutionary side branch that did not give rise to modern humans. He had been hoping for a creature that looked more like Sir Arthur Keith's ideal ancestor—one with a bigger brain and more modern teeth that would clearly fit in the genus *Homo*. (The definition of the genus *Homo* at the time required a brain of at least 750 cubic centimeters, which was larger than a gorilla's brain but less than the smallest known human brain.) He called those "true men," and the australopithecines "near-men." But his disappointment gave way to excitement as he saw other humanlike characteristics in the skull. As he and Mary excavated the skull, it became clear that it was a robust australopithecine with enormous jaws and teeth, which would prompt one colleague to dub it "Nutcracker man." Interestingly, Louis pronounced Zinj a hominid on the basis of traits in its skull and teeth—not on any direct evidence from the skeleton for upright walking.

Louis called it "the connecting link between the South African nearmen and true man as we know him." The robust skull is recognized today as a member of *Australopithecus boisei* (also known *Paranthropus boisei*). More members of the same species, including a partial skeleton, have been found in Ethiopia and Kenya, and today *A. boisei* is considered more of a cousin whose own lineage went extinct than a direct ancestor of humans. Along with the australopithecines in South Africa, Zinj reinforced the hypothesis that Africa was the birthplace of humanity.

The discoveries also offered a new view of human evolution, showing that the earliest sign of becoming human was not a big brain, as Sir Arthur Keith had predicted (and, interestingly, as Louis had believed, as well). The Taung baby's species and Zinj, as well as more robust forms of australopithecines from South Africa, showed that one of the first steps toward becoming human was walking upright. The big brain did not come until later, less than 2 million years ago, when it began to expand in *Homo erectus,* reaching its largest size in Neandertals and modern humans.

■

The recognition of Africa as the birthplace of humans came at the same time as a revolution in the dating of fossil sites in the 1950s and 1960s. Louis had claimed that Zinj lived more than 600,000 years ago, and geologists thought at the time that the Pleistocene—the Age of Man—was about 1 million years ago (that date of 1 million years had doubled since the 1920s). Thus, Louis and Mary were stunned when, in 1961, a team of geologists at the University of California in Berkeley announced a date that was much older.

Berkeley geologists Jack Evernden and Garniss Curtis had traveled to Olduvai in 1957 and 1959 and collected samples of dirt at the base of the layer of sediments where the skull of Zinj was found. Researchers are unable to date ancient fossils directly, since they lack the radioactive elements necessary for radiometric dating. Instead, geochronologists collect samples of sediment laid down directly above and below a fossil, if possible, to bracket the age of the fossil. Back in their lab at Berkeley, Evernden and Curtis applied a sophisticated new method they had helped develop with physicists at Berkeley in the 1950s to the sediments from Olduvai. The method, known as potassium-argon dating, gave the age of sediments by using the radioactive decay of elements found naturally in rocks and soil. The method capitalized on a century of research that showed that unstable elements in rocks, such as radioactive uranium, potassium, and argon, decay gradually over time, emitting radiation and producing daughter elements that are stable and, therefore, not radioactive. The decay happens at a constant rate: the unstable version, or isotope, of potassium in volcanic sediments—called potassium-40—steadily decays, or ticks over into a stable isotope of gas—argon-40—at a known rate.

In their lab, Curtis and Evernden heated the samples of sediment until they emitted the gas trapped inside. The pair used sophisticated new instruments to isolate and count atoms of potassium and argon gas as they were released from these sediments. The ratio of potassium-40 to argon-40 told them precisely how old these sediments were—the more argon, the more time had passed since the sediment was laid down at the fossil site.

The same methodology is also used with other isotopes, but the preferred method at ancient fossil sites in Africa today is a new version of the potassium-argon dating where researchers study the decay of two isotopes of argon. They use lasers to heat and count individual crystals of the isotopes of argon-40 and argon-39. The quality of the dates, however, depends on the quality of the sample of volcanic sediments, which must be rich in potassium or argon. Where fossils are found in soil without volcanic crystals, such as in Chad, researchers use other, less precise methods—such as dating sediments by the identification of extinct animals whose age is known.

In 1961, Curtis and Evernden were the first to use radiometric methods to date a fossil of an early human ancestor. Their potassium-argon dates for Zinj still hold; in this case, they dated the skull to about 1.75 million years. "It was four times older than previously thought at that time," said the eminent Berkeley anthropologist Sherwood Washburn. "It changed the conception of the rate of evolution regarding fossil man."

The new age fit Louis Leakey's notion that humans had been evolving in Africa far longer than previously believed. At about the same time, Evernden and Curtis also used the new method to date volcanic rock beds of known age in Italy that were a benchmark for the beginning of the Pleistocene. They showed, with others, that the Pleistocene had to be recalibrated to extend back to 2 million years, a date that still holds today.

■

For the Leakeys, nearly thirty years of work was justified, and fame unlike any they had know would soon follow. Louis embarked on a lecture tour of the United States and England, and entered into a long-standing arrangement with *National*

Geographic, which started supporting the Leakeys' efforts, paying them salaries and buying them tents and equipment. They began serious excavations at Olduvai immediately—work that soon led to the discovery of fossils that fit the criteria for the genus *Homo*. Louis called them *Homo habilis*, or "handyman," because they were found alongside stone tools that he felt certain they had used. With this discovery, Leakey had a fossil that fit better his notions of an early human ancestor—and he proposed that *Homo habilis* was on the true line to humans, while all the australopithecines, including Zinj, were not. He held this view until his death in 1972.

But success brought problems at home. "*Zinjanthropus* had come into our lives," Mary wrote later. "Though we were not immediately, aware of it, the whole nature of our research operation at Olduvai was about to be altered drastically, and we ourselves were going to be profoundly affected." The discovery of Zinj would have a "snowball effect" that would propel Louis and Mary down separate paths and eventually lead to the breakdown of their marriage. Where they had worked closely together before, Louis and Mary now traveled and worked separately more and more.

They were no longer alone, however, in their search for early humans in Africa. The French, who already had been working in Ethiopia, would expand their presence there, and a new generation of young researchers from Britain and the United States, unaccustomed to the ways of the British colonies, would push aggressively to get their own toehold on the prime fossil beds of Africa. For the first time, Louis would have serious competition in eastern Africa.

From *The First Human: The Race to Discover Our Earliest Ancestors*, Doubleday, April 2006, pp. 25–45. Copyright © 2006 by Ann Gibbons. Reprinted by permission of Doubleday, a division of Random House, Inc.

Hunting the First Hominid

PAT SHIPMAN

S elf-centeredly, human beings have always taken an exceptional interest in their origins. Each discovery of a new species of hominid—both our human ancestors and the near-relatives arising after the split from the gorilla-chimp lineage—is reported with great fanfare, even though the First Hominid remains elusive. We hope that, when our earliest ancestor is finally captured, it will reveal the fundamental adaptations that make us *us*.

There is no shortage of ideas about the essential nature of the human species and the basic adaptations of our kind. Some say hominids are fundamentally thinkers; others favor tool-makers or talkers; still others argue that hunting, scavenging or bipedal walking made hominids special. Knowing what the First Hominid looked like would add some meat to a soup flavored with speculation and prejudice.

Genetic and molecular studies provide one sort of insight, showing what sort of *stuff* we are made of, and that it is only slightly different stuff from that which makes up the apes (gorillas, chimps, orangutans and gibbons). From the molecular differences among the genes of humans and apes, geneticists estimate the time when each of the various ape and hominid lineages diverged from the common stem. The result is a sort of hairy Y diagram, with multiple branches instead of simply two as is usual on a Y. Each terminus represents a living species; each branching point or node represents the appearance of some new evolutionary trait, such as new molecules, new genes, new shapes or new proportions of limb, skull and tooth. Unfortunately, this way of diagramming the results tends to lull us into thinking (falsely) that all the evolutionary changes occurred at those nodes, and none along the branches.

Such studies omit a crucial part of our history: the extinct species. Only the fossil record contains evidence, in context, of the precise pathway that evolution took. In this technological age, when sophisticated instrumentation and gee-whiz algorithms seem downright necessary, the most basic information about the evolution of our lineage still comes from branches of science that operate in rather old-fashioned ways. The primary discoveries in paleontology (the study of old things), pale oanthropology (the study of old humans) and archaeology (the study of the old stuff that old humans leave around) still rely on the efforts of handful of investigators who slog around on foot or excavate in desolate landscapes. Fancy equipment can't replace eyes and brains, although instrumentation plays a crucial role in the dating an analysis of fossil remains.

Finding the evolutionary origin of hominids is little like stalking big game. Paleoanthropologist struggle to establish when and where their quarry was last seen and what it was like—hoping to follow its tracks backward in time. (*Why* hominids or any other group arose is such a metaphysical question that most paleoanthropologists run away screaming when it is asked.)

When the first hominid slinked through the underbush has been estimated from molecular clock and confirmed by radiometric dates. These line of evidence converge on a period between 5 million and 7 million years ago as the time when primitive, perhaps vaguely apelike species evolved some definitive new adaptation that transformed it into the First Hominid. But like the point of inflection on a line graph, the first species in any new lineage is only readily apparent after the fact. The emergence of the first hominid was probably not obvious in prospect but only now, in retrospect—in the context of the entire evolutionary record of the hominids—when the long-term evolutionary trends can be seen.

Where this dangerous creature once lived has to be Africa, since both our closest living relatives (chimpanzees and gorillas) and all early hominids (older than about 2 million years ago) are African.

What to Look For?

What happened, exactly, and to *whom*, remain to be discovered. Two newly discovered fossil species have each been proposed to be the First Hominid. This circumstance raises a significant issue: How would we know the First Hominid if we saw it?

Making a list of key features that differentiate apes from people is not difficult, but misleading. It is absurd to expect that all of these differences arose simultaneously during a single evolutionary event represented by the final fork on the hairy Y diagram. The first ape on the gorilla-chimp lineage was neither a gorilla nor a chimpanzee, for modem gorillas and chimps have had at least 5 million years to evolve in isolation before arriving at their modem form. In the same way the First Hominid on our lineage was not a human and did not possess all of the characteristics of modem humans.

Using a hairy Y diagram, we can limit the number of contenders for the essential or basal hominid adaptation:

—Hominids might be essentially bipeds. All known hominids are bipedal; no apes are.

—Wishful thinking aside, hominids are not simply brainy apes. Alas, until about 2 million years ago no hominid had a brain larger than an ape's relative to its body size.

—Hominids might be apelike creatures that have lost their sexual dimorphism. Sexual dimorphism is exhibited as male-female differences that are not related to reproduction. For example, male orangs are typically much larger than females and have longer canine teeth that hone sharper with wear. The fossil record shows that hominids lost their dental sexual dimorphism first, since all known hominids have small and flat-wearing canines. In contrast, sexual dimorphism in body size persisted in hominids until about 2 million years ago, when the genus *Homo* first appeared.

—Thick dental enamel may be a key hominid trait. All hominids have thick enamel, whereas all fossil and living apes (except those in the orangutan lineage) have thin enamel. Because the fossil record of apes is so poor, we do not know whether the primitive condition for apes and hominids was thick or thin enamel. Indeed, how enamel thickness is to be measured and evaluated has generated many pages of debate.

—Hominids are hand-graspers or manipulators (from the Latin for hand, *manus*), whereas apes are foot-graspers. These differences are reflected in the sharp contrasts in the hand and foot anatomy of apes and humans. Apes have divergent big toes and long, curved toe bones for holding onto branches; their thumbs are short and cannot be opposed to the other fingers for skillful manipulation. Human beings are the opposite, with long, opposable thumbs and big toes that are closely aligned with the remaining short, straight toes. Human feet are nearly useless for grasping but are well adapted to bipedalism. An intermediate condition occurs in early hominids such as Lucy (the best-known individual of *Australopithecus afarensis*), who had opposable thumbs and numerous adaptations to bipedalism, and yet retained rather long and curved toes. Lucy and probably other types of *Australopithecus* were walkers, hand-graspers *and* somewhat compromised foot-graspers.

A description of our desired prey, then, might read like this:

> An ape-brained and small-canined creature, with dental enamel of unknown thickness. Large if male but smaller if female. May be spotted climbing adeptly in trees or walking bipedally on the ground. Last seen in Africa between 5 million and 7 million years ago.

From this description, can we identify the First Hominid? Well, no—not yet.

A Tale of Two Trophies

Yohannes Haile-Selassie of the University of California, Berkeley, described a likely contender from Ethiopia in July of 2001. His material comes from Ethiopian sediments between 5.2 million and 5.8 million years old and is called *Ardipithecus ramidus kadabba*, a new subspecies. *Ardipithecus* means "root ape" in the Mar language, and the species has been explicitly proposed to be a "root species" ancestral to all later hominids.

Haile-Selassie's specimens include more than 20 teeth, some associated with a mandible or lower jaw; substantial pieces of two left humeri, or upper arm bones; a partial ulna from the same forearm as one humerus; a partial clavicle or collarbone; a half of one finger bone; and a complete toe bone.

No ironclad evidence of bipedality in any *Ardipithecus* specimen has yet been published. In this collection, the only evidence about habitual patterns of locomotion comes from the single toe bone. Its weight-bearing surface faces downward as in bipeds, not inward as in apes. Any jury might be suspicious that *Ardipithecus* was bipedal, but none of the really telltale body parts—pelvis, complete femur, tibia, or ankle bones—has yet been recovered. The preserved bones of the arm, finger and shoulder closely resemble those of Lucy and may have been used in grasping and tree-climbing. *Ardipithecus* is as yet too poorly known for its relative brain size or sexual dimorphism in body size to be assessed.

The other candidate that has already been bagged is a 6 million-year-old find from the Tugen Hills of Kenya called *Orrorin tugenensis*. Its generic name is derived from the Tugen language and means "original man"—a claim as bold as "root ape." Found by a joint French-Kenyan team headed by Brigitte Senut of the Centre nationale de la recherche scientifique, the *Orrorin* fossils include a few teeth, some embedded in a jaw fragment; a partial humerus; a finger bone; and substantial parts of three femurs, or thigh bones.

The femurs, which might provide definitive evidence of bipedality, are incomplete. The sole evidence for bipedality lies in the head of the femur in *Orrorin*, which is proportionately larger than Lucy's. One reason to evolve a large-headed femur is to dissipate the forces produced by bipedalism. The team concludes that *Orrorin* evolved bipedalism separately from Lucy (and from other species of *Australopithecus*), making *Orrorin* the only known ancestor of *Homo*. *Australopithecus* is displaced to an extinct side-branch of the hominid lineage.

Their surprising conclusion is not universally accepted. Skeptics reply that the femoral differences between *Orrorin* and *Australopithecus* might disappear if *Orrorin*'s femur were compared with that of a large male individual rather than with the diminutive Lucy.

As in *Ardipithecus*, the bones from the upper limb of *Orrorin* show tree-climbing adaptations. Neither relative brain size nor body size dimorphism can be evaluated in *Orrorin*.

Where *Orrorin* and *Ardipithecus* differ are in their teeth. *Orrorin* appears to have thick enamel, like a hominid or an orangutan, and *Ardipithecus* seems to have thin enamel, like other apes. This dental comparison might resolve the question "Who is the First Hominid?" in favor of *Orrorin*, except that *Orrorin*'s canine teeth imply the opposite conclusion. *Orrorin*'s single known canine is sizable and pointed and wears like an ape's canine. In contrast, the several known canines of *Ardipithecus* are all small-crowned and flat-wearing, like a hominid's canines.

Puzzlingly, *Ardipithecus* and *Orrorin* show different mosaics of hominid and ape features. Both may be bipeds, although *Ardipithecus* is a biped in the manner of *Australopithecus* and *Orrorin* is not. If one of these two newly announced species is the First Hominid, then the other must be banished to the ape lineage. The situation is deliciously complex and confusing.

It is also humbling. We thought we knew an ape from a person; we thought we could even identify a man in an ape suit or an ape in a tuxedo for what they were. Humans have long prided themselves on being very different from apes—but pride goeth before a fall. In this case, embarrassingly, we can't tell the ape from the hominid even though we have teeth, jaws, and arm and leg bones.

Paleoanthropologists must seriously reconsider the defining attributes of apes and hominids while we wait for new fossils. In the meantime, we should ponder our complicity, too, for we have been guilty of expecting evolution to be much simpler than it ever is.

PAT SHIPMAN is an adjunct professor of anthropology at the Pennsylvania State University. Address: Department of Anthropology, 315 Carpenter Building, Pennsylvania State University, University Park, PA 16801. Internet:pls10@psu.edu

Digital Ancestors Walk Again

Commonplace hospital gear opens up a new way of reconstructing forerunners of *Homo sapiens*

CARL ZIMMER

In the past, most of the big news about human evolution came from remote dig sites in places like Africa or Indonesia. In the future, the big news will come from familiar sites closer to home: hospitals. That's because hospitals are equipped with powerful new scanning machines primarily used to identify tumors, ballooning blood vessels, bone fractures, and a wide range of disorders in people. Those same scanners also make it possible for paleoanthropologists to look inside the fossils of ancient hominids and see things that until now have been shrouded in mystery.

Take brains, for example. The evolution of the human brain is one of the most important questions in the story of our origins. But when our ancestors died, their brains quickly rotted away. Fossilized skulls offer the only clues. Until recently, if a team of researchers found an intact braincase, they were limited in what they could learn unless they cut the fossil open. Because hominid skulls are rare, few would dare take such a radical step.

Now paleoanthropologists can put a hominid skull in a computed-tomography, or CT, scanner and create a virtual skull that they can split apart any way they want. If they remove that digital skull altogether, they leave behind the outlines of a virtual brain. In 2005 a virtual brain of the one known skull of *Homo floresiensis*—the three-foot-tall hominid discovered on the Indonesian island of Flores—provided evidence in the ongoing debate about whether the creature represents a separate species or was a human pygmy with a birth defect. The size and shape of the virtual brain lends credence to the separate species theory. Moreover, the brain was not just a simpler version of a human brain. Some regions were smaller than ours, but others were unusually large for such a small hominid, hinting that *Homo floresiensis* might have been capable of abstract thought and could make complicated plans.

Most hominid fossils are in much worse shape than the skull of *Homo floresiensis*. Over thousands of years, they have disin-

tegrated. Reconstructing a skull from bone chips used to be like assembling a three-dimensional puzzle with most of the pieces missing. Debates flare up over reconstructions. Was this hominid tall or short? Was that fossil a single individual, or a mélange of several? When they try new reconstructions, paleoanthropologists often wind up damaging the fossils as they cut through the glue and varnish that held pieces together. And when fossils are particularly smashed up, paleoanthropologists simply don't dare reconstruct them.

CT scans make it much easier to put these puzzles back together. Researchers can create virtual bone fragments and then use sophisticated mathematical software to find the best way to assemble them. In some cases, they can make the scans without even removing the fossils from the rock that encases them. This new method has already changed the way scientists think about Neanderthals. A Swiss research team has produced a virtual series of young Neanderthal skulls and compared their development with that of modern human children. It turns out that Neanderthal children are as different from modern humans as adult Neanderthals are—which suggests that Neanderthals did belong to a separate species and did not give rise to living Europeans.

As the use of CT scans expands, paleoanthropologists are developing new avenues for uncovering clues to our past. They are discovering signs of healed wounds, of toothless old hominids who must have been cared for by others. Some researchers are even producing full-length virtual skeletons to which they can attach virtual muscles and make the ancient hominids walk again. Most significantly, CT scans can liberate hominid fossils from museum drawers. Once a research team makes a scan, they can post the data on a Web site for other researchers to analyze, bringing a precious hominid fossil to new sets of eyes and new sets of questions.

Scavenger Hunt

As paleoanthropologists close in on their quarry, it may turn out to be a different beast from what they imaged

PAT SHIPMAN

In both textbooks and films, ancestral humans (hominids) have been portrayed as hunters. Small-brained, big-browed, upright, and usually mildly furry, early hominid males gaze with keen eyes across the gold savanna, searching for prey. Skillfully wielding a few crude stone tools, they kill and dismember everything from small gazelles to elephants, while females care for young and gather roots, tubers, and berries. The food is shared by group members at temporary camps. This familiar image of Man the Hunter has been bolstered by the finding of stone tools in association with fossil animal bones. But the role of hunting in early hominid life cannot be determined in the absence of more direct evidence.

I discovered one means of testing the hunting hypothesis almost by accident. In 1978, I began documenting the microscopic damage produced on bones by different events. I hoped to develop a diagnostic key for identifying the post-mortem history of specific fossil bones, useful for understanding how fossil assemblages were formed. Using a scanning electron microscope (SEM) because of its excellent resolution and superb depth of field, I inspected high-fidelity replicas of modern bones that had been subjected to known events or conditions. (I had to use replicas, rather than real bones, because specimens must fit into the SEM's small vacuum chamber.) I soon established that such common events as weathering, root etching, sedimentary abrasion, and carnivore chewing produced microscopically distinctive features.

In 1980, my SEM study took an unexpected turn. Richard Potts (now of Yale University), Henry Bunn (now of the University of Wisconsin at Madison), and I almost simultaneously found what appeared to be stone-tool cut marks on fossils from Olduvai Gorge, Tanzania, and Koobi Fora, Kenya. We were working almost side by side at the National Museums of Kenya, in Nairobi, where the fossils are stored. The possibility of cut marks was exciting, since both sites preserve some of the oldest known archaeological materials. Potts and I returned to the United States, manufactured some stone tools, and started "butchering" bones and joints begged from our local butchers. Under the SEM, replicas of these cut marks looked very different from replicas of carnivore tooth scratches, regardless of the species of carnivore or the type of tool involved. By comparing the marks on the fossils

with our hundreds of modern bones of known history, we were able to demonstrate convincingly that hominids using stone tools had processed carcasses of many different animals nearly two million years ago. For the first time, there was a firm link between stone tools and at least some of the early fossil animal bones.

This initial discovery persuaded some paleoanthropologists that the hominid hunter scenario was correct. Potts and I were not so sure. Our study had shown that many of the cut-marked fossils also bore carnivore tooth marks and that some of the cut marks were in places we hadn't expected—on bones that bore little meat in life. More work was needed.

In addition to more data about the Olduvai cut marks and tooth marks, I needed specific information about the patterns of cut marks left by known hunters performing typical activities associated with hunting. If similar patterns occurred on the fossils, then the early hominids probably behaved similarly to more modern hunters; if the patterns were different, then the behavior was probably also different. Three activities related to hunting occur often enough in peoples around the world and leave consistent enough traces to be used for such a test.

First, human hunters systematically disarticulate their kills, unless the animals are small enough to be eaten on the spot. Disarticulation leaves cut marks in a predictable pattern on the skeleton. Such marks cluster near the major joints of the limbs: shoulder, elbow, carpal joint (wrist), hip, knee, and hock (ankle). Taking a carcass apart at the joints is much easier than breaking or cutting through bones. Disarticulation enables hunters to carry food back to a central place or camp, so that they can share it with others or cook it or even store it by placing portions in trees, away from the reach of carnivores. If early hominids were hunters who transported and shared their kills, disarticulation marks would occur near joints in frequencies comparable to those produced by modern human hunters.

Second, human hunters often butcher carcasses, in the sense of removing meat from the bones. Butchery marks are usually found on the shafts of bones from the upper part of the front or hind limb, since this is where the big muscle masses lie. Butchery may be carried out at the kill site—especially if the animal is very large and its bones very heavy—or it may take place at the base camp, during the process of sharing food with others. Compared with disarticulation, butchery leaves relatively few marks. It is hard for a hunter to locate an animal's joints without leaving cut marks on the bone. In contrast, it is easier to cut the meat away from the midshaft of the bone without making such marks. If early hominids shared their food, however, there ought to be a number of cut marks located on the midshaft of some fossil bones.

Finally, human hunters often remove skin or tendons from carcasses, to be used for clothing, bags, thongs, and so on. Hide or tendon must be separated from the bones in many areas where there is little flesh, such as the lower limb bones of pigs, giraffes, antelopes, and zebras. In such cases, it is difficult to cut the skin without leaving a cut mark on the bone. Therefore, one expects to find many more cut marks on such bones than on the flesh-covered bones of the upper part of the limbs.

Unfortunately, although accounts of butchery and disarticulation by modern human hunters are remarkably consistent, quantitative studies are rare. Further, virtually all modern hunter-gatherers use metal tools, which leave more cut marks than stone tools. For these reasons I hesitated to compare the fossil evidence with data on modern hunters. Fortunately, Diane Gifford of the University of California, Santa Cruz, and her colleagues had recently completed a quantitative study of marks and damage on thousands of antelope bones processed by Neolithic (Stone Age) hunters in Kenya some 2,300 years ago. The data from Prolonged Drift, as the site is called, were perfect for comparison with the Olduvai material.

Assisted by my technician, Jennie Rose, I carefully inspected more than 2,500 antelope bones from Bed I at Olduvai Gorge, which is dated to between 1.9 and 1.7 million years ago. We made high-fidelity replicas of every mark that we thought might be either a cut mark or a carnivore tooth mark. Back in the United States, we used the SEM to make positive identifications of the marks. (The replication and SEM inspection was time consuming, but necessary: only about half of the marks were correctly identified by eye or by light microscope.) I then compared the patterns of cut mark and tooth mark distributions on Olduvai fossils with those made by Stone Age hunters at Prolonged Drift.

By their location, I identified marks caused either by disarticulation or meat removal and then compared their frequencies with those from Prolonged Drift. More than 90 percent of the Neolithic marks in these two categories were from disarticulation, but to my surprise, only about 45 percent of the corresponding Olduvai cut marks were from disarticulation. This difference is too great to have occurred by chance; the Olduvai bones did not show the predicted pattern. In fact, the Olduvai cut marks attributable to meat removal and disarticulation showed essentially the same pattern of distribution as the carnivore tooth marks. Apparently, the early hominids were not regularly disarticulating carcasses. This finding casts serious doubt on the idea that early hominids carried their kills back to camp to share with others, since both transport and sharing are difficult unless carcasses are cut up.

When I looked for cut marks attributable to skinning or tendon removal, a more modern pattern emerged. On both the Neolithic and Olduvai bones, nearly 75 percent of all cut marks occurred on bones that bore little meat; these cut marks probably came from skinning. Carnivore tooth marks were much less common on such bones. Hominids were using carcasses as a source of skin and tendon. This made it seem more surprising that they disarticulated carcasses so rarely.

A third line of evidence provided the most tantalizing clue. Occasionally, sets of overlapping marks occur on the Olduvai fossils. Sometimes, these sets include both cut marks and carnivore tooth marks. Still more rarely, I could see under the SEM which mark had been made first, because its features were overlaid by those of the later mark, in much the same way as old tire tracks on a dirt road are obscured by fresh ones. Although only thirteen such sets of marks were found, in eight cases the hominids made the cut marks after the carnivores made their tooth marks. This finding suggested a new hypothesis. Instead of hunting for prey and leaving the remains behind for carnivores to scavenge, perhaps hominids were scavenging from the carnivores. This might explain the hominids' apparently unsystematic use of carcasses: they took what they could get, be it skin, tendon, or meat.

Man the Scavenger is not nearly as attractive an image as Man the Hunter, but it is worth examining. Actually, although hunting and scavenging are different ecological strategies, many mammals do both. The only pure scavengers alive in Africa today are vultures; not one of the modern African mammalian carnivores is a pure scavenger. Even spotted hyenas, which have massive, bone-crushing teeth well adapted for eating the bones left behind by others, only scavenge about 33 percent of their food. Other carnivores that scavenge when there are enough carcasses around include lions, leopards, striped hyenas, and jackals. Long-term behavioral studies suggest that these carnivores scavenge when they can and kill when they must. There are only two nearly pure predators, or hunters—the cheetah and the wild dog—that rarely, if ever, scavenge.

What are the costs and benefits of scavenging compared with those of predation? First of all, the scavenger avoids the task of making sure its meal is dead: a predator has already endured the energetically costly business of chasing or stalking animal after animal until one is killed. But while scavenging may be cheap, it's risky. Predators rarely give up their prey to scavengers without defending it. In such disputes, the larger animal, whether a scavenger or a predator, usually wins, although smaller animals in a pack may defeat a lone, larger animal. Both predators and scavengers suffer the dangers inherent in fighting for possession of a carcass. Smaller scavengers such as jackals or striped hyenas avoid disputes to some extent by specializing in darting in and removing a piece of a carcass without trying to take possession of the whole thing. These two strategies can be characterized as that of the bully or that of the sneak: bullies need to be large to be successful, sneaks need to be small and quick.

Because carcasses are almost always much rarer than live prey, the major cost peculiar to scavenging is that scavengers must survey much larger areas than predators to find food. They can travel slowly, since their "prey" is already dead, but endurance is important. Many predators specialize in speed at the expense of endurance, while scavengers do the opposite.

The more committed predators among the East African carnivores (wild dogs and cheetahs) can achieve great top speeds when running, although not for long. Perhaps as a consequence, these "pure" hunters enjoy a much higher success rate in hunting (about three-fourths of their chases end in kills) than any of the scavenger-hunters do (less than half of their chases are successful). Wild dogs and cheetahs are efficient hunters, but they are neither big enough nor efficient enough in their locomotion to make good scavengers. In fact, the cheetah's teeth are so specialized for meat slicing that they probably cannot withstand the stresses of bone crunching and carcass dismembering carried out by scavengers. Other carnivores are less successful at hunting, but have specializations of size, endurance, or (in the case of the hyenas) dentition that make successful scavenging possible. The small carnivores seem to have a somewhat higher hunting success rate than the large ones, which balances out their difficulties in asserting possession of carcasses.

In addition to endurance, scavengers need an efficient means of locating carcasses, which, unlike live animals, don't move or make noises. Vultures, for example, solve both problems by flying. The soaring, gliding flight of vultures expends much less energy than walking or cantering as performed by the part-time mammalian scavengers. Flight enables vultures to maintain a foraging radius two to three times larger than that of spotted hyenas, while providing a better vantage point. This explains why vultures can scavenge all of their food in the same habitat in which it is impossible for any mammal to be a pure scavenger. (In fact, many mammals learn where carcasses are located from the presence of vultures.)

Since mammals can't succeed as full-time scavengers, they must have another source of food to provide the bulk of their diet. The large carnivores rely on hunting large animals to obtain food when scavenging doesn't work. Their size enables them to defend a carcass against others. Since the small carnivores—jackals and striped hyenas—often can't defend carcasses successfully, most of their diet is composed of fruit and insects. When they do hunt, they usually prey on very small animals, such as rats or hares, that can be consumed in their entirety before the larger competitors arrive.

The ancient habitat associated with the fossils of Olduvai and Koobi Fora would have supported many herbivores and carnivores. Among the latter were two species of large sabertoothed cats, whose teeth show extreme adaptations for meat slicing. These were predators with primary access to carcasses. Since their teeth were unsuitable for bone crushing, the sabertoothed cats must have left behind many bones covered with scraps of meat, skin, and tendon. Were early hominids among the scavengers that exploited such carcasses?

All three hominid species that were present in Bed I times (*Homo habilis, Australopithecus africanus, A. robustus*) were adapted for habitual, upright bipedalism. Many anatomists see evidence that these hominids were agile tree climbers as well. Although upright bipedalism is a notoriously peculiar mode of locomotion, the adaptive value of which has been argued for years (See Matt Cartmill's article, "Four Legs Good, Two Legs Bad," *Natural History*, November 1983), there are three general points of agreement.

First, bipedal running is neither fast nor efficient compared to quadrupedal gaits. However, at moderate speeds of 2.5 to 3.5 miles per hour, bipedal *walking* is more energetically efficient than quadrupedal walking. Thus, bipedal walking is an excellent means of covering large areas slowly, making it an unlikely adaptation for a hunter but an appropriate and useful adaptation for a scavenger. Second, bipedalism elevates the head, thus improving the hominid's ability to spot items on the ground—an advantage both to scavengers and to those trying to avoid becoming a carcass. Combining bipedalism with agile tree climbing improves the vantage point still further. Third, bipedalism frees the hands from locomotive duties, making it possible to carry items. What would early hominids have carried? Meat makes a nutritious, easy-to-carry package; the problem is that carrying meat attracts scavengers. Richard Potts suggests that carrying stone tools or unworked stones for toolmaking to caches would be a more efficient and less dangerous activity under many circumstances.

In short, bipedalism is compatible with a scavenging strategy. I am tempted to argue that bipedalism evolved because it provided a substantial advantage to scavenging hominids. But I doubt hominids could scavenge effectively without tools, and bipedalism predates the oldest known stone tools by more than a million years.

Is there evidence that, like modern mammalian scavengers, early hominids had an alternative food source, such as either hunting or eating fruits and insects? My husband, Alan Walker, has shown that the microscopic wear on an animal's teeth reflects its diet. Early hominid teeth wear more like that of chimpanzees and other modern fruit eaters than that of carnivores. Apparently, early hominids ate mostly fruit, as the smaller, modern scavengers do. This accords with the estimated body weight of early hominids, which was only about forty to eighty pounds—less than that of any of the modern carnivores that combine scavenging and hunting but comparable to the striped hyena, which eats fruits and insects as well as meat.

Would early hominids have been able to compete for carcasses with other carnivores? They were too small to use a bully strategy, but if they scavenged in groups, a combined bully-sneak strategy might have been possible. Perhaps they were able to drive off a primary predator long enough to grab some meat, skin, or marrow-filled bone before relinquishing the carcass. The effectiveness of this strategy would have been vastly improved by using tools to remove meat or parts of limbs, a task at which hominid teeth are poor. As agile climbers, early hominids may have retreated into the trees to eat their scavenged trophies, thus avoiding competition from large terrestrial carnivores.

In sum, the evidence on cut marks, tooth wear, and bipedalism, together with our knowledge of scavenger adaptation in general, is consistent with the hypothesis that two million years ago hominids were scavengers rather than accomplished hunters. Animal carcasses, which contributed relatively little to the hominid diet, were not systematically cut up and transported for sharing at base camps. Man the Hunter may not have appeared until 1.5 to 0.7 million years ago, when we do see a shift toward omnivory, with a greater proportion of meat in the diet. This more heroic ancestor may have been *Homo erectus*, equipped with Acheulean-style stone tools and, increasingly, fire. If we wish to look further back, we may have to become accustomed to a less flattering image of our heritage.

PAT SHIPMAN is an assistant professor in the Department of Cell Biology and Anatomy at The Johns Hopkins University School of Medicine.

The Scavenging of "Peking Man"

New evidence shows that a venerable cave was neither hearth nor home.

NOEL T. BOAZ AND RUSSELL L. CIOCHON

China is filled with archaeological wonders, but few can rival the Peking Man Site at Zhoukoudian, which has been inscribed on UNESCO's World Heritage List. Located about thirty miles southwest of Beijing, the town of Zhoukoudian boasts several attractions, including ruins of Buddhist monasteries dating from the Ming Dynasty (1368–1644). But the town's main claim to fame is Longgushan, or Dragon Bone Hill, the site of the cave that yielded the first (and still the largest) cache of fossils of *Homo erectus pekinensis*, historically known as Peking man—a human relative who walked upright and whose thick skull bones and beetling brow housed a brain three-quarters the size of *H. sapiens*'s.

The remains of about forty-five individuals—more than half of them women and children—along with thousands of stone stools, debris from tool manufacturing, and thousands of animal bones, were contained within the hundred-foot-thick deposits that once completely filled the original cave. The task of excavation, initiated in 1921, was not completed until 1982. Some evidence unearthed at the site suggested that these creatures, who lived from about 600,000 to 300,000 years ago, had mastered the use of fire and practiced cannibalism. But despite years of excavation and analysis, little is certain about what occurred here long ago. In the past two years we have visited the cave site, reexamined the fossils, and carried out new tests in an effort to sort out the facts.

To most of the early excavators, such as anatomist Davidson Black, paleontologist Pierre Teilhard de Chardin, and archaeologist Henri Breuil, the likely scenario was that these particular early humans lived in the cave where their bones and stone tools were found and that the animal bones were the remains of meals, proof of their hunting expertise. Excavation exposed ash in horizontal patches within the deposits or in vertical patches along the cave's walls; these looked very much like the residue of hearths built up over time.

A more sensational view, first advanced by Breuil in 1929, was that the cave contained evidence of cannibalism. If the animal bones at the site were leftovers from the cave dwellers' hunting forays, he argued, why not the human bones as well? And skulls were conspicuous among the remains, suggesting to him that these might be the trophies of headhunters. Perhaps, Breuil even proposed, the dull-witted *H. erectus* had been prey

to a contemporary, advanced cousin, some ancestral form of *H. sapiens*. Most paleoanthropologists rejected this final twist, but the cannibalism hypothesis received considerable support.

In the late 1930s Franz Weidenreich, an eminent German paleoanthropologist working at Peking Union Medical College, described the *H. erectus* remains in scientific detail. A trained anatomist and medical doctor, he concluded that some of the skulls showed signs of trauma, including scars and fresh injuries from attacks with both blunt and sharp instruments, such as clubs and stone tools. Most convincing to him and others was the systematic destruction of the skulls, apparently at the hands of humans who had decapitated the victims and then broken open the skull bases to retrieve the brains. Weidenreich also believed that the large longitudinal splits seen, for example, in some of the thighbones could only have been caused by humans and were probably made in an effort to extract the marrow.

Others held dissenting views. Chinese paleoanthropologist Pei Wenzhong, who codirected the early Zhoukoudian excavations, disagreed with Breuil and suggested in 1929 that the skulls had been chewed by hyenas. Some Western scientists also had doubts. In 1939 German paleontologist Helmuth Zapfe published his findings on the way hyenas at the Vienna zoo fed on cow bones. Echoing Pei's earlier observations, of which he was aware, Zapfe convincingly argued that many of the bones found at sites like Longgushan closely resembled modern bones broken up by hyenas. In fact, a new term, taphonomy, was coined shortly thereafter for the field Zapfe pioneered: the study of how, after death, animal and plant remains become modified, moved, buried, and fossilized. Franz Weidenreich soon revised his prior interpretation of several *H. erectus* bones whose condition he had attributed to human cannibalistic activity, but he continued to argue that the long-bone splinters and broken skull bases must have resulted from human action.

Following disruptions in fieldwork during World War II (including the loss of all the *H. erectus* fossils collected at Longgushan up to that time, leaving only the casts that had been made of them), Chinese paleoanthropologists resumed investigation of the site. While rejecting the idea of cannibalism, they continued to look upon the cave as a shelter used by early humans equipped with stone tools and fire, as reflected in the title of paleoanthropologist Jia Lampo's book *The Cave Home of Peking Man*, published in 1975.

About this time, Western scientists began to appreciate and develop the field of taphonomy. A few scholars, notably U.S. archaeologist Lewis R. Binford, then reexamined the Longgushan evidence, but only from a distance, concluding that the burning of accumulated bat or bird guano may have accounted for the ash in the cave. With the founding in 1993 of the Zhoukoudian International Paleoanthropological Research Center at Beijing's Institute of Vertebrate Paleontology and Paleoanthropology, a new era of multidisciplinary and international research at Longgushan began. At the institute, we have been able to collaborate with paleontologists Xu Qinqi and Liu Jinyi and with other scholars in a reassessment of the excavations.

It looked as if H. erectus had smashed open the skulls to cannibalize the brains.

One of taphonomy's maxims is that the most common animals at a fossil site and/or the animals whose remains there are the most complete are most likely the ones to have inhabited the area in life. Standing in the Beijing institute amid row after row of museum cases filled with mammal fossils from the cave, we were immediately struck by how few belonged to *H. erectus*—perhaps only 0.5 percent. This suggests that most of the time, this species did not live in the cave. Furthermore, none of the *H. erectus* skeletons is complete. There is a dearth of limb bones, especially of forearms, hands, lower leg bones, and feet—indicating to us that these individuals died somewhere else and that their partial remains were subsequently brought to the cave. But how?

The answer was suggested by the remains of the most common and complete animal skeletons in the cave deposit: those of the giant hyena, *Pachycrocuta brevirostris*. Had *H. erectus*, instead of being the mighty hunters of anthropological lore, simply met the same ignominious fate as the deer and other prey species in the cave? This possibility, which had been raised much earlier by Pei and Zapfe, drew backing from subsequent studies by others. In 1970, for example, British paleontologist Anthony J. Sutcliffe reported finding a modern hyena den in Kenya that contained a number of human bones, including skulls, which the animals had apparently obtained from a nearby hospital cemetery. In the same year, South African zoologist C. K. Brain published the findings of his extensive feeding experiments with captive carnivores, akin to those of Zapfe three decades earlier. One of Brain's conclusions was that carnivores tend to chew up and destroy the ends of the extremities, leaving, in the case of primates, very little of the hands and feet.

To test the giant hyena hypothesis, we examined all the fossil casts and the few actual fossils of *H. erectus* from Longgushan. We looked for both carnivore bite marks and the shallow, V-shaped straight cuts that would be left by stone tools (although we realized that cut marks would probably not be detectable on the casts). We also analyzed each sample's fracture patterns. Breaks at right angles indicate damage long after death, when the bone is fossilized or fossilizing; fractures in fresh bone tend to be irregu-

lar, following natural structural lines. Breakage due to crushing by cave rocks is usually massive, and the fracture marks characteristically match rock fragments pushed into the bone.

We were surprised by our findings. Two-thirds of Longgushan's *H. erectus* fossils display what we are convinced are one or more of the following kinds of damage: puncture marks from a carnivore's large, pointed front teeth, most likely the canines of a hyena; long, scraping bite marks, typified by U-shaped grooves along the bone; and fracture patterns comparable to those created by modern hyenas when they chew bone. Moreover, we feel that the longitudinal splitting of large bones—a feature that Weidenreich considered evidence of human activity—can also be attributed to a hyena, especially one the size of the extinct *Pachycrocuta*, the largest hyena known, whose preferred prey was giant elk and woolly rhinoceros. One of the *H. erectus* bones, part of a femur, even reveals telltale surface etchings from stomach acid, indicating it was swallowed and then disgorged.

The pattern of damage on some of the skulls sheds light on how hyenas may have handled them. Bite marks on the brow ridge above the eyes indicate that this protrusion had been grasped and bitten by an animal in the course of chewing off the face. Most animals' facial bones are quite thin, and modern hyenas frequently attack or bite the face first; similarly, their ancient predecessors would likely have discovered this vulnerable region in *H. erectus*. Practically no such facial bones, whose structure is known to us from discoveries at other sites, have been found in the Longgushan cave.

The rest of the skull is a pretty tough nut to crack, however, even for *Pachycrocuta*, since it consists of bones half again as thick as those of a modern human, with massive mounds called tori above the eyes and ears and around the back of the skull. Puncture marks and elongated bite marks around the skulls reveal that the hyenas gnawed at and grappled with them, probably in an effort to crack open the cranium and consume the tasty, lipid-rich brain. We concluded that the hyenas probably succeeded best by chewing through the face, gaining a purchase on the bone surrounding the foramen magnum (the opening in the cranium where the spinal cord enters), and then gnawing away until the skull vault cracked apart or the opening was large enough to expose the brain. This is how we believe the skull bases were destroyed—not by the actions of cannibalistic *H. erectus*.

Two-thirds of the fossils show bite marks or fractures inflicted by carnivores.

We know from geological studies of the cave that the animal bones found there could not have been washed in by rains or carried in by streams: the sediments in which the bones are found are either very fine-grained—indicating gradual deposition by wind or slow-moving water—or they contain angular, sharp-edged shards that would not have survived in a stream or flood. Some of the bones may have belonged to animals that died inside the cave during the course of living in it or frequenting it. Other bones were probably brought in and chewed on by hyenas and other carnivores.

Cut marks we observed on several mammal bones from the cave suggest that early humans did sometimes make use of Longgushan, even if they were not responsible for accumulating most of the bones. Stone tools left near the cave entrance also attest to their presence. Given its long history, the cave may have served a variety of occupants or at times have been configured as several separate, smaller shelters. Another possibility is that, in a form of time-sharing, early humans ventured partway into the cave during the day to scavenge on what the hyenas had not eaten and to find temporary shelter. They may not have realized that the animals, which roamed at twilight and at night, were sleeping in the dark recesses a couple of hundred feet away.

What about the ash in the cave, which has been taken as evidence that *H. erectus* used fire? Recently published work by geochemist Steve Weiner and his team at the Weizmann Institute of Science in Israel suggests that the fires were not from hearths. In detailed studies of the ash levels, they discovered no silica-rich layers, which would be left by the burning of wood. Wood (as well as grass and leaves) contains silica particles known as phytoliths—heat-resistant residues that are ubiquitous in archaeological hearth sites. The results indicate that fire was present in the cave but that its controlled use in hearths was not part of the story.

Still, a human hand may somehow be implicated in these fires. One possibility we are exploring in the next phase of our research is that Longgushan was a place where *Pachycrocuta* and *H. erectus* confronted each other as the early humans sought to snatch some of the meat brought back to the cave by the large hyenas. *Pachycrocuta* would have had the home court advantage, but *H. erectus*, perhaps using fire to hold the carnivore at bay, could have quickly sliced off slivers of meat. Although today we might turn up our noses at such carrion, it may have been a dependable and highly prized source of food during the Ice Age.

Reprinted with permission from *Natural History*, March 2001, pp. 46-51. © 2001 by Natural History Magazine, Inc.

Erectus Rising

Oh No. Not This. The Hominids Are Acting Up Again...

JAMES SHREEVE

Just when it seemed that the recent monumental fuss over the origins of modern human beings was beginning to quiet down, an ancient ancestor is once more running wild. Trampling on theories. Appearing in odd places, way ahead of schedule. Demanding new explanations. And shamelessly flaunting its contempt for conventional wisdom in the public press.

The uppity ancestor this time is *Homo erectus*—alias Java man, alias Peking man, alias a mouthful of formal names known only to the paleontological cognoscenti. Whatever you call it, *erectus* has traditionally been a quiet, average sort of hominid: low of brow, thick of bone, endowed with a brain larger than that of previous hominids but smaller than those that followed, a face less apelike and projecting than that of its ancestors but decidedly more simian than its descendants'. In most scenarios of human evolution, *erectus*'s role was essentially to mark time—a million and a half years of it—between its obscure, presumed origins in East Africa just under 2 million years ago and its much more recent evolution into something deserving the name *sapiens*.

Erectus accomplished only two noteworthy deeds during its long tenure on Earth. First, some 1.5 million years ago, it developed what is known as the Acheulean stone tool culture, a technology exemplified by large, carefully crafted tear-shaped hand axes that were much more advanced than the bashed rocks that had passed for tools in the hands of earlier hominids. Then, half a million years later, and aided by those Acheulean tools, the species carved its way out of Africa and established a human presence in other parts of the Old World. But most of the time, *Homo erectus* merely existed, banging out the same stone tools millennium after millennium, over a time span that one archeologist has called "a period of unimaginable monotony."

Asian and African fossils were lumped into one far-flung taxon, a creature not quite like us but human enough to be welcomed into our genus: Homo erectus.

Or so read the old script. These days, *erectus* has begun to ad-lib a more vigorous, controversial identity for itself. Research within the past year has revealed that rather than being 1 million years old, several *erectus* fossils from Southeast Asia are in fact almost 2 million years old. That is as old as the oldest African members of the species, and it would mean that *erectus* emerged from its home continent much earlier than has been thought—in fact, almost immediately after it first appeared. There's also a jawbone, found in 1991 near the Georgian city of Tbilisi, that resembles *erectus* fossils from Africa and may be as old as 1.8 million years, though that age is still in doubt. These new dates—and the debates they've engendered—have shaken *Homo erectus* out of its interpretive stupor, bringing into sharp relief just how little agreement there is on the rise and demise of the last human species on Earth, save one.

"Everything now is in flux," says Carl Swisher of the Berkeley Geochronology Center, one of the prime movers behind the redating of *erectus* outside Africa. "It's all a mess."

The focal point for the flux is the locale where the species was first found: Java. The rich but frustration-soaked history of paleoanthropology on that tropical island began just over 100 years ago, when a young Dutch anatomy professor named Eugène Dubois conceived the idée fixe that the "missing link" between ape and man was to be found in the jungled remoteness of the Dutch East Indies. Dubois had never left Holland, much less traveled to the Dutch East Indies, and his pick for the spot on Earth where humankind first arose owed as much to a large part of the Indonesian archipelago's being a Dutch colony as it did to any scientific evidence. He nevertheless found this missing link—the top of an oddly thick skull with massive browridges—in 1891 on the banks of the Solo River, near a community called Trinil in central Java. About a year later a thighbone that Dubois thought might belong to the same individual was found nearby; it looked so much like a modern human thighbone that Dubois assumed this ancient primate had walked upright. He christened the creature *Pithecanthropus erectus*—"erect ape-man"—and returned home in triumph.

Finding the fossil proved to be the easy part. Though Dubois won popular acclaim, neither he nor his "Java man" received the full approbation of the anatomists of the day, who considered his ape-man either merely an ape or merely a man. In an apparent pique, Dubois cloistered away the fossils for a quarter-century, refusing others the chance to view his prized possessions. Later, other similarly primitive human remains began to turn up in China and East Africa. All shared a collection of anatomical traits, including a long, low braincase with prominent browridges and a flattened forehead; a sharp angle to the back of the skull when viewed in profile; and a deep, robustly built jaw showing no hint of a chin. Though initially given separate regional names, the fossils were eventually lumped together into one far-flung taxon, a creature not quite like us but human enough to be welcomed into our genus: *Homo erectus*.

Over the decades the most generous source of new *erectus* fossils has been the sites on or near the Solo River in Java. The harvest continues: two more skulls, including one of the most complete *erectus* skulls yet known, were found at a famous fossil site called Sangiran just in the past year. Though the Javan yield of ancient humans has been rich, something has always been missing—the crucial element of time. Unless the age of a fossil can be determined, it hangs in limbo, its importance and place in the larger scheme of human evolution forever undercut with doubt. Until researchers can devise better methods for dating bone directly—right now there are no techniques that can reliably date fossilized, calcified bone more than 50,000 years old—a specimen's age has to be inferred from the geology that surrounds it. Unfortunately, most of the discoveries made on the densely populated and cultivated island of Java have been made not by trained excavators but by sharp-eyed local farmers who spot the bones as they wash out with the annual rains and later sell them. As a result, the original location of many a prized specimen, and thus all hopes of knowing its age, are a matter of memory and word of mouth.

Despite the problems, scientists continue to try to pin down dates for Java's fossils. Most have come up with an upper limit of around 1 million years. Along with the dates for the Peking man skulls found in China and the Acheulean tools from Europe, the Javan evidence has come to be seen as confirmation that *erectus* first left Africa at about that time.

By the early 1970s most paleontologists were firmly wedded to the idea that Africa was the only human-inhabited part of the world until one million years ago.

There are those, however, who have wondered about these dates for quite some time. Chief among them is Garniss Curtis, the founder of the Berkeley Geochronology Center. In 1971 Curtis, who was then at the University of California at Berkeley, attempted to determine the age of a child's skull from a site called Mojokerto, in eastern Java, by using the potassium-argon method to date volcanic minerals in the sediments from which the skull was purportedly removed. Potassium-argon dating had been in use since the 1950s, and Curtis had been enormously successful with it in dating ancient African hominids—including Louis Leakey's famous hominid finds at Olduvai Gorge in Tanzania. The method takes advantage of the fact that a radioactive isotope of potassium found in volcanic ash slowly and predictably decays over time into argon gas, which becomes trapped in the crystalline structure of the mineral. The amount of argon contained in a given sample, measured against the amount of the potassium isotope, serves as a kind of clock that tells how much time has passed since a volcano exploded and its ash fell to earth and buried the bone in question.

Applying the technique to the volcanic pumice associated with the skull from Mojokerto, Curtis got an extraordinary age of 1.9 million years. The wildly anomalous date was all too easy to dismiss, however. Unlike the ash deposits of East Africa, the volcanic pumices in Java are poor in potassium. Also, not unexpectedly, a heavy veil of uncertainty obscured the collector's memories of precisely where he had found the fossil some 35 years earlier. Besides, most paleontologists were by this time firmly wedded to the idea that Africa was the only human-inhabited part of the world until 1 million years ago. Curtis's date was thus deemed wrong for the most stubbornly cherished of reasons: because it couldn't possibly be right.

In 1992 Curtis—under the auspices of the Institute for Human Origins in Berkeley—returned to Java with his colleague Carl Swisher. This time he was backed up by far more sensitive equipment and a powerful refinement in the dating technique. In conventional potassium-argon dating, several grams' worth of volcanic crystals gleaned from a site are needed to run a single experiment. While the bulk of these crystals are probably from the eruption that covered the fossil, there's always the possibility that other materials, from volcanoes millions of years older, have gotten mixed in and will thus make the fossil appear to be much older than it actually is. The potassium-argon method also requires that the researcher divide the sample of crystals in two. One half is dissolved in acid and passed through a flame; the wavelengths of light emitted tell how much potassium is in the sample. The other half is used to measure the amount of argon gas that's released when the crystals are heated. This two-step process further increases the chance of error, simply by giving the experiment twice as much opportunity to go wrong.

The refined technique, called argon-argon dating, neatly sidesteps most of these difficulties. The volcanic crystals are first placed in a reactor and bombarded with neutrons; when one of these neutrons penetrates the potassium nucleus, it displaces a proton, converting the potassium into an isotope of argon that doesn't occur in nature. Then the artificially created argon and the naturally occurring argon are measured in a single experiment. Because the equipment used to measure the isotopes can look for both types of argon at the same time, there's no need to divide the sample, and so the argon-argon method can produce clear results from tiny amounts of material.

In some cases—when the volcanic material is fairly rich in potassium—all the atoms of argon from a single volcanic crystal can be quick-released by the heat from a laser beam and then counted. By doing a number of such single-crystal experiments, the researchers can easily pick out and discard any data from older, contaminant crystals. But even when the researchers are forced to sample more than one potassium-poor crystal to get any reading at all—as was the case at Mojokerto—the argon-argon method can still produce a highly reliable age. In this case, the researchers carefully heat a few crystals at a time to higher and higher temperatures, using a precisely controlled laser. If all the crystals in a sample are the same age, then the amount of argon released at each temperature will be the same. But if contaminants are mixed in, or if severe weathering has altered the crystal's chemical composition, the argon measurements will be erratic, and the researchers will know to throw out the results.

Curtis and Swisher knew that in the argon-argon step-heating method they had the technical means to date the potassium-poor deposits at Mojokerto accurately. But they had no way to prove that those deposits were the ones in which the skull had been buried: all they had was the word of the local man who had found it. Then, during a visit to the museum in the regional capital, where the fossil was being housed, Swisher noticed something odd. The hardened sediments that filled the inside of the fossil's braincase looked black. But back at the site, the deposits of volcanic pumice that had supposedly sheltered the infant's skull were whitish in color. How could a skull come to be filled with black sediments if it had been buried in white ones? Was it possible that the site and the skull had nothing to do with each other after all? Swisher suspected something was wrong. He borrowed a penknife, picked up the precious skull, and nicked off a bit of the matrix inside.

"I almost got kicked out of the country at that point," he says. "These fossils in Java are like the crown jewels."

Luckily, his impulsiveness paid off. The knife's nick revealed white pumice under a thin skin of dark pigment: years earlier, someone had apparently painted the surface of the hardened sediments black. Since there were no other deposits within miles of the purported site that contained a white pumice visually or chemically resembling the matrix in the skull, its tie to the site was suddenly much stronger. Curtis and Swisher returned to Berkeley with pumice from that site and within a few weeks proclaimed the fossil to be 1.8 million years old, give or take some 40,000 years. At the same time, the geochronologists ran tests on pumice from the lower part of the Sangiran area, where *erectus* facial and cranial bone fragments had been found. The tests yielded an age of around 1.6 million years. Both numbers obviously shatter the 1-million-year barrier for *erectus* outside Africa, and they are a stunning vindication of Curtis's work at Mojokerto 20 years ago. "That was very rewarding," he says, "after having been told what a fool I was by my colleagues."

While no one takes Curtis or Swisher for a fool now, some of their colleagues won't be fully convinced by the new dates until the matrix inside the Mojokerto skull itself can be tested. Even then, the possibility will remain that the skull may have drifted down over the years into deposits containing older volcanic crystals that have nothing to do with its original burial site, or that it was carried by a river to another, older site. But Swisher contends that the chance of such an occurrence is remote: it would have to have happened at both Mojokerto and Sangiran for the fossils' ages to be refuted. "I feel really good about the dates," he says. "But it has taken me a while to understand their implications."

The implications that can be spun out from the Javan dates depend on how one chooses to interpret the body of fossil evidence commonly embraced under the name *Homo erectus*. The earliest African fossils traditionally attributed to *erectus* are two nearly complete skulls from the site of Koobi Fora in Kenya, dated between 1.8 and 1.7 million years old. In the conventional view, these early specimens evolved from a more primitive, smaller-brained ancestor called *Homo habilis*, well represented by bones from Koobi Fora, Olduvai Gorge, and sites in South Africa.

If this conventional view is correct, then the new dates mean that *erectus* must have migrated out of Africa very soon after it evolved, quickly reaching deep into the farthest corner of Southeast Asia. This is certainly possible: at the time, Indonesia was connected to Asia by lower sea levels—thus providing an overland route from Africa—and Java is just 10,000 to 15,000 miles from Kenya, depending on the route. Even if *erectus* traveled just one mile a year, it would still take no more than 15,000 years to reach Java—a negligible amount of evolutionary time.

If *erectus* did indeed reach Asia almost a million years earlier than thought, then other, more controversial theories become much more plausible. Although many anthropologists believe that the African and Asian *erectus* fossils all represent a single species, other investigators have recently argued that the two groups are too different to be so casually lumped together. According to paleoanthropologist Ian Tattersall of the American Museum of Natural History in New York, the African skulls traditionally assigned to *erectus* often lack many of the specialized traits that were originally used to define the species in Asia, including the long, low cranial structure, thick skull bones, and robustly built faces. In his view, the African group deserves to be placed in a separate species, which he calls *Homo ergaster*.

Most anthropologists believe that the only way to distinguish between species in the fossil record is to look at the similarities and differences between bones; the age of the fossil should not play a part. But age is often hard to ignore, and Tattersall believes that the new evidence for what he sees as two distinct populations living at the same time in widely separate parts of the Old World is highly suggestive. "The new dates help confirm that these were indeed two different species," he says. "In my view, *erectus* is a separate variant that evolved only in Asia."

Other investigators still contend that the differences between the African and Asian forms of *erectus* are too minimal to merit

Meanwhile, In Siberia...

The presence of *Homo erectus* in Asia twice as long ago as previously thought has some people asking whether the human lineage might have originated in Asia instead of Africa. This long-dormant theory runs contrary to all current thinking about human evolution and lacks an important element: evidence. Although the new Javan dates do place the species in Asia at around the same time it evolved in Africa, all confirmed specimens of other, earlier hominids—the first members of the genus *Homo*, for instance, and the australopithecines, like Lucy—have been found exclusively in Africa. Given such an overwhelming argument, most investigators continue to believe that the hominid line began in Africa.

Most, but not all. Some have begun to cock an ear to the claims of Russian archeologist Yuri Mochanov. For over a decade Mochanov has been excavating a huge site on the Lena River in eastern Siberia—far from Africa, Java, or anywhere else on Earth an ancient hominid bone has ever turned up. Though he hasn't found any hominid fossils in Siberia, he stubbornly believes he's uncovered the next best thing: a trove of some 4,000 stone artifacts—crudely made flaked tools, but tools nonetheless—that he maintains are at least 2 million years old, and possibly 3 million. This, he says, would mean that the human lineage arose not in tropical Africa but in the cold northern latitudes of Asia.

"For evolutionary progress to occur, there had to be the appearance of new conditions: winter, snow, and, accompanying them, hunger," writes Mochanov. "[The ancestral primates] had to learn to walk on the ground, to change their carriage, and to become accustomed to meat—that is, to become 'clever animals of prey.'" And to become clever animals of prey, they'd need tools.

Although he is a well-respected investigator, Mochanov has been unable to convince either Western anthropologists or his Russian colleagues of the age of his site. Until recently the chipped rocks he was holding up as human artifacts were simply dismissed as stones broken by natural processes, or else his estimate of the age of the site was thought to be wincingly wrong. After all, no other signs of human occupation of Siberia appear until some 35,000 years ago.

But after a lecture swing through the United States earlier this year—in which he brought more data and a few prime examples of the tools for people to examine and pass around—many archeologists concede that it is difficult to explain the particular pattern of breakage of the rocks by any known natural process. "Everything I have heard or seen about the context of these things suggests that they are most likely tools," says anthropologist Rick Potts of the Smithsonian Institution, which was host to Mochanov last January.

They're even willing to concede that the site might be considerably older than they'd thought, though not nearly as old as Mochanov estimates. (To date the site, Mochanov compared the tools with artifacts found early in Africa; he also employed an arcane dating technique little known outside Russia.)

Preliminary results from an experimental dating technique performed on soil samples from the site by Michael Waters of Texas A&M and Steve Forman of Ohio State suggest that the layer of sediment bearing the artifacts is some 400,000 years old. That's a long way from 2 million, certainly, but it's still vastly older than anything else found in Siberia—and the site is 1,500 miles farther north than the famous Peking man site in China, previously considered the most northerly home of *erectus*.

"If this does turn out to be 400,000 years old, it's very exciting," says Waters. "If people were able to cope and survive in such a rigorous Arctic environment at such an early time, we would have to completely change our perception of the evolution of human adaptation."

"I have no problem with hominids being almost anywhere at that age—they were certainly traveling around," says Potts. "But the environment is the critical thing. If it was really cold up there"—temperatures in the region now often reach −50 degrees in deep winter—"we'd all have to scratch our heads over how these early hominids were making it in Siberia. There is no evidence that Neanderthals, who were better equipped for cold than anyone, were living in such climates. But who knows? Maybe a population got trapped up there, went extinct, and Mochanov managed to find it." He shrugs. "But that's just arm waving."

—J. S

placing them in separate species. But if Tattersall is right, his theory raises the question of who the original emigrant out of Africa really was. *Homo ergaster* may have been the one to make the trek, evolving into *erectus* once it was established in Asia. Or perhaps a population of some even more primitive, as-yet-unidentified common ancestor ventured forth, giving rise to *erectus* in Asia while a sister population evolved into *ergaster* on the home continent.

Furthermore, no matter who left Africa first, there's the question of what precipitated the migration, a question made even more confounding by the new dates. The old explanation, that the primal human expansion across the hem of the Old World was triggered by the sophisticated Acheulean tools, is no longer tenable with these dates, simply because the tools had not yet been invented when the earliest populations would have moved out. In hindsight, that notion seems a bit shopworn anyway. Acheulean tools first appear in Africa around 1.5 million years ago, and soon after at a site in the nearby Middle East. But they've never been found in the Far East, in spite of the abundant fossil evidence for *Homo erectus* in the region.

Until now, that absence has best been explained by the "bamboo line." According to paleoanthropologist Geoffrey Pope of William Paterson College in New Jersey, *erectus* populations venturing from Africa into the Far East found the land rich in bamboo, a raw material more easily worked into cutting and butchering tools than recalcitrant stone. Sensibly, they abandoned their less efficient stone industry for one based on the pliable plant, which leaves no trace of itself in the archeological record. This is still a viable theory, but the new dates from Java add an even simpler dimension to it: there are no Acheulean tools in the Far East because the first wave of *erectus* to leave Africa didn't have any to bring with them.

So what *did* fuel the quick-step migration out of Africa? Some researchers say the crucial development was not cultural but physical. Earlier hominids like *Homo habilis* were small-bodied creatures with more apelike limb proportions, notes paleoanthropologist Bernard Wood of the University of Liverpool, while African *erectus* was built along more modern lines. Tall, relatively slender, with long legs better able to range over distance and a body better able to dissipate heat, the species was endowed with the physiology needed to free it from the tropical shaded woodlands of Africa that sheltered earlier hominids. In fact, the larger-bodied *erectus* would have required a bigger feeding range to sustain itself, so it makes perfect sense that the expansion out of Africa should begin soon after the species appeared. "Until now, one was always having to account for what kept *erectus* in Africa so long after it evolved," says Wood. "So rather than raising a problem, in some ways the new dates in Java solve one."

Of course, if those dates are right, the accepted time frame for human evolution outside the home continent is nearly doubled, and that has implications for the ongoing debate over the origins of modern human beings. There are two opposing theories. The "out of Africa" hypothesis says that *Homo sapiens* evolved from *erectus* in Africa, and then—sometime in the last 100,000 years—spread out and replaced the more archaic residents of Eurasia. The "multiregional continuity" hypothesis says that modern humans evolved from *erectus* stock in various parts of the Old World, more or less simultaneously and independently. According to this scenario, living peoples outside Africa should look for their most recent ancestors not in African fossils but in the anatomy of ancient fossils within their own region of origin.

As it happens, the multiregionalists have long claimed that the best evidence for their theory lies in Australia, which is generally thought to have become inhabited around 50,000 years ago, by humans crossing over from Indonesia. There are certain facial and cranial characteristics in modern Australian aborigines, the multiregionalists say, that can be traced all the way back to the earliest specimens of *erectus* at Sangiran—characteristics that differ from and precede those of any more recent, *Homo sapiens* arrival from Africa. But if the new Javan dates are right, then these unique characteristics, and thus the aborigines' Asian *erectus* ancestors, must have been evolving separately from the rest of humankind for almost 2 million years. Many anthropologists, already skeptical of the multiregionalists' potential 1-million-year-long isolation for Asian *erectus*, find a 2-million-year-long isolation exceedingly difficult to swallow. "Can anyone seriously propose that the lineage of Australian aborigines could go back that far?" wonders paleoanthropologist Chris Stringer of the Natural History Museum in London, a leading advocate of the out-of-Africa theory.

The multiregionalists counter that they've never argued for *complete* isolation—that there's always been some flow of genes between populations, enough interbreeding to ensure that clearly beneficial *sapiens* characteristics would quickly be conferred on peoples throughout the Old World. "Just as genes flow now from Johannesburg to Beijing and from Melbourne to Paris, they have been flowing that way ever since humanity evolved," says Alan Thorne of the Australian National University in Canberra, an outspoken multiregionalist.

Stanford archeologist Richard Klein, another out-of-Africa supporter, believes the evidence actually *does* point to just such a long, deep isolation of Asian populations from African ones. The fossil record, he says, shows that while archaic forms of *Homo sapiens* were developing in Africa, *erectus* was remaining much the same in Asia. In fact, if some *erectus* fossils from a site called Ngandong in Java turn out to be as young as 100,000 years, as some researchers believe, then *erectus* was still alive on Java at the same time that fully modern human beings were living in Africa and the Middle East. Even more important, Klein says, is the cultural evidence. That Acheulean tools never reached East Asia, even after their invention in Africa, could mean the inventors never reached East Asia either. "You could argue that the new dates show that until very recently there was a long biological and cultural division between Asia on one hand, and Africa and Europe on the other," says Klein. In other words, there must have been two separate lineages of *erectus*, and since there aren't two separate lineages of modern humans, one of those must have gone extinct: presumably the Asian lineage, hastened into oblivion by the arrival of the more culturally adept, tool-laden *Homo sapiens*.

Naturally this argument is anathema to the multiregionalists. But this tenacious debate is unlikely to be resolved without basketfuls of new fossils, new ways of interpreting old ones—and new dates. In Berkeley, Curtis and Swisher are already busy applying the argon-argon method to the Ngandong fossils, which could represent some of the last surviving *Homo erectus* populations on Earth. They also hope to work their radiometric magic on a key *erectus* skull from Olduvai Gorge. In the meantime, at least one thing has become clear: *Homo erectus*, for so long the humdrum hominid, is just as fascinating, contentious, and elusive a character as any other in the human evolutionary story.

Further Reading

Eugène Dubois & the Ape-Man from Java. Bert Theunissen. Kluwer Academic, 1989. When a Dutch army surgeon, determined to prove Darwin right, traveled to Java in search of the missing link between apes and humans, he inadvertently opened a paleontological Pandora's box. This is Dubois's story, the story of the discovery of *Homo erectus*.

His book, *The Neandertal Enigma: Solving the Mystery of Modern Human Origins*, was published in 1995, and he is at work on a novel that a reliable source calls "a murder thriller about the species question."

JAMES SHREEVE is the coauthor, with anthropologist Donald Johanson, of *Lucy's Child: The Discovery of a Human Ancestor*.

Reprinted with permission of the author from *Discover* magazine, September 1994, pp. 80–84, 86, 88–89. © 1994 by James Shreeve.

UNIT 5
Late Hominid Evolution

Unit Selections

Key Points to Consider

- What evidence is there for hard times among the Neanderthals?

- Are Neanderthals part of our ancestry? Explain.

- If the Neanderthals were so smart, why did they disappear?

- When did language ability arise in our ancestry and why?

- When, where and how did modern humans evolve?

- Where does *Homo* floresiensis fit in the hominid family tree and why?

Student Website
www.mhcls.com/online

Internet References
Further information regarding these websites may be found in this book's preface or online.

Human Prehistory
http://users.hol.gr/~dilos/prehis.htm

The most important aspect of human evolution is also the most difficult to decipher from the fossil evidence: our development as sentient, social beings, capable of communicating by means of language.

We detect hints of incipient humanity in the form of crudely chipped tools, the telltale signs of a home base, or the artistic achievements of ornaments and cave art. Yet none of these indicators of a distinctly hominid way of life can provide us with the nuances of the everyday lives of these creatures, their social relations, or their supernatural beliefs, if any. Most of what remains is the rubble of bones and stones from which we interpret what we can of their lifestyle, thought processes, and ability to communicate. Our ability to glean from the fossil record is not completely without hope, however. In fact, informed speculation is what makes possible such essays as "Hard Times Among the Neanderthals" by Erik Trinkaus, "Rethinking Neanderthals" by Joe Alper and "The Gift *of* Gab" by Matt Cartmill. Each is a fine example of careful, systematic, and thought-provoking work that is based upon an increased understanding of hominid fossil sites as well as the more general environmental circumstances in which our predecessors lived.

Beyond the technological and anatomical adaptations, questions have arisen as to how our hominid forebears organized themselves socially and whether modern-day human behavior is inherited as a legacy of our evolutionary past or is a learned product of contemporary circumstances. Attempts to address these questions have given rise to the technique referred to as the "ethnographic analogy." This is a method whereby anthropologists use "ethnographies" or field studies of modern-day hunters and gatherers whose lives we take to be the best approximations we have to what life might have been like for our ancestors. Granted, these contemporary foragers have been living under conditions of environmental and social change just as industrial peoples have. Nevertheless, it seems that, at least in some aspects of their lives, they have not changed as much as we have. So, if we are to make any enlightened assessments of prehistoric behavior patterns, we are better off looking at them than at ourselves.

As if to show that controversial interpretations of the evidence are not limited to the earlier hominid period, in this unit we also see how long-held beliefs about the rise of modern human behavior are being threatened by new fossil evidence as well as new archeological evidence (See "We Are All Africans" by Pat Shipman and "A Cave Full of Clues" by Fredric Heeren). There may have even been a species of hominid living contemporaneously alongside our own ancestors 13,000 years ago (as described in "The Littlest Human"). For some scientists, these revelations fit in quite comfortably with previously held positions; for others it seems that reputations, as well as theories, are at stake.

Hard Times Among the Neanderthals

Although life was difficult, these prehistoric people may not have been as exclusively brutish as usually supposed

Erik Trinkaus

Throughout the century that followed the discovery in 1856 of the first recognized human fossil remains in the Neander Valley (*Neanderthal* in German) near Düsseldorf, Germany, the field of human paleontology has been beset with controversies. This has been especially true of interpretations of the Neanderthals, those frequently maligned people who occupied Europe and the Near East from about 100,000 years ago until the appearance of anatomically modern humans about 35,000 years ago.

During the last two decades, however, a number of fossil discoveries, new analyses of previously known remains, and more sophisticated models for interpreting subtle anatomical differences have led to a reevaluation of the Neanderthals and their place in human evolution.

This recent work has shown that the often quoted reconstruction of the Neanderthals as semierect, lumbering caricatures of humanity is inaccurate. It was based on faulty anatomical interpretations that were reinforced by the intellectual biases of the turn of the century. Detailed comparisons of Neanderthal skeletal remains with those of modern humans have shown that there is nothing in Neanderthal anatomy that conclusively indicates locomotor, manipulative, intellectual, or linguistic abilities inferior to those of modern humans. Neanderthals have therefore been added to the same species as ourselves—*Homo sapiens*—although they are usually placed in their own subspecies, *Homo sapiens neanderthalensis*.

Despite these revisions, it is apparent that there are significant anatomical differences between the Neanderthals and present-day humans. If we are to understand the Neanderthals, we must formulate hypotheses as to why they evolved from earlier humans about 100,000 years ago in Europe and the Near East, and why they were suddenly replaced about 35,000 years ago by peoples largely indistinguishable from ourselves. We must determine, therefore, the behavioral significance of the anatomical differences between the Neanderthals and other human groups, since it is patterns of successful behavior that dictate the direction of natural selection for a species.

In the past, behavioral reconstructions of the Neanderthals and other prehistoric humans have been based largely on archeological data. Research has now reached the stage at which behavioral interpretations from the archeological record can be significantly supplemented by analyses of the fossils themselves. These analyses promise to tell us a considerable amount about the ways of the Neanderthals and may eventually help us to determine their evolutionary fate.

One of the most characteristic features of the Neanderthals is the exaggerated massiveness of their trunk and limb bones. All of the preserved bones suggest a strength seldom attained by modern humans. Furthermore, not only is this robustness present among the adult males, as one might expect, but it is also evident in the adult females, adolescents, and even children. The bones themselves reflect this hardiness in several ways.

First, the muscle and ligament attachment areas are consistently enlarged and strongly marked. This implies large, highly developed muscles and ligaments capable of generating and sustaining great mechanical stress. Secondly, since the skeleton must be capable of supporting these levels of stress, which are frequently several times as great as body weight, the enlarged attachments for muscles and ligaments are associated with arm and leg bone shafts that have been reinforced. The shafts of all of the arm and leg bones are modified tubular structures that have to absorb stress from bending and twisting without fracturing. When the habitual load on a bone increases, the bone responds by laying down more bone in those areas under the greatest stress.

In addition, musculature and body momentum generate large forces across the joints. The cartilage, which covers joint surfaces, can be relatively easily overworked to the point where it degenerates, as is indicated by the prevalence of arthritis in joints subjected to significant wear and tear over the years. When the surface area of a joint is increased, the force per unit area of cartilage is reduced, decreasing the pressure on the cartilage.

Most of the robustness of Neanderthal arm bones is seen in muscle and ligament attachments. All of the muscles that go from the trunk or the shoulder blade to the upper end of the arm show massive development. This applies in particular to the muscles responsible for powerful downward movements of the arm and, to a lesser extent, to muscles that stabilize the shoulder during vigorous movements.

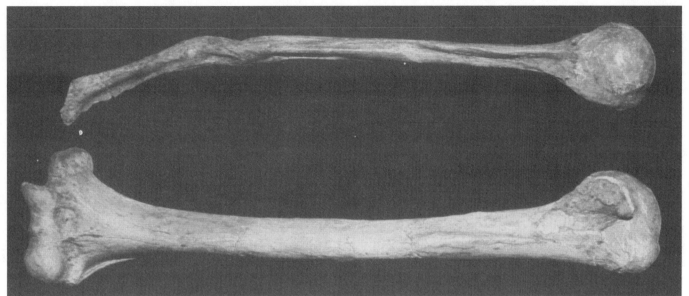

Diagonal lines on these two arm bones from Shanidar 1 are healed fractures. The bottom bone is normal. That on the top is atrophied and has a pathological tip, caused by either amputation or an improperly healed elbow fracture.

Virtually every major muscle or ligament attachment on the hand bones is clearly marked by a large roughened area or a crest, especially the muscles used in grasping objects. In fact, Neanderthal hand bones frequently have clear bony crests, where on modern human ones it is barely possible to discern the attachment of the muscle on the dried bone.

In addition, the flattened areas on the ends of the fingers, which provide support for the nail and the pulp of the finger tip, are enormous among the Neanderthals. These areas on the thumb and the index and middle fingers are usually two to three times as large as those of similarly sized modern human hands. The overall impression is one of arms to rival those of the mightiest blacksmith.

Neanderthal legs are equally massive; their strength is best illustrated in the development of the shafts of the leg bones. Modern human thigh and shin bones possess characteristic shaft shapes adapted to the habitual levels and directions of the stresses acting upon them. The shaft shapes of the Neanderthals are similar to those in modern humans, but the cross-sectional areas of the shafts are much greater. This implies significantly higher levels of stress.

Further evidence of the massiveness of Neanderthal lower limbs is provided by the dimensions of their knee and ankle joints. All of these are larger than in modern humans, especially with respect to the overall lengths of the bones.

The development of their limb bones suggests that the Neanderthals frequently generated high levels of mechanical stress in their limbs. Since most mechanical stress in the body is produced by body momentum and muscular contraction, it appears that the Neanderthals led extremely active lives. It is hard to conceive of what could have required such exertion, especially since the maintenance of vigorous muscular activity would have required considerable expenditure of energy. That level of

energy expenditure would undoubtedly have been maladaptive had it not been necessary for survival.

The available evidence from the archeological material associated with the Neanderthals is equivocal on this matter. Most of the archeological evidence at Middle Paleolithic sites concerns stone tool technology and hunting activities. After relatively little change in technology during the Middle Paleolithic (from about 100,000 years to 35,000 years before the present), the advent of the Upper Paleolithic appears to have brought significant technological advances. This transition about 35,000 years ago is approximately coincident with the replacement of the Neanderthals by the earliest anatomically modern humans. However, the evidence for a significant change in hunting patterns is not evident in the animal remains left behind. Yet even if a correlation between the robustness of body build and the level of hunting efficiency could be demonstrated, it would only explain the ruggedness of the Neanderthal males. Since hunting is exclusively or at least predominantly a male activity among humans, and since Neanderthal females were in all respects as strongly built as the males, an alternative explanation is required for the females.

Some insight into why the Neanderthals consistently possessed such massiveness is provided by a series of partial skeletons of Neanderthals from the Shanidar Cave in northern Iraq. These fossils were excavated between 1953 and 1960 by anthropologist Ralph Solecki of Columbia University and have been studied principally by T. Dale Stewart, an anthropologist at the Smithsonian Institution, and myself. The most remarkable aspect of these skeletons is the number of healed injuries they contain. Four of the six reasonably complete adult skeletons show evidence of trauma during life.

The identification of traumatic injury in human fossil remains has plagued paleontologists for years. There has been a

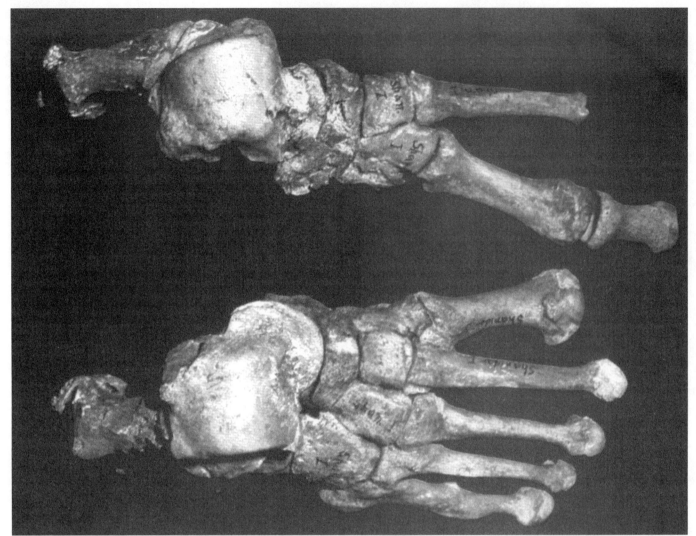

The ankle and big toe of Shanidar 1's bottom foot show evidence of arthritis, which suggests an injury to those parts. The top foot is normal though incomplete.

tendency to consider any form of damage to a fossil as conclusive evidence of prehistoric violence between humans if it resembles the breakage patterns caused by a direct blow with a heavy object. Hence a jaw with the teeth pushed in or a skull with a depressed fracture of the vault would be construed to indicate blows to the head.

The central problem with these interpretations is that they ignore the possibility of damage after death. Bone is relatively fragile, especially as compared with the rock and other sediment in which it is buried during fossilization. Therefore when several feet of sediment caused compression around fossil remains, the fossils will almost always break. In fact, among the innumerable cases of suggested violence between humans cited over the years, there are only a few exceptional examples that cannot be readily explained as the result of natural geologic forces acting after the death and burial of the individual.

One of these examples is the trauma of the left ninth rib of the skeleton of Shanidar 3, a partially healed wound inflicted by

a sharp object. The implement cut obliquely across the top of the ninth rib and probably pierced the underlying lung. Shanidar 3 almost certainly suffered a collapsed left lung and died several days or weeks later, probably as a result of secondary complications. This is deduced from the presence of bony spurs and increased density of the bone around the cut.

The position of the wound on the rib, the angle of the incision, and the cleanness of the cut make it highly unlikely that the injury was accidentally inflicted. In fact, the incision is almost exactly what would have resulted if Shanidar 3 had been stabbed in the side by a right-handed adversary in face-to-face conflict. This would therefore provide conclusive evidence of violence between humans, the *only* evidence so far found of such violence among the Neanderthals.

In most cases, however, it is impossible to determine from fossilized remains the cause of an individual's death. The instances that can be positively identified as prehistoric traumatic injury are those in which the injury was inflicted prior to death and some

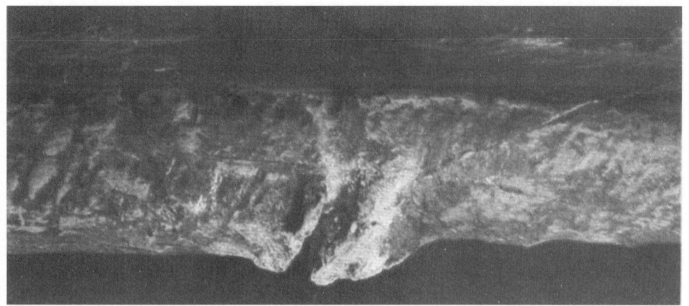

The scar on the left ninth rib of Shanidar 3 is a partially healed wound inflicted by a sharp object. This wound is one of the few examples of trauma caused by violence.

healing took place. Shortly after an injury to bone, whether a cut or a fracture, the damaged bone tissue is resorbed by the body and new bone tissue is laid down around the injured area. As long as irritation persists, new bone is deposited, creating a bulge or spurs of irregular bone extending into the soft tissue. If the irritation ceases, the bone will slowly re-form so as to approximate its previous, normal condition. However, except for superficial injuries or those sustained during early childhood, some trace of damage persists for the life of the individual.

In terms of trauma, the most impressive of the Shanidar Neanderthals is the first adult discovered, known as Shanidar 1. This individual suffered a number of injuries, some of which may be related. On the right forehead there are scars from minor surface injuries, probably superficial scalp cuts. The outside of the left eye socket sustained a major blow that partially collapsed that part of the bony cavity, giving it a flat rather than a rounded contour. This injury possibly caused loss of sight in the left eye and pathological alterations of the right side of the body.

Shanidar 1's left arm is largely preserved and fully normal. The right arm, however, consists of a highly atrophied but otherwise normal collarbone and shoulder blade and a highly abnormal upper arm bone shaft. That shaft is atrophied to a fraction of the diameter of the left one but retains most of its original length. Furthermore, the lower end of the right arm bone has a healed fracture of the atrophied shaft and an irregular, pathological tip. The arm was apparently either intentionally amputated just above the elbow or fractured at the elbow and never healed.

This abnormal condition of the right arm does not appear to be a congenital malformation, since the length of the bone is close to the estimated length of the normal left upper arm bone. If, however, the injury to the left eye socket also affected the left side of the brain, directly or indirectly, by disrupting the blood supply to part of the brain, the result could have been partial pa-

ralysis of the right side. Motor and sensory control areas for the right side are located on the left side of the brain, slightly behind the left eye socket. This would explain the atrophy of the whole right arm since loss of nervous stimulation will rapidly lead to atrophy of the affected muscles and bone.

The abnormality of the right arm of Shanidar 1 is paralleled to a lesser extent in the right foot. The right ankle joint shows extensive arthritic degeneration, and one of the major joints of the inner arch of the right foot has been completely reworked by arthritis. The left foot, however, is totally free of pathology. Arthritis from normal stress usually affects both lower limbs equally; this degeneration therefore suggests that the arthritis in the right foot is a secondary result of an injury, perhaps a sprain, that would not otherwise be evident on skeletal remains. This conclusion is supported by a healed fracture of the right fifth instep bone, which makes up a major portion of the outer arch of the foot. These foot pathologies may be tied into the damage to the left side of the skull; partial paralysis of the right side would certainly weaken the leg and make it more susceptible to injury.

The trauma evident on the other Shanidar Neanderthals is relatively minor by comparison. Shanidar 3, the individual who died of the rib wound, suffered debilitating arthritis of the right ankle and neighboring foot joints, but lacks any evidence of pathology on the left foot; this suggests a superficial injury similar to the one sustained by Shanidar 1. Shanidar 4 had a healed broken rib. Shanidar 5 received a transverse blow across the left forehead that left a large scar on the bone but does not appear to have affected the brain.

None of these injuries necessarily provides evidence of deliberate violence among the Neanderthals; all of them could have been accidentally self-inflicted or accidentally caused by another individual. In either case, the impression gained of the Shanidar Neanderthals is of a group of invalids. The crucial variable, however, appears to be age. All four of these individ-

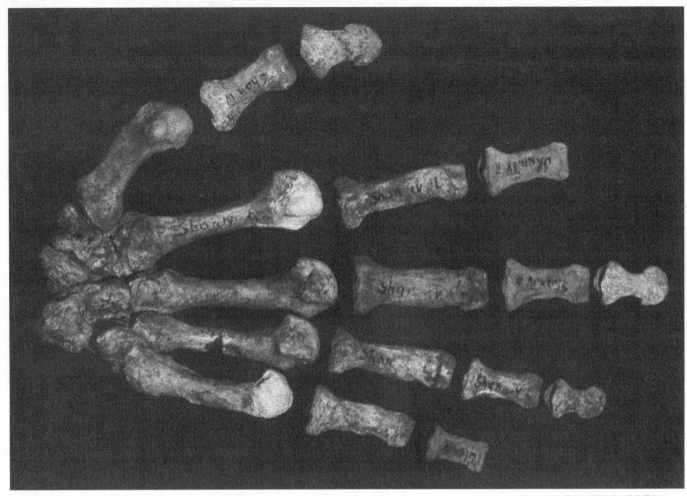

Photograph by Erik Trinkaus

The right hand of Shanidar 4 demonstrates the enlarged finger tips and strong muscle markings characteristic of Neanderthal hands.

uals died at relatively advanced ages, probably between 40 and 60 years (estimating the age at death for Neanderthals beyond the age of 25 is extremely difficult); they therefore had considerable time to accumulate the scars of past injuries. Shanidar 2 and 6, the other reasonably complete Shanidar adults, lack evidence of trauma, but they both died young, probably before reaching 30.

Other Neanderthal remains, all from Europe, exhibit the same pattern. Every fairly complete skeleton of an elderly adult shows evidence of traumatic injuries. The original male skeleton from the Neander Valley had a fracture just below the elbow of the left arm, which probably limited movement of that arm for life. The "old man" from La Chapelle-aux-Saints, France, on whom most traditional reconstructions of the Neanderthals have been based, suffered a broken rib. La Ferrassi 1, the old adult male from La Ferrassie, France, sustained a severe injury to the right hip, which may have impaired his mobility.

In addition, several younger specimens and ones of uncertain age show traces of trauma. La Quina 5, the young adult female from La Quina, France, was wounded on her right upper arm. A young adult from Sala, Czechoslovakia, was superficially wounded on the right forehead just above the brow. And an individual of unknown age and sex from the site of Krapina, Yugoslavia, suffered a broken forearm, in which the bones never reunited after the fracture.

The evidence suggests several things. First, life for the Neanderthals was rigorous. If they lived through childhood and early adulthood, they did so bearing the scars of a harsh and dangerous life. Furthermore, this incident of trauma correlates with the massiveness of the Neanderthals; a life style that so consistently involved injury would have required considerable strength and fortitude for survival.

There is, however, another, more optimistic side to this. The presence of so many injuries in a prehistoric human group, many of which were debilitating and sustained years before death, shows that individuals were taken care of long after their economic usefulness to the social group had ceased. It is perhaps no accident that among the Neanderthals, for the first time in human history, people lived to a comparatively old age. We also find among the Neanderthals the first intentional burials of the dead, some of which involved offerings. Despite the hardships of their life style, the Neanderthals apparently had a deep-seated respect and concern for each other.

Taken together, these different pieces of information paint a picture of life among the Neanderthals that, while harsh and dangerous, was not without personal security. Certainly the hardships the Neanderthals endured were beyond those commonly experienced in the prehistoric record of human caring and respect as well as of violence between individuals. Perhaps for these reasons, despite their physical appearance, the Neanderthals should be considered the first modern humans.

Reprinted with permission of the author from *Natural History,* December 1978. © 1978 by Erik Trinkaus.

Rethinking Neanderthals

**Research suggests the so-called brutes fashioned tools,
buried their dead, maybe cared for the sick and even conversed.
But why, if they were so smart, did they disappear?**

JOE ALPER

Bruno Maureille unlocks the gate in a chain-link fence, and we walk into the fossil bed past a pile of limestone rubble, the detritus of an earlier dig. We're 280 miles southwest of Paris, in rolling farm country dotted with long-haired cattle and etched by meandering streams. Maureille, an anthropologist at the University of Bordeaux, oversees the excavation of this storied site called Los Pradelles, where for three decades researchers have been uncovering, fleck by fleck, the remains of humanity's most notorious relatives, the Neanderthals.

We clamber 15 feet down a steep embankment into a swimming pool-size pit. Two hollows in the surrounding limestone indicate where shelters once stood. I'm just marveling at the idea that Neanderthals lived here about 50,000 years ago when Maureille, inspecting a long ledge that a student has been painstakingly chipping away, interrupts my reverie and calls me over. He points to a whitish object resembling a snapped pencil that's embedded in the ledge. "Butchered reindeer bone," he says. "And here's a tool, probably used to cut meat from one of these bones." The tool, or lithic, is shaped like a hand-size D.

All around the pit, I now see, are other lithics and fossilized bones. The place, Maureille says, was probably a butchery where Neanderthals in small numbers processed the results of what appear to have been very successful hunts. That finding alone is significant, because for a long time paleoanthropologists have viewed Neanderthals as too dull and too clumsy to use efficient tools, never mind organize a hunt and divvy, up the game. Fact is, this site, along with others across Europe and in Asia, is helping overturn the familiar conception of Neanderthals as dumb brutes. Recent studies suggest they were imaginative enough to carve artful objects and perhaps clever enough to invent a language.

Neanderthals, traditionally designated *Homo sapiens neanderthalensis*, were not only "human" but also, it turns out, more "modern" than scientists previously allowed. "In the minds of the European anthropologists who first studied them, Neanderthals were the embodiment of primitive humans, subhumans if you will," says Fred H. Smith, a physical anthropologist at Loy-ola University in Chicago who has been studying Neanderthal DNA. "They were believed to be scavengers who made primitive tools and were incapable of language or symbolic thought." Now, he says, researchers believe that Neanderthals "were highly intelligent, able to adapt to a wide variety of ecological zones, and capable of developing highly functional tools to help them do so. They were quite accomplished."

Contrary to the view that Neanderthals were evolutionary failures—they died out about 28,000 years ago—they actually had quite a run. "If you take success to mean the ability to survive in hostile, changing environments, then Neanderthals were a great success," says archaeologist John Shea of the State University of New York at Stony Brook. "They lived 250,000 years or more in the harshest climates experienced by primates, not just humans." In contrast, we modern humans have only been around for 100,000 years or so and moved into colder, temperate regions only in the past 40,000 years.

Though the fossil evidence is not definitive, Neanderthals appear to have descended from an earlier human species, *Homo erectus*, between 500,000 to 300,000 years ago. Neanderthals shared many features with their ancestors—a prominent brow, weak chin, sloping skull and large nose—but were as big-brained as the anatomically modern humans that later colonized Europe, *Homo sapiens*. At the same time, Neanderthals were stocky, a build that would have conserved heat efficiently. From musculature marks on Neanderthal fossils and the heft of arm and leg bones, researchers conclude they were also incredibly strong. Yet their hands were remarkably like those of modern humans; a study published this past March in *Nature* shows that Neanderthals, contrary to previous thinking, could touch index finger and thumb, which would have given them considerable dexterity.

Neanderthal fossils suggest that they must have endured a lot of pain. "When you look at adult Neanderthal fossils, particularly the bones of the arms and skull, you see [evidence of] fractures," says Erik Trinkaus, an anthropologist at Washington University in St. Louis. "I've yet to see an adult Neanderthal skeleton that doesn't have at least one fracture, and in adults in their 30s, it's common to see multiple healed fractures." (That they suffered so many broken bones suggests they hunted large animals up close, probably stabbing prey with heavy spears—a risky tactic.) In addition, fossil evidence indicates that Neanderthals suffered from a wide range of ailments, including pneumonia and malnourishment. Still, they persevered, in some cases living to the ripe old age of 45 or so.

Perhaps surprisingly, Neanderthals must also have been caring: to survive disabling injury or illness requires the help of fellow clan members, paleoanthropologists say. A telling example came from an Iraqi cave known as Shanidar, 250 miles north of Baghdad, near the border with Turkey and Iran. There, archaeologist Ralph Solecki discovered nine nearly complete Neanderthal skeletons in the late 1950s. One belonged to a 40- to 45-year-old male with several major fractures. A blow to the left side of his head had crushed an eye socket and almost certainly blinded him. The bones of his right shoulder and upper arm appeared shriveled, most likely the result of a trauma that led to the amputation of his right forearm. His right foot and lower right leg had also been broken while he was alive. Abnormal wear in his right knee, ankle and foot shows that he suffered from injury-induced arthritis that would have made walking painful, if not impossible. Researchers don't know how he was injured but believe that he could not have survived long without a hand from his fellow man.

"This was really the first demonstration that Neanderthals behaved in what we think of as a fundamentally human way," says Trinkaus, who in the 1970s helped reconstruct and catalog the Shanidar fossil collection in Baghdad. (One of the skeletons is held by the Smithsonian Institution's National Museum of Natural History.) "The result was that those of us studying Neanderthals started thinking about these people in terms of their behavior and not just their anatomy."

NEANDERTHALS INHABITED a vast area roughly from present-day England east to Uzbekistan and south nearly to the Red Sea. Their time spanned periods in which glaciers advanced and retreated again and again, But the Neanderthals adjusted. When the glaciers moved in and edible plants became scarcer, they relied more heavily on large, hoofed animals for food, hunting the reindeer and wild horses that grazed the steppes and tundra.

There are hints of cannibalism: deer and neanderthal bones at the same site bear identical scrape marks.

Paleoanthropologists have no idea how many Neanderthals existed (crude estimates are in the many thousands), but archaeologists have found more fossils from Neanderthals than from any extinct human species. The first Neanderthal fossil was uncovered in Belgium in 1830, though nobody accurately identified it for more than a century. In 1848, the Forbes Quarry in Gibraltar yielded one of the most complete Neanderthal skulls ever found, but it, too, went unidentified, for 15 years. The name Neanderthal arose after quarrymen in Germany's Neander Valley found a cranium and several long bones in 1856; they gave the specimens to a local naturalist, Johann Karl Fuhlrott, who soon recognized them as the legacy of a previously unknown type of human. Over the years, France, the Iberian Peninsula, southern Italy and the Levant have yielded abundances of Neanderthal remains, and those finds are being supplemented by newly opened excavations in Ukraine and Georgia. "It seems that everywhere we look, we're finding Neanderthal remains," says Loyola's Smith. "It's an exciting time to be studying Neanderthals."

Clues to some Neanderthal ways of life come from chemical analyses of fossilized bones, which confirm that Neanderthals were meat eaters. Microscopic studies hint at cannibalism; fossilized deer and Neanderthal bones found at the same site bear identical scrape marks, as though the same tool removed the muscle from both animals.

The arrangement of fossilized Neanderthal skeletons in the ground demonstrates to many archaeologists that Neanderthals buried their dead. "They might not have done so with elaborate ritual, since there has never been solid evidence that they included symbolic objects in graves, but it is clear that they did not just dump their dead with the rest of the trash to be picked over by hyenas and other scavengers," says archaeologist Francesco d'Errico of the University of Bordeaux.

Paleoanthropologists generally agree that Neanderthals lived in groups of 10 to 15, counting children. That assessment is based on a few lines of evidence, including the limited remains at burial sites and the modest size of rock shelters. Also, Neanderthals were top predators, and some top predators, such as lions and wolves, live in small groups.

Steven Kuhn, an archaeologist at the University of Arizona, says experts "can infer quite a bit about who Neanderthal was by studying tools in conjunction with the other artifacts they left behind." For instance, recovered stone tools are typically fashioned from nearby sources of flint or quartz, indicating to some researchers that a Neanderthal group did not necessarily range far.

The typical Neanderthal tool kit contained a variety of implements, including large spear points and knives that would have been hafted, or set in wooden handles. Other tools were suitable for cutting meat, cracking open bones (to get at fat-rich marrow) or scraping hides (useful for clothing, blankets or shelter). Yet other stone tools were used for woodworking; among the very few wooden artifacts associated with Neanderthal sites are objects that resemble spears, plates and pegs.

I get a feel for Neanderthal handiwork in Maureille's office, where plastic milk crates are stacked three high in front of his desk. They're stuffed with plastic bags full of olive and tan flints from Les Pradelles. With his encouragement, I take a palm-size, D-shaped flint out of a bag. Its surface is scarred as though by chipping, and the flat side has a thin edge. I readily imagine I could scrape a hide with it or whittle a stick. The piece, Maureille says, is about 60,000 years old. "As you can see from the number of lithics we've found," he adds, referring to the crates piling up in his office, "Neanderthals were prolific and accomplished toolmakers."

AMONG THE NEW APPROACHES to Neanderthal study is what might be called paleo-mimicry, in which researchers themselves fashion tools to test their ideas. "What we do is make our own tools out of flint, use them as a Neanderthal might have, and then look at the fine detail of the cutting edges with a high-powered microscope," explains Michael Bisson, chairman of anthropology at McGill University in Montreal. "A tool used to work wood will have one kind of wear pattern that differs from that seen when a tool is used to cut meat from a bone, and we can see those different patterns on the implements recovered from Neanderthal sites." Similarly, tools used to scrape hide show few microscopic scars, their edges having been smoothed by repeated rubbing against skin, just as stropping a straight razor will hone its edge. As Kuhn, who has also tried to duplicate Neanderthal handicraft, says: "There is no evidence of really fine, precise work, but they were skilled in what they did."

Based on the consistent form and quality of the tools found at sites across Europe and western Asia, it appears likely that Neanderthal was able to pass along his toolmaking techniques to others. "Each Neanderthal or Neanderthal group did not have to reinvent the wheel when it came to their technologies," says Bisson.

The kinds of tools that Neanderthals began making about 200,000 years ago are known as Mousterian, after the site in France where thousands of artifacts were first found. Neanderthals struck off pieces from a rock "core" to make an implement, but the "flaking" process was not random; they evidently examined a core much as a diamond cutter analyzes a rough gemstone today, trying to strike just the spot that would yield "flakes," for knives or spear points, requiring little sharpening or shaping.

Around 40,000 years ago, Neanderthals innovated again. In what passes for the blink of an eye in paleoanthropology, some Neanderthals were suddenly making long, thin stone blades and hafting more tools. Excavations in southwest France and northern Spain have uncovered Neanderthal tools betraying a more refined technique involving, Kuhn speculates, the use of soft hammers made of antler or bone.

What happened? According to the conventional wisdom, there was a culture clash. In the early 20th century, when researchers first discovered those "improved" lithics—called Châtelperronian and Uluzzian, depending on where they were found—they saw the relics as evidence that modern humans, Homo sapiens or Cro-Magnon, had arrived in Neanderthal territory. That's because the tools resembled those unequivocally associated with anatomically modern humans, who began colonizing western Europe 38,000 years ago. And early efforts to assign a date to those Neanderthal lithics yielded time frames consistent with the arrival of modern humans.

But more recent discoveries and studies, including tests that showed the lithics to be older than previously believed, have prompted d'Errico and others to argue that Neanderthals advanced on their own. "They could respond to some change in their environment that required them to improve their technology," he says. "They could behave like modern humans."

Meanwhile, these "late" Neanderthals also discovered ornamentation, says d'Errico and his archaeologist colleague João Zilhão of the University of Lisbon. Their evidence includes items made of bone, ivory and animal teeth marked with grooves and perforations. The researchers and others have also found dozens of pieces of sharpened manganese dioxide—black crayons, essentially—that Neanderthals probably used to color animal skins or even their own. In his office at the University of Bordeaux, d'Errico hands me a chunk of manganese dioxide. It feels silky, like soapstone. "Toward the end of their time on earth," he says, "Neanderthals were using technology as advanced as that of contemporary anatomically modern humans and were using symbolism in much the same way."

As Neanderthals retreated, modern humans were right on their heels. The two may have mated—or tried to.

Generally, anthropologists and archaeologists today proffer two scenarios for how Neanderthals became increasingly resourceful in the days be fore they vanished. On the one hand, it may be that Neanderthals picked up a few new technologies from invading humans in an effort to copy their cousins. On the other, Neanderthals learned to innovate in parallel with anatomically modern human beings, our ancestors,

MOST RESEARCHERS AGREE that Neanderthals were skilled hunters and craftsmen who made tools, used fire, buried their dead (at least on occasion), cared for their sick and injured and even had a few symbolic notions. Likewise, most researchers believe that Neanderthals probably had some facility, for language, at least as we usually think of it. It's not far-fetched to think that language skills developed when Neanderthal groups mingled and exchanged mates; such interactions may have been necessary for survival, some researchers speculate, because Neanderthal groups were too small to sustain the species. "You need to have a breeding population of at least 250 adults, so some kind of exchange had to take place," says archaeologist Ofer Bar-Yosef of Harvard University. "We see this type of behavior in all hunter-gatherer cultures, which is essentially what Neanderthals had."

But if Neanderthals were so smart, why did they go extinct? "That's a question we'll never really have an answer to," says Clive Finlayson, who runs the Gibraltar Museum, "though it doesn't stop any of us from putting forth some pretty elaborate scenarios." Many researchers are loath even to speculate on the cause of Neanderthals' demise, but Finlayson suggests that a combination of climate change and the cumulative effect of repeated population busts eventually did them in. "I think it's the culmination of 100,000 years of climate hitting Neanderthals hard, their population diving during the cold years, rebounding some during warm years, then diving further when it got cold again," Finlayson says.

As Neanderthals retreated into present-day southern Spain and parts of Croatia toward the end of their time, modern human beings were right on their heels. Some researchers, like Smith, believe that Neanderthals and Cro-Magnon humans probably mated, if only in limited numbers. The question of whether Neanderthals and modern humans bred might be resolved within a decade by scientists studying DNA samples from Neanderthal and Cro-Magnon fossils.

But others argue that any encounter was likely to be hostile. "Brotherly love is not the way I'd describe any interaction between different groups of humans," Shea says. In fact, he speculates that modern humans were superior warriors and wiped out the Neanderthals. "Modern humans are very competitive and really good at using projectile weapons to kill from a distance," he says, adding they also probably worked together better in large groups, providing a battlefield edge.

In the end, Neanderthals, though handy, big-brained, brawny and persistent, went the way of every human species but one. "There have been a great many experiments at being human preceding us and none of them made it, so we should not think poorly of Neanderthal just because they went extinct," says Rick Potts, head of the Smithsonian's Human Origins Program. "Given that Neanderthal possessed the very traits that we think guarantee our success should make us pause about our place here on earth."

JOE ALPER, a freelance writer in Louisville, Colorado, is a frequent contributor to *Science* magazine. This is his first article for SMITHSONIAN. Paris-based **ERIC SANDER** photographed "Master Class" in the October 2002 issue. **STAN FELLOWS'** illustrations appeared in "Hamilton Takes Command," in the January issue.

Caveful of Clues About Early Humans

Interbreeding with Neanderthals among theories being explored

FREDRIC HEEREN

"Field research" projects often require scientists to endure discomfort and danger to get where they need to be, but not many can trump this summer's expedition to what may be the world's most inaccessible human fossil site, a cave in the foothills of Romania's Carpathian Mountains.

For the seven-member team, the hazards of reaching the site, accessible only by diving through frigid underwater passages, were worth it. Their finds may help answer some of the most hotly debated questions about early humans: Did they make love or war with Neanderthals? Were Neanderthals intellectually inferior to our human ancestors?

This may be asking a lot of the scanty fossil remains of three individuals who lived 35,000 years ago, but their age makes them the earliest modern humans ever found in Europe. The uniqueness of the site, which was discovered in 2002, was motivation enough for the specially trained team to devote a month of cold and dangerous underground journeys to reach and excavate the site known as Pestera cu Oase—Cave with Bones.

The team included a Portuguese shipwreck diver and archaeologist, a French Neanderthal specialist, a Romanian cave biologist, and the three Romanian adventurers who discovered the human fossils while exploring submerged caves.

At the start of each day's nine-hour excursion underground, team members stepped into a frigid mountain river that flows into a cave, their helmet-mounted lights piercing the perpetual fog of the cave's 100 percent humidity. As the equipment-laden crew sloshed past stalagmites, the cave narrowed and the air temperature plunged from the 90s to the upper 40s Fahrenheit.

Further in, the ceiling lowered until they were forced, first, to swim on their backs and, finally, don their diving masks and enter a narrow, 80-foot-long underwater passage called "the sump." Underwater visibility was about three feet.

Lead diver Stefan Milota warned newcomers: "Do you know how long it takes to die in the sump? Twenty, thirty seconds and you're gone."

When Milota first told biologist Oana Moldovan in 2002 that he had found a human jaw in a closed-off chamber, Moldovan wanted to see for herself—even though she had to learn to dive to do it.

"I had to see the cave," said Moldovan, now the team's project coordinator. "So I was very motivated. But I was very scared. My first time, it was horrible."

Surfacing inside another chamber, the divers peeled off their wetsuits and changed into warm clothing. The next step was climbing "the pit," a series of underground cliff faces that the cavers scaled in dizzying climbs up a succession of ladders they had carried in earlier.

Finally, to reach the gallery of bones, they passed through "the gate," an opening that Milota had first spotted when he felt warmer air emerging. He and his explorer friends widened it just enough for the thinnest of them to squeeze through. Each day, the cavers had to plunge head-and-arms first at a slight uphill angle, then wriggle and rest, wriggle and rest, to cover the final, winding 10 feet.

Inside the final gallery, there was room for only three workers at a time because the rest of the floor was covered with thousands of fragile fossils. Most belonged to a cave bear species that became extinct 10,000 years ago—animals almost twice as big as today's bears.

The original entrance caved in long ago, sealing off the galleries from the outside. After two labs independently yielded radiocarbon dates of about 35,000 years for the jaw, or mandible, that Milota had found, more scientists took interest. In a 2003 expedition, they found a full face and an ear region of a skull from two more individuals, with puzzling traits that suggested a mix of Neanderthal and human features, something scientists had thought impossible.

Anthropologist Erik Trinkaus of Washington University in St. Louis and Joao Zilhao of Cidade University in Lisbon joined this summer's excavation to look for more specimens and to try to find out how the human remains got into the cave. Because they turned up no sign of torches, charcoal or tools, they concluded that the human remains had washed in through fissures.

The biggest payoff of the summer was the discovery of more fragments of the three individuals found earlier, which added to the evidence of hybrid traits.

Trinkaus said the Oase fossils show features of modern humans: projecting chin, no brow ridge, a high and rounded brain case. But they also have clear archaic features that place them outside the range of variation for modern humans: a huge face, a large crest of bone behind the ear and enormous teeth that get even larger toward the back.

Trinkaus made a CT scan of the face to measure the unerupted teeth. "To find wisdom teeth that big," he said, "you have to go back 500,000 years."

The team considered whether early humans might have interbred with other hominids with Neanderthal-like features, but "in this time period," said Trinkaus, "the only archaic humans those modern humans could have interbred with were Neanderthals." The mosaic of Neanderthal and modern traits remind Trinkaus and Zilhao of similar traits they found in a 25,000-year-old fossil of a child in Portugal.

Researchers pondering why the Neanderthals died out have speculated that early humans might have killed them off, and Zilhao said the signs of interbreeding do not exclude that possibility. "We know that even when people fight, the winner might kill the males and keep the females from the other side," he said.

The signs of interbreeding challenge the standard wisdom that Neanderthals were a distinct, less intelligent species.

"If you look at the archaeological evidence," argued Trinkaus, "which includes things like burials, there is very little difference between what we find associated with Neanderthals and what we find associated with early modern humans—from the same time period."

Richard Klein of Stanford University thinks this holds true only until about 50,000 years ago, when modern human behavior changed dramatically. "There could have been interbreeding," Klein conceded. "But all the genetic evidence we have suggests that, if it occurred, it was remarkably rare."

Six years ago, Zilhao and Francesco d'Errico of the University of Bordeaux published evidence that Neanderthals independently invented and used personal ornamentation. Zilhao said these finds have changed the view that Neanderthals were an inferior species.

Klein said the picture is changing, but not in that direction. The real question today, he said, is "whether modern humans fully replaced the Neanderthals or simply swamped them" genetically, with greater numbers. "And it may never be possible to say."

The Gift *of* Gab

Grooves and holes in fossil skulls may reveal when our ancestors began to speak. The big question, though, is what drove them to it?

MATT CARTMILL

People can talk. Other animals can't. They can all communicate in one way or another—to lure mates, at the very least—but their whinnies and wiggles don't do the jobs that language does. The birds and beasts can use their signals to attract, threaten, or alert each other, but they can't ask questions, strike bargains, tell stories, or lay out a plan of action.

Those skills make *Homo sapiens* a uniquely successful, powerful, and dangerous mammal. Other creatures' signals carry only a few limited kinds of information about what's happening at the moment, but language lets us tell each other in limitless detail about what used to be or will be or might be. Language lets us get vast numbers of big, smart fellow primates all working together on a single task—building the Great Wall of China or fighting World War II or flying to the moon. It lets us construct and communicate the gorgeous fantasies of literature and the profound fables of myth. It lets us cheat death by pouring out our knowledge, dreams, and memories into younger people's minds. And it does powerful things for us inside our own minds because we do a lot of our thinking by talking silently to ourselves. Without language, we would be only a sort of upright chimpanzee with funny feet and clever hands. With it, we are the self-possessed masters of the planet.

How did such a marvelous adaptation get started? And if it's so marvelous, why hasn't any other species come up with anything similar? These may be the most important questions we face in studying human evolution. They are also the least understood. But in the past few years, linguists and anthropologists have been making some breakthroughs, and we are now beginning to have a glimmering of some answers.

COULD NEANDERTHALS talk? They seem to have had nimble tongues, but some scientists think the geometry of their throats prevented them from making many clear vowel sounds.

We can reasonably assume that by at least 30,000 years ago people were talking—at any rate, they were producing carvings, rock paintings, and jewelry, as well as ceremonial graves containing various goods. These tokens of art and religion are high-level forms of symbolic behavior, and they imply that the everyday symbol-handling machinery of human language must have been in place then as well.

Language surely goes back further than that, but archeologists don't agree on just how far. Some think that earlier, more basic human behaviors—hunting in groups, tending fires, making tools—also demanded language. Others think these activities are possible without speech. Chimpanzees, after all, hunt communally, and with human guidance they can learn to tend fires and chip flint.

Paleontologists have pored over the fossil bones of our ancient relatives in search of evidence for speech abilities. Because the most crucial organ for language is the brain, they have looked for signs in the impressions left by the brain on the inner surfaces of fossil skulls, particularly impressions made by parts of the brain called speech areas because damage to them can impair a person's ability to talk or understand language. Unfortunately, it turns out that you can't tell whether a fossil hominid was able to talk simply by looking at brain impressions on the inside of its skull. For one thing, the fit between the brain and the bony braincase is loose in people and other large mammals, and so the impressions we derive from fossil skulls are disappointingly fuzzy. Moreover, we now know that language functions are not tightly localized but spread across many parts of the brain.

Faced with these obstacles, researchers have turned from the brain to other organs used in speech, such as the throat and tongue. Some have measured the fossil skulls and jaws of early hominids, tried to reconstruct the shape of their vocal tracts, and then applied the laws of acoustics to them to see whether they might have been capable of producing human speech.

All mammals produce their vocal noises by contracting muscles that compress the rib cage. The air in the lungs is driven out through the windpipe to the larynx, where it flows between the vocal cords. More like flaps than cords, these structures vibrate in the breeze, producing a buzzing sound that becomes the

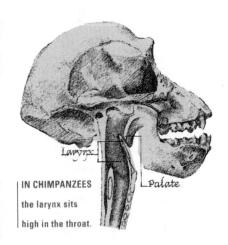

IN CHIMPANZEES the larynx sits high in the throat.

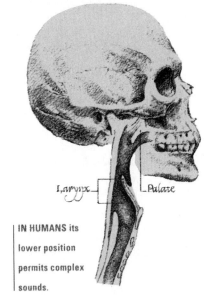

IN HUMANS its lower position permits complex sounds.

ILLUSTRATIONS BY DUGALD STERMER

voice. The human difference lies in what happens to the air after it gets past the vocal cords.

In people, the larynx lies well below the back of the tongue, and most of the air goes out through the mouth when we talk. We make only a few sounds by exhaling through the nose—for instance, nasal consonants like *m* or *n*, or the so-called nasal vowels in words like the French *bon* and *vin*. But in most mammals, including apes, the larynx sticks farther up behind the tongue, into the back of the nose, and most of the exhaled air passes out through the nostrils. Nonhuman mammals make mostly nasal sounds as a result.

At some point in human evolution the larynx must have descended from its previous heights, and this change had some serious drawbacks. It put the opening of the windpipe squarely in the path of descending food, making it dangerously easy for us to choke to death if a chunk of meat goes down the wrong way—something that rarely happens to a dog or a cat. Why has evolution exposed us to this danger?

Some scientists think that the benefits outweighed the risks, because lowering the larynx improved the quality of our vowels

and made speech easier to understand. The differences between vowels are produced mainly by changing the size and shape of the airway between the tongue and the roof of the mouth. When the front of the tongue almost touches the palate, you get the *ee* sound in *beet*; when the tongue is humped up high in the back (and the lips are rounded), you get the *oo* sound in *boot*, and so on. We are actually born with a somewhat apelike throat, including a flat tongue and a larynx lying high up in the neck, and this arrangement makes a child's vowels sound less clearly separated from each other than an adult's.

Philip Lieberman of Brown University thinks that an apelike throat persisted for some time in our hominid ancestors. His studies of fossil jaws and skulls persuade him that a more modern throat didn't evolve until some 500,000 years ago, and that some evolutionary lines in the genus *Homo* never did acquire modern vocal organs. Lieberman concludes that the Neanderthals, who lived in Europe until perhaps 25,000 years ago, belonged to a dead-end lineage that never developed our range of vowels, and that their speech—if they had any at all—would have been harder to understand than ours. Apparently, being easily understood wasn't terribly important to them—not important enough, at any rate, to outweigh the risk of inhaling a chunk of steak into a lowered larynx. This suggests that vocal communication wasn't as central to their lives as it is to ours.

Many paleoanthropologists, especially those who like to see Neanderthals as a separate species, accept this story. Others have their doubts. But the study of other parts of the skeleton in fossil hominids supports some of Lieberman's conclusions. During the 1980s a nearly complete skeleton of a young *Homo* male was recovered from 1.5-million-year-old deposits in northern Kenya. Examining the vertebrae attached to the boy's rib cage, the English anatomist Ann MacLarnon discovered that his spinal cord was proportionately thinner in this region than it is in people today. Since that part of the cord controls most of the muscles that drive air in and out of the lungs, MacLarnon concluded that the youth may not have had the kind of precise neural control over breathing movements that is needed for speech.

This year my colleague Richard Kay, his student Michelle Balow, and I were able to offer some insights from yet another part of the hominid body. The tongue's movements are controlled almost solely by a nerve called the hypoglossal. In its course from the brain to the tongue, this nerve passes through a hole in the skull, and Kay, Balow, and I found that this bony canal is relatively big in modern humans—about twice as big in cross section as that of a like-size chimpanzee. Our larger canal presumably reflects a bigger hypoglossal nerve, giving us the precise control over tongue movements that we need for speech.

We also measured this hole in the skulls of a number of fossil hominids. Australopithecines have small canals like those of apes, suggesting that they couldn't talk. But later *Homo* skulls, beginning with a 400,000-year-old skull from Zambia, all have big, humanlike hypoglossal canals. These are also the skulls that were the first to house brains as big as our own. On these counts our work supports Lieberman's ideas. We disagree only on the matter of Neanderthals. While he claims their throats couldn't have produced human speech, we find that their skulls also had human-size canals for the hypoglossal nerve, suggesting that they could indeed talk.

THE VERDICT IS STILL out on the language abilities of Neanderthals. I tend to think they must have had fully human language. After all, they had brains larger than those of most humans.

In short, several lines of evidence suggest that neither the australopithecines nor the early, small-brained species of *Homo* could talk. Only around half a million years ago did the first big-brained *Homo* evolve language. The verdict is still out on the language abilities of Neanderthals. I tend to think that they must have had fully human language. After all, they had brains larger than those of most modern humans, made elegant stone tools, and knew how to use fire. But if Lieberman and his friends are right about those vowels, Neanderthals may have sounded something like the Swedish chef on *The Muppet Show*.

We are beginning to get some idea of when human language originated, but the fossils can't tell us how it got started, or what the intermediate stages between animal calls and human language might have been like. When trying to understand the origin of a trait that doesn't fossilize, it's sometimes useful to look for similar but simpler versions of it in other creatures living today. With luck, you can find a series of forms that suggest how simple primitive makeshifts could have evolved into more complex and elegant versions. This is how Darwin attacked the problem of the evolution of the eye. Earlier biologists had pointed to the human eye as an example of a marvelously perfect organ that must have been specially created all at once in its final form by God. But Darwin pointed out that animal eyes exist in all stages of complexity, from simple skin cells that can detect only the difference between light and darkness, to pits lined with such cells, and so on all the way to the eyes of people and other vertebrates. This series, he argued, shows how the human eye could have evolved from simpler precursors by gradual stages.

Can we look to other animals to find simpler precursors of language? It seems unlikely. Scientists have sought experimental evidence of language in dolphins and chimpanzees, thus far without success. But even if we had no experimental studies, common sense would tell us that the other animals can't have languages like ours. If they had, we would be in big trouble because they would organize against us. They don't. Outside of Gary Larson's *Far Side* cartoons and George Orwell's *Animal Farm*, farmers don't have to watch their backs when they visit the cowshed. There are no conspiracies among cows, or even among dolphins and chimpanzees. Unlike human slaves or prisoners, they never plot rebellions against their oppressors.

Even if language as a whole has no parallels in animal communication, might some of its peculiar properties be foreshadowed among the beasts around us? If so, that might tell us

something about how and in what order these properties were acquired. One such property is reference. Most of the units of human languages refer to things—to individuals (like *Fido*), or to types of objects (*dog*), actions (*sit*), or properties (*furry*). Animal signals don't have this kind of referential meaning. Instead, they have what is called instrumental meaning: this is, they act as stimuli that trigger desired responses from others. A frog's mating croak doesn't *refer* to sex. Its purpose is to get some, not to talk about it. People, too, have signals of this purely animal sort—for example, weeping, laughing, and screaming—but these stand outside language. They have powerful meanings for us but not the kind of meaning that words have.

Some animal signals have a focused meaning that looks a bit like reference. For example, vervet monkeys give different warning calls for different predators. When they hear the "leopard" call, vervets climb trees and anxiously look down; when they hear the "eagle" call, they hide in low bushes or look up. But although the vervets' leopard call is in some sense about leopards, it isn't a word for leopard. Like a frog's croak or human weeping, its meaning is strictly instrumental; it's a stimulus that elicits an automatic response. All a vervet can "say" with it is "*Eeek!* A leopard!"—not "I really hate leopard!" or "No leopards here, thank goodness" or "A leopard ate Alice yesterday."

AUSTRALOPITHECUS africanus and other early hominids couldn't speak.

In these English sentences, such referential words as *leopard* work their magic through an accompanying framework of non-referential, grammatical words, which set up an empty web of meaning that the referential symbols fill in. When Lewis Carroll tells us in "Jabberwocky" that "the slithy toves did gyre and gimble in the wabe," we have no idea what he is talking about, but we do know certain things—for instance, that all this happened in the past and that there was more than one tove but only one wabe. We know these things because of the grammatical structure of the sentence, a structure that linguists call syntax. Again, there's nothing much like it in any animal signals.

But if there aren't any intermediate stages between animal calls and human speech, then how could language evolve? What was there for it to evolve from? Until recently, linguists have shrugged off these questions—or else concluded that language didn't evolve at all, but just sprang into existence by accident, through some glorious random mutation. This theory drives Darwinians crazy, but the linguists have been content with it because it fits neatly into some key ideas in modern linguistics.

Forty years ago most linguists thought that people learn to talk through the same sort of behavior reinforcement used in training an animal to do tricks: when children use a word correctly or produce a grammatical sentence, they are rewarded. This picture was swept away in the late 1950s by the revolutionary ideas of Noam Chomsky. Chomsky argued that the structures of syntax lie in unconscious linguistic patterns—so-called deep structures—that are very different from the surface

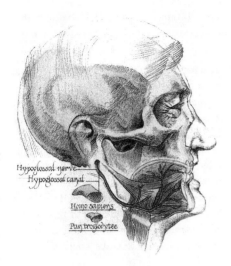

THE TONGUE-CONTROLLING

hypoglossal nerve is larger in

humans than in chimps.

strings of words that come out of our mouths. Two sentences that look different on the surface (for instance, "A leopard ate Alice" and "Alice was eaten by a leopard") can mean the same thing because they derive from a single deep structure. Conversely, two sentences with different deep structures and different meanings can look exactly the same on the surface (for example, "Fleeing leopards can be dangerous"). Any models of language learning based strictly on the observable behaviors of language, Chomsky insisted, can't account for these deep-lying patterns of meaning.

Chomsky concluded that the deepest structures of language are innate, not learned. We are all born with the same fundamental grammar hard-wired into our brains, and we are preprogrammed to pick up the additional rules of the local language, just as baby ducks are hard-wired to follow the first big animal they see when they hatch. Chomsky could see no evidence of other animals' possessing this innate syntax machinery. He concluded that we can't learn anything about the origins of language by studying other animals and they can't learn language from us. If language learning were just a matter of proper training, Chomsky reasoned, we ought to be able to teach English to lab rats, or at least to apes.

As we have seen, apes aren't built to talk. But they can be trained to use sign language or to point to word-symbols on a keyboard. Starting in the 1960s, several experimenters trained chimpanzees and other great apes to use such signs to ask for things and answer questions to get rewards. Linguists, however, were unimpressed. They said that the apes' signs had a purely instrumental meaning: the animals were just doing tricks to get a treat. And there was no trace of syntax in the random-looking jumble of signs the apes produced; an ape that signed "You give me cookie please" one minute might sign "Me cookie please you cookie eat give" the next.

Duane Rumbaugh and Sue Savage-Rumbaugh set to work with chimpanzees at the Yerkes Regional Primate Research Center in Atlanta to try to answer the linguists' criticisms. After many years of mixed results, Sue made a surprising break-through with a young bonobo (or pygmy chimp) named Kanzi. Kanzi had watched his mother, Matata, try to learn signs with little success. When Sue gave up on her and started with Kanzi, she was astonished to discover that he already knew the meaning of 12 of the keyboard symbols. Apparently, he had learned them without any training or rewards. In the years that followed, he learned new symbols quickly and used them referentially, both to answer questions and to "talk" about things that he intended to do or had already done. Still more amazingly, he had a considerable understanding of spoken English—including its syntax. He grasped such grammatical niceties as case structures ("Can you throw a potato to the turtle?") and if-then implication ("You can have some cereal if you give Austin your monster mash to play with"). Upon hearing such sentences, Kanzi behaved appropriately 72 percent of the time—more than a 30-month-old human child given the same tests.

Kanzi is a primatologist's dream and a linguist's nightmare. His language-learning abilities seem inexplicable. He didn't need any rewards to learn language, as the old behaviorists would have predicted; but he also defies the Chomskyan model, which can't explain why a speechless ape would have an innate tendency to learn English. It looks as though some animals can develop linguistic abilities for reasons unrelated to language itself.

BRAIN ENLARGEMENT in hominids may have been the result of evolutionary pressures that favored intelligence. As a side effect, human evolution crossed a threshold at which language became possible.

Neuroscientist William Calvin of the University of Washington and linguist Derek Bickerton of the University of Hawaii have a suggestion as to what those reasons might be. In their forthcoming book, *Lingua ex Machina*, they argue that the ability to create symbols—signs that refer to things—is potentially present in any animal that can learn to interpret natural signs, such as a trail of footprints. Syntax, meanwhile, emerges from the abstract thought required for a social life. In apes and some other mammals with complex and subtle social relationships, individuals make alliances and act altruistically towards others, with the implicit understanding that their favors will be returned. To succeed in such societies, animals need to choose trustworthy allies and to detect and punish cheaters who take but never give anything in return. This demands fitting a shifting constellation of individuals into an abstract mental model of social roles (debtors, creditors, allies, and so on) connected by social expectations ("If you scratch my back, I'll scratch yours"). Calvin and Bickerton believe that such abstract models of social obligation furnished the basic pattern for the deep structures of syntax.

These foreshadowings of symbols and syntax, they propose, laid the groundwork for language in a lot of social animals but didn't create language itself. That had to wait until our ancestors evolved brains big enough to handle the large-scale operations needed to generate and process complex strings of signs. Calvin and Bickerton suggest that brain enlargement in our ancestry was the result of evolutionary pressures that favored intelligence and motor coordination for making tools and throwing weapons. As a side effect of these selection pressures, which had nothing to do with communication, human evolution crossed a threshold at which language became possible. Big-brained, nonhuman animals like Kanzi remain just on the verge of language.

FOSSILS HINT that language dawned 500,000 years ago.

This story reconciles natural selection with the linguists' insistence that you can't evolve language out of an animal communication system. It is also consistent with what we know about language from the fossil record. The earliest hominids with modern-size brains also seem to be the first ones with modern-size hypoglossal canals. Lieberman thinks that these are also the first hominids with modern vocal tracts. It may be no coincidence that all three of these changes seem to show up together around half a million years ago. If Calvin and Bickerton are right, the enlargement of the brain may have abruptly brought language into being at this time, which would have placed new selection pressures on the evolving throat and tongue.

This account may be wrong in some of its details, but the story in its broad outlines solves so many puzzles and ties up so many loose ends that something like it must surely be correct. It also promises to resolve our conflicting views of the boundary between people and animals. To some people, it seems obvious that human beings are utterly different from any beasts. To others, it's just as obvious that many other animals are essentially like us, only with fewer smarts and more fur. Each party finds the other's view of humanity alien and threatening. The story of language origins sketched above suggests that both parties are right: the human difference is real and profound, but it is rooted in aspects of psychology and biology that we share with our close animal relatives. If the growing consensus on the origins of language can join these disparate truths together, it will be a big step forward in the study of human evolution.

MATT CARTMILL is a professor at Duke, where he teaches anatomy and anthropology and studies animal locomotion. Cartmill is also the author of numerous articles and books on the evolution of people and other animals, including an award-winning book on hunting, *A View to a Death in the Morning.* He is the president of the American Association of Physical Anthropologists.

We Are All Africans

Pat Shipman

Nice hypothesis, you may say, but *is it true?* The heart of the scientific method lies in that skeptic's simple question. A merely plausible explanation of what we observe will not suffice, nor will a hypothesis bolstered only by some expert's endorsement. Modern science is wonderfully egalitarian, and it demands proof that all can see: measurements, objects or evidence of some kind. Scientific hypotheses must be both well-defined and firmly supported, preferably by several different types of data.

If a hypothesis meets these requirements, does this mean it is true? No, according to the philosopher Karl Popper, who believed that hypotheses could never be proven true—only false. Frederick Nietzsche claimed that the great charm of hypotheses was that they were refutable. While Nietzsche may be correct in the abstract, my experience suggests that scientists do not gleefully bury hypotheses in which they have invested many years of research.

In the study of human evolution, dueling hypotheses are commonplace because of long-standing, frequently erupting feuds about the interpretation of the fossil record. The field is so argumentative, in part, because the theories reflect directly on the nature and origin of humans. There is immense room for giving and taking offense when the subject is oneself. Too, the primary data of paleoanthropology—fossilized remains of our ancestors and near-relatives—are rare and difficult to obtain. Hence, it is not a simple matter to collect more evidence to clarify or support hypotheses. Fossil hunting requires tremendous knowledge and effort, good organizational skills, substantial grants, and a huge dollop of luck. New theories are, sadly, easier to come by than new primary evidence.

Thus it is a joyous occasion when my paleoanthropologist colleagues appear to resolve one of the most bitterly debated questions in the discipline: the issue of when, where and how modern humans evolved. For simplicity, I use "modern humans" or "recent humans" to denote the species to which I (and all the readers of this magazine) belong, but the formal term is either "anatomically modern *Homo sapiens*" or "*Homo sapiens sapiens*." Humans who lived in the past and did not have modern anatomy are often referred to as archaic or primitive.

Opposable Theories

For decades, paleoanthropologists have argued over two competing theories about the origin of our kind. The older notion, which owes its crude beginnings to Charles Darwin, is the *Out of Africa hypothesis*. This theory maintains that modern humans evolved in Africa and then spread around the world. Boiled down to its essence, the hypothesis states that modern humans are both relatively recent (100,000 to 200,000 years old) and African in origin. A major prediction of this hypothesis is that the earliest remains of modern humans will be found in Africa, dated to an appropriate time period.

The rival *Multiregional hypothesis* argues that modern humans evolved in many locations around the world from a precursor species, *Homo erectus*, approximately one to two million years ago. According to this school of thought, these regional populations evolved along parallel paths and reached modernity at roughly the same time. Because the populations were largely isolated from one another, they developed distinctive regional features, which people recognize today as "racial" differences. The Multiregional hypothesis predicts that the fossilized remains of the earliest modern humans will be found all over the Old World and that these scattered fossils will all date from about the same time. Furthermore, the theory requires these early populations to show anatomical and genetic continuity with the current inhabitants of the same region. For example, Multiregionalists believe that Neandertals, an archaic human form, are most closely related to modern indigenous Europeans.

Unfortunately for adherents of the Multiregional hypothesis, recent results are weighing heavily against them. Three very different strains of evidence have converged to offer convincing support for the rival theory.

Evidence for "Eve"

In April of this year, Sarah Tishkoff of the University of Maryland and a team of coworkers reported genetic analyses of more than 600 living Tanzanians from 14 different tribes and four linguistic groups. They analyzed mitochondrial DNA (mtDNA)—the tool of choice for tracing ancestry because it is inherited only through the mother as part of the ovum. The number of mutations that have accumulated in mtDNA is a rough measure of the time that has passed since that lineage first appeared. The owner of the first modern human mtDNA (by definition, a woman) is often referred to as "Eve," although many women of that time are likely to have shared similar mtDNA.

Tishkoff and her colleagues chose to investigate East African peoples for specific reasons. The number of linguistic and cultural differences is unusually high in the region, as is the variation in physical appearance—East Africans are tall or short, darker-skinned or lighter-skinned, round-faced or narrow-faced, and so on. This observation suggested that the genetic composition of the population is highly diverse, and as expected, the team found substantial variation in the mtDNA. In fact, members of five of the lineages showed an exceptionally high number of mutations compared with other populations, indicating that these East African lineages are of great antiquity. Identified by tribal affiliation, these are: the Sandawe, who speak a "click" language related to that of the Bushmen of the Kalahari desert; the Burunge and Gorowaa, who migrated to Tanzania from Ethiopia within the last five thousand years; and the Maasai and the Datog, who probably originated in the Sudan. The efforts of the University of Maryland group reflect a substantially larger database and more certain geographic origins for its subjects than earlier mtDNA studies. Further, the work by Tishkoff's team reveals that these five East African populations have even older origins than the !Kung San of southern Africa, who previously had the oldest known mtDNA.

"These samples showed really deep, old lineages with lots of genetic diversity," Tishkoff says. "They are the oldest lineages identified to date. And that fact makes it highly likely that 'Eve' was an East or Northeast African. My guess is that the region of Ethiopia or the Sudan is where modern humans originated."

By assuming that mtDNA mutates at a constant rate, Tishkoff's team estimated that the oldest lineages in their study originated 170,000 years ago, although she cautions that the method only gives an approximate date. Nonetheless, this finding is neatly congruent with new fossil evidence.

Idaltu Means "Elder"

This past June, an international team headed by Tim White and F. Clark Howell of the University of California at Berkeley, and Berhane Asfaw of the Rift Valley Research Service in Addis Ababa announced the discovery of three fossilized human skulls in the Herto Bouri area of Ethiopia. Volcanic layers immediately above and below the layer were dated to 154,000 and 160,000 years using radioisotopes, meaning that the owners of the skulls lived sometime between those dates.

The most remarkable of the three specimens is an adult male cranium: With the exception of a few missing teeth and some damage on the left side of the skull, the fossil is complete. There is also part of another male skull and an immature cranium from a six- or seven-year-old child.

Once these specimens were cleaned and pieced together, the team was able to make some telling observations. Like modern humans, the owners of these skulls had small faces tucked under capacious braincases, making the facial profile vertical. The cranial volume of the most complete specimen, designated BOU-VP-16/1, is 1,450 cubic centimeters—large even for modern humans. The braincase of the other adult skull may have been even bigger. Although the African Herto skulls are longer and more robust than those of recent humans, the team considers the Herto specimens to be the earliest modern *Homo sapien*s yet found—direct ancestors of people living today. In an unknowing echo of Tishkoff's genetic findings, Tim White concludes, "We are all, in this sense, Africans."

Because the discoverers of the Herto skulls were unable to find convincing links between these fossils and archaic humans from any single geographic region, they put the three specimens into a new subspecies, *Homo sapiens idaltu*. The subspecies name *idaltu* comes from the Afar language of Ethiopia. It means "elder."

Even paleoanthropologists who were not associated with the finds overwhelmingly agree that the Herto skulls are the earliest securely dated modern humans yet found, meshing with the Out of Africa hypothesis. The Herto fossils also fit neatly into an African succession: Older skulls from the region include *Homo erectus* fossils from Daka, dated to about 1 million years ago, and the archaic Bodo skull, estimated to be about 500,000 years old. Meanwhile, fossils from Omo Kibish, also in Ethiopia, are more recent than the Herto skulls, according to a reanalysis of those remains. For a long time, the Omo Kibish specimens were regarded as ambiguous: They were fragmentary, making their anatomy less clear, and the site was originally dated using older, less reliable methods. However, a recent relocation of the site turned up new pieces that glued onto specimens found in 1967, and the site was redated to about 125,000 years using modern techniques. Further evidence comes from the Qafzeh site in Israel—on a plausible route from Africa—where there is a 92,000-year-old modern human skull.

These findings establish the earliest modern humans in Africa, but they do not exclude the simultaneous evolution of modern man in other parts of the world, as suggested by the Multiregional hypothesis. The most pertinent test of Multiregionalism focuses on Neandertals, which are a uniquely European form of primitive humans. According to Multiregionalists, Neandertals (which lived between about 200,000 and 27,000 years ago) are a transitional form that connects European *Homo erectus* to modern *Homo sapiens sapiens*. Could the Herto skulls simply be the regional, African equivalent of Neandertals?

"No," says co-leader Berhane Asfaw definitively. "The Herto skulls show that people in Africa had already developed the anatomy of modern humans while European Neandertals were still quite different." Indeed, the Herto skulls, though robust, lack many of the diagnostic anatomical features of Neandertal skulls. Asfaw states, "We can conclusively say that Neandertals had nothing to do with modern humans based on these skulls and on the genetic evidence."

Neandertals as European Ancestors?

The genetic evidence to which he refers has accumulated over the last six years, but the most dramatic advance came in 1999, when a team led by Svante Pääbo of the Max Planck Institute for Evolutionary Anthropology in Leipzig, Germany, became the first to extract mtDNA from the original Neandertal specimen. His group's success was a spectacular tour de force of meticulous technique and solid research design. The ancient mtDNA was compared with mtDNA from more than 2,000 people living in various regions around the world and differed from each of the modern regional groups by an average of 27 mutations (out of a possible 379 that were examined). Contrary to the predictions of the Multiregional hypothesis, the mtDNA of Neandertals was not closer to that of the modern Europeans. The work was a strong blow to the theory that humans evolved in several places simultaneously.

Multiregionalists Milford Wolpoff of the University of Michigan and Alan Thorne, then of Australian National University in Canberra, challenged the conclusions. They urged an investigation of mtDNA from additional Neandertals, in case the single individual used by Pääbo's team was particularly unusual. They also suggested that Neandertal mtDNA might be closer to samples from the fossilized remains of early modern humans in Europe than from living Europeans.

Since these criticisms were levied, several teams have carried out additional studies of mtDNA from Neandertals and fossilized modern humans. All have shown that Neandertal samples differ significantly from modern mtDNA, which is indistinguishable from fossilized modern human mtDNA. Giorgio Bertorelle of the University of Ferrara in Italy led one of these teams, which published important results last May. Bertorelle's team compared mtDNA from two early modern humans (Cro-Magnons) from Italy, dated to 23,000 and 24,720 years old, with four mtDNA sequences of Neandertals from 42,000 to 29,000 years ago. The chronological proximity of the Neandertal and modern fossils was key because it increased the likelihood that Neandertal mtDNA would strongly resemble the early modern human mtDNA—if the former evolved into the latter, as the Multiregional hypothesis states.

Bertorelle and colleagues found that the Cro-Magnon mtDNA was unlike the Neandertal samples, differing from them at 22 and 28 sites out of 360. Instead, the Cro-Magnon mtDNA sequences fell squarely within the range of variation of living humans. One of the Italian Cro-Magnons had a sequence shared by 359 (14 percent) of 2,566 modern samples in Europe and the Near East, and the other differs by only one mutation.

"The early modern humans had sequences that living individuals still have," concluded Bertorelle, "they [have] … nothing to do with Neandertal sequences." He and Asfaw might chime in unison that Neandertals cannot represent a regional European transition from *Homo erectus* to modern *Homo sapiens*

The identity between Cro-Magnon and modern human mtDNA sequences in this study and others is striking, and it has caused some researchers to worry about the possibility of mtDNA contamination from researchers or others who have handled the fossils. Although contamination is a major problem in such studies, Bertorelle asserts spiritedly that his data are clean, stating that his group performed nine different tests to check for contamination and followed the most stringent procedures and methodology. He also points out the irony of questioning the validity of the mtDNA of a prehistoric human only because it is identical to that of modern humans.

Compelling Congruity

The Out of Africa hypothesis has become compelling because these different studies have all yielded congruent answers. Tishkoff's work points to East Africa in general, and Ethiopia/Sudan in particular, as the region where the oldest modern human lineages are found—and probably evolved. Studies of ancient mtDNA by groups led by Pääbo, Bertorelle and others emphasize the genetic discrepancies between Neandertals and modern humans and demonstrate that some early anatomically modern fossils were also genetically modern—undermining the Multiregional hypothesis.

Despite the power of these genetic studies, only the fossils can tell us what our ancestors actually looked like, what they actually did and where they actually lived. It is singularly satisfying that the White-Howell-Asfaw team has discovered fossilized human remains from the right place (Ethiopia) and time (about 160,000 years ago) that also have the right (modern human) anatomy. The authors of the Out of Africa hypothesis are celebrating.

I don't expect that the subscribers of the Multiregional hypothesis will be waving a white flag of surrender, although they have lost the great majority of their supporters. At least one of the theory's most ardent proponents, Wolpoff, is still steadfast in defense of the hypothesis he has so long espoused. While it remains possible that new findings will shift the balance in favor of the Multiregional viewpoint, the consilience of such evidence creates a powerful testament. It would take many new fossils and many new genetic studies to resculpt this intellectual landscape.

PAT SHIPMAN is an adjunct professor of anthropology at the Pennsylvania State University. Address: Department of Anthropology, 315 Carpenter Building, Pennsylvania State University, University Park, PA 16801. pls10@psu.edu

Bibliography

Caramelli, D., C. Lalueza-Fox, C. Vernesi, M. Lari, A. Casoli, F. Mallegni, B. Chiarelli, I. Dupanloup, J. Bertranpetit, G. Barbujani and G. Bertorelle. 2003. "Evidence for a genetic discontinuity between Neandertals and 24,000-year-old anatomically modern Europeans." *Proceedings of the National Academy of Science of the United States of America* 100:6593-6597.

Barbujani, G., and G. Bertorelle. 2003. "Were Cro-Magnons too like us for DNA to tell?" *Nature* 424:127.

Krings, M., A. Stone, R. W. Schmitz, H. Krainitzki, M. Stoneking and S. Pääbo. 1997. "Neandertal DNA sequences and the origin of modern humans." *Cell* 90:19-30.

Tishkoff, S. A., K. Gonder, J. Hirbo, H. Mortensen, K. Powell, A. Knight and J. Mountain. 2003. "The genetic diversity of linguistically diverse Tanzanian populations: A multilocus analysis." *American Journal of Physical Anthropology Supplement* 36:208-209.

White, T. D., B. Asfaw, D. DaGusta, H. Gilbert, G. D. Richards, G. Suwa and F. C. Howell. 2003. "Pleistocene *Homo sapiens* from Middle Awash, Ethiopia." *Nature* 423:742-747.

The Littlest Human

A spectacular find in Indonesia reveals that a strikingly different hominid shared the earth with our kind in the not so distant past

KATE WONG

O n the island of Flores in Indonesia, villagers have long told tales of a diminutive, upright-walking creature with a lopsided gait, a voracious appetite, and soft, murmuring speech.

They call it *ebu gogo*, "the grandmother who eats anything." Scientists' best guess was that macaque monkeys inspired the *ebu gogo* lore. But last October, an alluring alternative came to light. A team of Australian and Indonesian researchers excavating a cave on Flores unveiled the remains of a lilliputian human—one that stood barely a meter tall—whose kind lived as recently as 13,000 years ago.

The announcement electrified the paleoanthropology community. *Homo sapiens* was supposed to have had the planet to itself for the past 25 millennia, free from the company of other humans following the apparent demise of the Neandertals in Europe and *Homo erectus* in Asia. Furthermore, hominids this tiny were known only from fossils of australopithecines (Lucy and the like) that lived nearly three million years ago—long before the emergence of *H. sapiens*. No one would have predicted that our own species had a contemporary as small and primitive-looking as the little Floresian. Neither would anyone have guessed that a creature with a skull the size of a grapefruit might have possessed cognitive capabilities comparable to those of anatomically modern humans.

Isle of Intrigue

This is not the first time Flores has yielded surprises. In 1998 archaeologists led by Michael J. Morwood of the University of New England in Armidale, Australia, reported having discovered crude stone artifacts some 840,000 years old in the Soa Basin of central Flores. Although no human remains turned up with the tools, the implication was that *H. erectus*, the only hominid known to have lived in Southeast Asia during that time, had crossed the deep waters separating Flores from Java. To the team, the find showed *H. erectus* to be a seafarer, which was startling because elsewhere *H. erectus* had left behind little material culture to suggest that it was anywhere near capable of making watercraft. Indeed, the earliest accepted date for boat-building was 40,000 to 60,000 years ago, when modern humans colonized Australia. (The other early fauna

Overview/Mini Humans

- Conventional wisdom holds that *Homo sapiens* has been the sole human species on the earth for the past 25,000 years. Remains discovered on the Indonesian island of Flares have upended that view.
- The bones are said to belong to a dwarf species of *Homo* that lived as recently as 13,000 years ago.
- Although the hominid is as small in body and brain as the earliest humans, it appears to have made sophisticated stone tools, raising questions about the relation between brain size and intelligence.
- The find is controversial, however—some experts wonder whether the discoverers have correctly diagnosed the bones and whether anatomically modern humans might have made those advanced artifacts.

on Flores probably got there by swimming or accidentally drifting over on flotsam. Humans are not strong enough swimmers to have managed that voyage, but skeptics say they may have drifted across on natural rafts).

Hoping to document subsequent chapters of human occupation of the island, Morwood and Radien P. Soejono of the Indonesian Center for Archaeology in Jakarta turned their attention to a large limestone cave called Liang Bua located in western Flores. Indonesian archaeologists had been excavating the cave intermittently since the 1970s, depending on funding availability, but workers had penetrated only the uppermost deposits. Morwood and Soejono set their sights on reaching bedrock and began digging in July 2001. Before long, their team's efforts turned up abundant stone tools and bones of a pygmy version of an extinct elephant relative known as *Stegodon*. But it was not until nearly the end of the third season of fieldwork that diagnostic hominid material in the form of an isolated tooth surfaced. Morwood brought a cast of the tooth back to Armidale to show to his department colleague Peter Brown. "It was clear that while the premolar was broadly humanlike, it wasn't from a modern human," Brown recollects. Seven days later Morwood received word that the Indonesians had recovered a skeleton. The Australians boarded the next plane to Jakarta.

Peculiar though the premolar was, nothing could have prepared them for the skeleton, which apart from the missing arms was largely complete. The pelvis anatomy revealed that the individual was bipedal and probably a female, and the tooth eruption and wear indicated that it was an adult. Yet it was only as tall as a modern three-year-old, and its brain was as small as the smallest australopithecine brain known. There were other primitive traits as well, including the broad pelvis and the long neck of the femur. In other respects, however, the specimen looked familiar. Its small teeth and narrow nose, the overall shape of the braincase and the thickness of the cranial bones all evoked *Homo*.

Brown spent the next three months analyzing the enigmatic skeleton, catalogued as LB1 and affectionately nicknamed the Hobbit by some of the team members, after the tiny beings in J.R.R. Tolkien's *The Lord of the Rings* books. The decision about how to classify it did not come easily. Impressed with the characteristics LB1 shared with early hominids such as the australopithecines, he initially proposed that it represented a new genus of human. On further consideration, however, the similarities to *Homo* proved more persuasive. Based on the 18,000-year age of LB1, one might have reasonably expected the bones to belong to *H. sapiens*, albeit a very petite representative. But when Brown and his colleagues considered the morphological characteristics of small-bodied modern humans—including normal ones, such as pygmies, and abnormal ones, such as pituitary dwarfs—LB1 did not seem to fit any of those descriptions. Pygmies have small bodies and large brains—the result of delayed growth during puberty, when the brain has already attained its full size. And individuals with genetic disorders that produce short stature and small brains have a range of distinctive features not seen in LB1 and rarely reach adulthood, Brown says. Conversely, he notes, the Flores skeleton exhibits archaic traits that have never been documented for abnormal small-bodied *H. sapiens*.

> DWARFS AND GIANTS tend to evolve on islands, with animals larger than rabbits shrinking and animals smaller than rabbits growing. The shifts appear to be adaptive responses to the limited food supplies available in such environments. *Stegodon*, an extinct proboscidean, colonized Flores several times, dwindling from elephant to water buffalo proportions. Some rats, in contrast, became rabbit-sized over time. *H. floresiensis* appears to have followed the island rule as well. It is thought to be a dwarfed descendant of *H. erectus*, which itself was nearly the size of a modern human.

What LB1 looks like most, the researchers concluded, is a miniature *H. erectus*. Describing the find in the journal *Nature*, they assigned LB1 as well as the isolated tooth and an arm bone from older deposits to a new species of human, *Homo floresiensis*. They further argued that it was a descendant of *H. erectus* that had become marooned on Flores and evolved in isolation into a dwarf species, much as the elephant-like *Stegodon* did.

Biologists have long recognized that mammals larger than rabbits tend to shrink on small islands, presumably as an adaptive response to the limited food supply. They have little to lose by doing so, because these environments harbor few predators. On Flores, the only sizable predators were the Komodo dragon and another, even larger monitor lizard. Animals smaller than rabbits, on the other hand, tend to attain brobdingnagian proportions—perhaps because bigger bodies are more energetically efficient than small ones. Liang Bua has yielded evidence of that as well, in the form of a rat as robust as a rabbit.

But attributing a hominid's bantam size to the so-called island rule was a first. Received paleoanthropological wisdom holds that culture has buffered us humans from many of the selective pressures that mold other creatures—we cope with cold, for example, by building fires and making clothes, rather than evolving a proper pelage. The discovery of a dwarf hominid species indicates that, under the right conditions, humans can in fact respond in the same, predictable way that other large mammals do when the going gets tough. Hints that *Homo* could deal with resource fluxes in this manner came earlier in 2004 from the discovery of a relatively petite *H. erectus* skull from Olorgesailie in Kenya, remarks Richard Potts of the Smithsonian Institution, whose team recovered the bones. "Getting small is one of the things *H. erectus* had in its biological tool kit," he says, and the Flores hominid seems to be an extreme instance of that.

> SHARED FEATURES between LB1 and members of our own genus led to the classification of the Flores hominid as *Homo*, despite its tiny brain size. Noting that the specimen most closely resembles *H. erectus*, the researchers posit that it is a new species, *H. floresiensis*, that dwarfed from a *H. erectus* ancestor. *H. floresiensis* differs from *H. sapiens* in having, among other characteristics, no chin, a relatively projecting face, a prominent brow and a low braincase.

Curiouser and Curiouser

H. FLORESIENSIS's teeny brain was perplexing. What the hominid reportedly managed to accomplish with such a modest organ was nothing less than astonishing. Big brains are a hallmark of human evolution. In the space of six million to seven million years, our ancestors more than tripled their cranial capacity, from some 360 cubic centimeters in *Sahelanthropus*, the earliest putative hominid, to a whopping 1,350 cubic centimeters on average in modern folks. Archaeological evidence indicates that behavioral complexity increased correspondingly. Experts were thus fairly certain that large brains are a prerequisite for advanced cultural practices. Yet whereas the pea-brained australopithecines left behind only crude stone tools at best (and most seem not to have done any stone working at all), the comparably gray-matter-impoverished *H. floresiensis* is said to have manufactured implements that exhibit a level of sophistication elsewhere associated exclusively with *H. sapiens*.

The bulk of the artifacts from Liang Bua are simple flake tools struck from volcanic rock and chert, no more advanced than the implements made by late australopithecines and early *Homo*. But mixed in among the pygmy *Stegodon* remains excavators found a fancier set of tools, one that included finely worked points, large blades, awls and small blades that may have been hafted for use as spears. To the team, this association suggests that *H. floresiensis* regularly hunted *Stegodon*. Many of the *Stegodon* bones are those of young individuals that one *H. floresiensis* might have been able to bring down alone. But some belonged to adults that weighed up to half a ton, the hunting and transport of which must have been a coordinated group activity—one that probably required language, surmises team member Richard G. ("Bert") Roberts of the University of Wollongong in Australia.

The discovery of charred animal remains in the cave suggests that cooking, too, was part of the cultural repertoire of *H. floresiensis*. That a hominid as cerebrally limited as this one might have had control of fire gives pause. Humans are not thought to have tamed flame until relatively late in our collective cognitive development: the earliest unequivocal evidence of fire use comes from 200,000-year-old hearths in Europe that were the handiwork of the large-brained Neandertals.

If the *H. floresiensis* discoverers are correct in their interpretation, theirs is one of the most important paleoanthropological finds in decades. Not only does it mean that another species of human coexisted with our ancestors just yesterday in geological terms, and that our genus is far more variable than expected, it raises all sorts of questions about brain size and intelligence. Perhaps it should come as no surprise, then, that controversy has accompanied their claims.

Classification Clash

It did not take long for alternative theories to surface. In a letter that ran in the October 31 edition of Australia's *Sunday Mail*, just three days after the publication of the *Nature* issue containing the initial reports, paleoanthropologist Maciej Henneberg of the University of Adelaide countered that a pathological condition known as microcephaly (from the Greek for "small brain") could explain LB1's unusual features. Individuals afflicted with the most severe congenital form of microcephaly, primordial microcephalic dwarfism, die in childhood. But those with milder forms, though mentally retarded, can survive into adulthood. Statistically comparing the head and face dimensions of LB1 with those of a 4,000-year-old skull from Crete that is known to have belonged to a microcephalic, Henneberg found no significant differences between the two. Furthermore, he argued, the isolated forearm bone found deeper in the deposit corresponds to a height of 151 to 162 centimeters—the stature of many modern women and some men, not that of a dwarf—suggesting that larger-bodied people, too, lived at Liang Bua. In Henneberg's view, these findings indicate that LB1 is more likely a microcephalic *H. sapiens* than a new branch of *Homo*.

Susan C. Antón of New York University disagrees with that assessment. "The facial morphology is completely different in microcephalic [modern] humans," and their body size is normal, not small, she says. Antón questions whether LB1 warrants a new species, however. "There's little in the shape that differentiates it from *Homo erectus*," she notes. One can argue that it's a new species, Antón allows, but the difference in shape between LB1 and *Homo erectus* is less striking than that between a Great Dane and a Chihuahua. The possibility exists that the LB1 specimen is a *H. erectus* individual with a pathological growth condition stemming from microcephaly or nutritional deprivation, she observes.

But some specialists say the Flores hominid's anatomy exhibits a more primitive pattern. According to Colin P. Groves of the Australian National University and David W. Cameron of the University of Sydney, the small brain, the long neck of the femur and other characteristics suggest an ancestor along the lines of *Homo habilis*, the earliest member of our genus, rather than the more advanced *H. erectus*. Milford H. Wolpoff of the University of Michigan at Ann Arbor wonders whether the Flores find might even represent an offshoot of *Australopithecus*. If LB1 is a descendant of *H. sapiens* or *H. erectus*, it is hard to imagine how natural selection left her with a brain that's even smaller than expected for her height, Wolpoff says. Granted, if she descended from *Australopithecus*, which had massive jaws and teeth, one has to account for her relatively delicate jaws and dainty dentition. That, however, is a lesser evolutionary conundrum than the one posed by her tiny brain, he asserts. After all, a shift in diet could explain the reduced chewing apparatus, but why would selection downsize intelligence?

Finding an australopithecine that lived outside of Africa—not to mention all the way over in Southeast Asia—18,000 years ago would be a first. Members of this group were thought to have died out in Africa one and a half million years ago, never having left their mother continent. Perhaps, researchers reasoned, hominids needed long, striding limbs, large brains and better technology

Awl

Blade

Centimeters

Point

© Mark Moore, Ph.D./University of New England, Armidale, Australia

ADVANCED IMPLEMENTS appear to have been the handiwork of *H. floresiensis*. Earlier hominids with brains similar in size to that of *H. floresiensis* made only simple flake tools at most. But in the same stratigraphic levels as the hominid remains at Liang Bua, researchers found a suite of sophisticated artifacts—including awls, blades and points—exhibiting a level of complexity previously thought to be the sole purview of *H. sapiens*.

before they could venture out into the rest of the Old World. But the recent discovery of 1.8 million-year-old *Homo* fossils at a site called Dmanisi in the Republic of Georgia refuted that explanation—the Georgian hominids were primitive and small and utilized tools like those australopithecines had made a million years before. Taking that into consideration, there is no a priori reason why australopithecines (or habilines, for that matter) could not have colonized other continents.

Troubling Tools

Yet if *Australopithecus* made it out of Africa and survived on Flores until quite recently, that would raise the question of why no other remains supporting that scenario have turned up in the region. According to Wolpoff, they may have: a handful of poorly studied Indonesian fossils discovered in the 1940s have been variously classified as *Australopithecus, Meganthropus* and, most recently, *H. erectus*. In light of the Flores find, he says, those remains deserve reexamination.

Many experts not involved in the discovery back Brown and Morwood's taxonomic decision, however. "Most of the differences [between the Flores hominid and known members of *Homo*], including apparent similarities to australopithecines, are almost certainly related to very small body mass," declares David R. Begun of the University of Toronto. That is, as the Flores people dwarfed from *H. erectus*, some of their anatomy simply converged on that of the likewise little australopithecines. Because LB1 shares some key derived features with *H. erectus* and some with other members of *Homo*, "the most straightforward option is to call it a new species of *Homo*," he remarks. "It's a fair and reasonable interpretation," *H. erectus* expert G. Philip Rightmire of Binghamton University agrees. "That was quite a little experiment in Indonesia."

Even more controversial than the position of the half-pint human on the family tree is the notion that it made those advanced-looking tools. Stanford University paleoanthropologist Richard Klein notes that the artifacts found near LB1 appear to include few, if any, of the sophisticated types found elsewhere in the cave. This brings up the possibility that the modern-looking tools were produced by modern humans, who could have occupied the cave at a different time. Further excavations are necessary to determine the stratigraphic relation between the implements and the hominid remains, Klein opines. Such efforts may turn up modern humans like us. The question then, he says, will be whether there were two species at the site or whether modern humans alone occupied Liang Bua—in which case LB1 was simply a modern who experienced a growth anomaly.

Stratigraphic concerns aside, the tools are too advanced and too large to make manufacture by a primitive, diminutive hominid likely, Groves contends. Although the Liang Bua implements allegedly date back as far as 94,000 years ago, which the team argues makes them too early to be the handiwork of *H. sapiens*, Groves points out that 67,000-year-old tools have turned up in Liujiang, China, and older indications of a modern human presence in the Far East might yet emerge. "*H. sapiens*, once it was out of Africa, didn't take long to spread into eastern Asia," he comments.

"At the moment there isn't enough evidence" to establish that *H. floresiensis* created the advanced tools, concurs Bernard Wood of George Washington University. But as a thought experiment, he says, "let's pretend that they did." In that case, "I don't have a clue about brain size and ability," he confesses. If a hominid with no more gray matter than a chimp has can create a material culture like this one, Wood contemplates, "why did it take people such a bloody long time to make tools" in the first place?

"If *Homo floresiensis* was capable of producing sophisticated tools, we have to say that brain size doesn't add up to much," Rightmire concludes. Of course, humans today exhibit considerable variation in gray matter volume, and great thinkers exist at both ends of the spectrum. French writer Jacques Anatole Francois Thibault (also known as Anatole France), who won the 1921 Nobel Prize for Literature, had a cranial capacity of only about 1,000 cubic centimeters; England's General Oliver Cromwell had more than twice that. "What that means is that once you get the brain to a certain size, size no longer matters, it's the organization of the brain," Potts states. At some point, he adds, "the internal wiring of the brain may allow competence even if the brain seems small."

LB1's brain is long gone, so how it was wired will remain a mystery. Clues to its organization may reside on the interior of the braincase, however. Paleontologists can sometimes obtain latex molds of the insides of fossil skulls and then create plaster endocasts that reveal the morphology of the organ. Because LBI's bones are too fragile to withstand standard casting procedures, Brown is working on creating a virtual endocast based on CT scans of the skull that he can then use to generate a physical endocast via stereolithography, a rapid prototyping technology.

"If it's a little miniature version of an adult human brain, I'll be really blown away," says paleoneurologist Dean Falk of the University of Florida. Then again, she muses, what happens if the convolutions look chimplike? Specialists have long wondered whether bigger brains fold differently simply because they are bigger or whether the reorganization reflects selection for increased cognition. "This specimen could conceivably answer that," Falk observes.

Return to the Lost World

Since submitting their technical papers to *Nature*, the Liang Bua excavators have reportedly recovered the remains of another five or so individuals, all of which fit the *H. floresiensis* profile. None are nearly so complete as LB1, whose long arms turned up during the most recent field season. But they did unearth a second lower jaw that they say is identical in size and shape to LB1's. Such duplicate bones will be critical to their case that they have a population of these tiny humans (as opposed to a bunch of scattered bones from one person). That should in turn dispel concerns that LB1 was a diseased individual.

Additional evidence may come from DNA: hair samples possibly from *H. floresiensis* are undergoing analysis at the University of Oxford, and the hominid teeth and bones may contain viable DNA as well. "Tropical environments are not the best for long-term preservation of DNA, so we're not holding our breath," Roberts remarks, "but there's certainly no harm in looking."

The future of the bones (and any DNA they contain) is uncertain, however. In late November, Teuku Jacob of the Gadjah Mada University in Yogyakarta, Java, who was not involved in the discovery or the analyses, had the delicate specimens transported from their repository at the Indonesian Center for Archaeology to his own laboratory with Soejono's assistance. Jacob, the dean of Indonesian paleoanthropology, thinks LB1 was a microcephalic and allegedly ordered the transfer of it and the new, as yet undescribed finds for examination and safekeeping, despite strong objections from other staff members at the center. At the time this article was going to press, the team was waiting for Jacob to make good on his promise to return the remains to Jakarta by January 1 of this year, but his reputation for restricting scientific access to fossils has prompted pundits to predict that the bones will never be studied again.

Efforts to piece together the *H. floresiensis* puzzle will proceed, however. For his part, Brown is eager to find the tiny hominid's large-bodied forebears. The possibilities are three-fold, he notes. Either the ancestor dwarfed on Flores (and was possibly the maker of the 840,000-year-old Soa Basin tools), or it dwindled on another island and later reached Flores, or the ancestor was small before it even arrived in Southeast Asia. In fact, in many ways, LB1 more closely resembles African *H. erectus* and the Georgian hominids than the geographically closer Javan *H. erectus*, he observes. But whether these similarities indicate that *H. floresiensis* arose from an earlier *H. erectus* foray into Southeast Asia than the one that produced Javan *H. erectus* or are merely coincidental results of the dwarfing process remains to be determined. Future excavations may connect the dots. The team plans to continue digging on Flores and Java and will next year begin work on other Indonesian islands, including Sulawesi to the north.

The hominid bones from Liang Bua now span the period from 95,000 to 13,000 years ago, suggesting to the team that the little Floresians perished along with the pygmy *Stegodon* because of a massive volcanic eruption in the area around 12,000 years ago, although they may have survived later farther east. If *H. erectus* persisted on nearby Java until 25,000 years ago, as some evidence suggests, and *H. sapiens* had arrived in the region by 40,000 years ago, three human species lived cheek by jowl in Southeast Asia for at least 15,000 years. And the discoverers of *H. floresiensis* predict that more will be found. The islands of Lombok and Sumbawa would have been natural stepping-stones for hominids traveling from Java or mainland Asia to Flores. Those that put down roots on these islands may well have set off on their own evolutionary trajectories.

Perhaps, it has been proposed, some of these offshoots of the *Homo* lineage survived until historic times. Maybe they still live in remote pockets of Southeast Asia's dense rain forests, awaiting (or avoiding) discovery. On Flores, oral histories hold that the *ebu gogo* was still in existence when Dutch colonists settled there in the 19th century. And Malay folklore describes another small, humanlike being known as the *orang pendek* that supposedly dwells on Sumatra to this day.

"Every country seems to have myths about these things," Brown reflects. "We've excavated a lot of sites around the world, and we've never found them. But then [in September 2003] we found LB1." Scientists may never know whether tales of the *ebu gogo* and *orang pendek* do in fact recount actual sightings of other hominid species, but the newfound possibility will no doubt spur efforts to find such creatures for generations to come.

KATE WONG is editorial director of ScientificAmerican.com.

UNIT 6

Human Diversity

Unit Selections

Key Points to Consider

- Discuss whether the human species can be subdivided into racial categories. Support your position.

- How and why did the concept of race develop?

- Why does skin color vary among humans?

- To what extent is height a barometer of the health of a society, and why?

Student Website

www.mhcls.com/online

Internet References

Further information regarding these websites may be found in this book's preface or online.

Hominid Evolution Survey
  http://www.geocities.com/SoHo/Atrium/1381/index.html
Human Genome Project Information
  http://www.ornl.gov/TechResources/Human_Genome/home.html
OMIM Home Page-Online Mendelian Inheritance in Man
  http://www3.ncbi.nlm.nih.gov/omim/
The Human Diversity Resource Page
  http://community-1.webtv.net/SoundBehavior/DIVERSITYFORSOUND/

The field of biological anthropology has come a long way since the days when one of its primary concerns was the classification of human beings according to racial type. Although human diversity is still a matter of major interest in terms of how and why we differ from one another, most anthropologists have concluded that human beings cannot be sorted into sharply distinct entities. Without denying the fact of human variation throughout the world, the prevailing view today is that the differences between us exist along geographical gradients, as differences in degree, rather than in terms of the separate and discrete reproductive entities perceived in the past.

One of the old ways of looking at human "races" was that each such group was a subspecies of humans that, if left reproductively isolated long enough, would eventually evolve into separate species. While this concept of subspecies, or racial varieties within a species, would seem to apply to some living creatures (such as the dog and wolf or the horse and zebra) and might even be relevant to hominid diversification in the past, the current consensus is that it does not apply today, at least not within the human species.

A more recent attempt to salvage the idea of human races has been to perceive them not so much as reproductively isolated entities but as so many clusters of gene frequencies, separable only by the fact that the proportions of traits (such as skin color, hair form, etc.) differ in each artificially constructed group. Some scientists in the area of forensic physical anthropology appreciate the practical value of this approach. (See "Does Race Exist? A Proponent's Perspective".) In a similar manner, our ability to reconstruct human prehistory is dependent upon an understanding of human variation (as in "Skin Deep" by Nina G. Jablonski and George Chaplin and "The Tall and the Short of It" by Barry Bogin).

Lest anyone think that anthropologists are "in denial" regarding the existence of human races and that some of the viewpoints expressed in this section are merely expressions of contemporary political correctness, it should be pointed out that serious, scholarly attempts to classify people in terms of precise, biological units have been going on now for 200 years and, so far, nothing of scientific value has come of them.

Complicating the matter, as Jonathan Marks elucidates in "Black, White, Other," is that there actually are two concepts of race: the strictly biological one, as described above, and the one of popular culture, which has been around since time immemorial. These "two constantly intersecting ways of thinking about the divisions between us," says Marks, have resulted not only in fuzzy thinking about racial biology, but they have also infected the way we think about people and, therefore, the way we treat each other in the social arena.

What we should recognize, claim most anthropologists, is that, despite the superficial physical and biological differences between us, when it comes to intelligence, all human beings are basically the same. The degrees of variation within our species may be accounted for by the subtle and changing selective forces experienced as one moves from one geographical area to another. However, no matter what the environmental pressures have been, the same intellectual demands have been made upon all of us. This is not to say, of course, that we do not vary from each other as individuals. Rather, what is being said is that when we look at these artificially created groups of people called "races," we find the same range of intellectual skills within each group. Indeed, even when we look at traits other than intelligence, we find much greater variation within each group than we find between groups.

It is time, therefore, to put the idea of human races to rest, at least as far as science is concerned. If such notions remain in the realm of social discourse, then so be it. That is where the problems associated with notions of race have to be solved anyway. At least, says Marks, in speaking for the anthropological community: "You may group humans into a small number of races if you want to, but you are denied biology as a support for it."

Skin Deep

Throughout the world, human skin color has evolved to be dark enough to prevent sunlight from destroying the nutrient folate but light enough to foster the production of vitamin D

NINA G. JABLONSKI AND GEORGE CHAPLIN

Among primates, only humans have a mostly naked skin that comes in different colors. Geographers and anthropologists have long recognized that the distribution of skin colors among indigenous populations is not random: darker peoples tend to be found nearer the equator, lighter ones closer to the poles. For years, the prevailing theory has been that darker skins evolved to protect against skin cancer. But a series of discoveries has led us to construct a new framework for understanding the evolutionary basis of variations in human skin color. Recent epidemiological and physiological evidence suggests to us that the worldwide pattern of human skin color is the product of natural selection acting to regulate the effects of the sun's ultraviolet (UV) radiation on key nutrients crucial to reproductive success.

From Hirsute to Hairless

The evolution of skin pigmentation is linked with that of hairlessness, and to comprehend both these stories, we need to page back in human history. Human beings have been evolving as an independent lineage of apes since at least seven million years ago, when our immediate ancestors diverged from those of our closest relatives, chimpanzees. Because chimpanzees have changed less over time than humans have, they can provide an idea of what human anatomy and physiology must have been like. Chimpanzees' skin is light in color and is covered by hair over most of their bodies. Young animals have pink faces, hands, and feet and become freckled or dark in these areas only as they are exposed to sun with age. The earliest humans almost certainly had a light skin covered with hair. Presumably hair loss occurred first, then skin color changed. But that leads to the question, When did we lose our hair?

The skeletons of ancient humans—such as the well-known skeleton of Lucy, which dates to about 3.2 million years ago—give us a good idea of the build and the way of life of our ancestors. The daily activities of Lucy and other hominids that lived before about three million years ago appear to have been similar to those of primates living on the open savannas of Africa today.

They probably spent much of their day foraging for food over three to four miles before retiring to the safety of trees to sleep.

By 1.6 million years ago, however, we see evidence that this pattern had begun to change dramatically. The famous skeleton of Turkana Boy—which belonged to the species *Homo ergaster*—is that of a long-legged, striding biped that probably walked long distances. These more active early humans faced the problem of staying cool and protecting their brains from overheating. Peter Wheeler of John Moores University in Liverpool, England, has shown that this was accomplished through an increase in the number of sweat glands on the surface of the body and a reduction in the covering of body hair. Once rid of most of their hair, early members of the genus *Homo* then encountered the challenge of protecting their skin from the damaging effects of sunlight, especially UV rays.

Built-in Sunscreen

In chimpanzees, the skin on the hairless parts of the body contains cells called melanocytes that are capable of synthesizing the dark-brown pigment melanin in response to exposure to UV radiation. When humans became mostly hairless, the ability of the skin to produce melanin assumed new importance. Melanin is nature's sunscreen: it is a large organic molecule that Overview/Skin Color Evolution serves the dual purpose of physically and chemically filtering the harmful effects of UV radiation; it absorbs UV rays, causing them to lose energy, and it neutralizes harmful chemicals called free radicals that form in the skin after damage by UV radiation.

Anthropologists and biologists have generally reasoned that high concentrations of melanin arose in the skin of peoples in tropical areas because it protected them against skin cancer. James E. Cleaver of the University of California at San Francisco, for instance, has shown that people with the disease xeroderma pigmentosum, in which melanocytes are destroyed by exposure to the sun, suffer from significantly higher than normal rates of squamous and basal cell carcinomas, which are usually easily treated. Malignant melanomas are more fre-

Overview/Skin Color Evolution

- After losing their hair as an adaptation for keeping cool, early hominids gained pigmented skins. Scientists initially thought that such pigmentation arose to protect against skin-cancer-causing ultraviolet [UV] radiation.
- Skin cancers tend to arise after reproductive age, however. An alternative theory suggests that dark skin might have evolved primarily to protect against the breakdown of folate, a nutrient essential for fertility and for fetal development.
- Skin that is too dark blocks the sunlight necessary for catalyzing the production of vitamin D, which is crucial for maternal and fetal bones. Accordingly, humans have evolved to be light enough to make sufficient vitamin B yet dark enough to protect their stores of folate.
- As a result of recent human migrations, many people now live in areas that receive more [or less] UV radiation than is appropriate for their skin color.

quently fatal, but they are rare (representing 4 percent of skin cancer diagnoses) and tend to strike only light-skinned people. But all skin cancers typically arise later in life, in most cases after the first reproductive years, so they could not have exerted enough evolutionary pressure for skin protection alone to account for darker skin colors. Accordingly, we began to ask what role melanin might play in human evolution.

The Folate Connection

In 1991 one of us (Jablonski) ran across what turned out to be a critical paper published in 1978 by Richard F. Branda and John W. Eaton, now at the University of Vermont and the University of Louisville, respectively. These investigators showed that light-skinned people who had been exposed to simulated strong sunlight had abnormally low levels of the essential B vitamin folate in their blood. The scientists also observed that subjecting human blood serum to the same conditions resulted in a 50-percent loss of folate content within one hour.

The significance of these findings to reproduction—and hence evolution—became clear when we learned of research being conducted on a major class of birth defects by our colleagues at the University of Western Australia. There Fiona J. Stanley and Carol Bower had established by the late 1980s that folate deficiency in pregnant women is related to an increased risk of neural tube defects such as spina bifida, in which the arches of the spinal vertebrae fail to close around the spinal cord. Many research groups throughout the world have since confirmed this correlation, and efforts to supplement foods with folate and to educate women about the importance of the nutrient have become widespread.

We discovered soon afterward that folate is important not only in preventing neural tube defects but also in a host of other processes. Because folate is essential for the synthesis of DNA in dividing cells, anything that involves rapid cell proliferation, such as spermatogenesis (the production of sperm cells), requires folate. Male rats and mice with chemically induced folate deficiency have impaired spermatogenesis and are infertile. Although no comparable studies of humans have been conducted, Wai Yee Wong and his colleagues at the University Medical Center of Nijmegen in the Netherlands have recently reported that folic acid treatment can boost the sperm counts of men with fertility problems.

Such observations led us to hypothesize that dark skin evolved to protect the body's folate stores from destruction. Our idea was supported by a report published in 1996 by Argentine pediatrician Pablo Lapunzina, who found that three young and otherwise healthy women whom he had attended gave birth to infants with neural tube defects after using sun beds to tan themselves in the early weeks of pregnancy. Our evidence about the breakdown of folate by UV radiation thus supplements what is already known about the harmful (skin-cancer-causing) effects of UV radiation on DNA.

Human Skin on the Move

The earliest members of *Homo sapiens*, or modern humans, evolved in Africa between 120,000 and 100,000 years ago and had darkly pigmented skin adapted to the conditions of UV radiation and heat that existed near the equator. As modern humans began to venture out of the tropics, however, they encountered environments in which they received significantly less UV radiation during the year. Under these conditions their high concentrations of natural sunscreen probably proved detrimental. Dark skin contains so much melanin that very little UV radiation, and specifically very little of the shorter-wavelength UVB radiation, can penetrate the skin. Although most of the effects of UVB are harmful, the rays perform one indispensable function: initiating the formation of vitamin D in the skin. Dark-skinned people living in the tropics generally receive sufficient UV radiation during the year for UVB to penetrate the skin and allow them to make vitamin D. Outside the tropics this is not the case. The solution, across evolutionary time, has been for migrants to northern latitudes to lose skin pigmentation.

The connection between the evolution of lightly pigmented skin and vitamin D synthesis was elaborated by W. Farnsworth Loomis of Brandeis University in 1967. He established the importance of vitamin D to reproductive success because of its role in enabling calcium absorption by the intestines, which in turn makes possible the normal development of the skeleton and the maintenance of a healthy immune system. Research led by Michael Holick of the Boston University School of Medicine has, over the past 20 years, further cemented the significance of vitamin D in development and immunity. His team also showed that not all sunlight contains enough UVB to stimulate vitamin D production. In Boston, for instance, which is located at about 42 degrees north latitude, human skin cells begin to produce vitamin D only after mid-March. In the wintertime there isn't enough UVB to do the job. We realized that this was another piece of evidence essential to the skin color story.

During the course of our research in the early 1990s, we searched in vain to find sources of data on actual UV radiation levels at the earth's surface. We were rewarded in 1996, when we contacted Elizabeth Weatherhead of the Cooperative Institute for Research in Environmental Sciences at the University of Colorado at Boulder. She shared with us a database of measurements of UV radiation at the earth's surface taken by NASA's Total Ozone Mapping Spectrophotometer satellite between 1978 and 1993. We were then able to model the distribution of UV radiation on the earth and relate the satellite data to the amount of UVB necessary to produce vitamin D.

We found that the earth's surface could be divided into three vitamin D zones: one comprising the tropics, one the subtropics and temperate regions, and the last the circumpolar regions north and south of about 45 degrees latitude. In the first, the dosage of UVB throughout the year is high enough that humans have ample opportunity to synthesize vitamin D all year. In the second, at least one month during the year has insufficient UVB radiation, and in the third area not enough UVB arrives on average during the entire year to prompt vitamin D synthesis. This distribution could explain why indigenous peoples in the tropics generally have dark skin, whereas people in the subtropics and temperate regions are lighter-skinned but have the ability to tan, and those who live in regions near the poles tend to be very light skinned and burn easily.

One of the most interesting aspects of this investigation was the examination of groups that did not precisely fit the predicted skin-color pattern. An example is the Inuit people of Alaska and northern Canada. The Inuit exhibit skin color that is somewhat darker than would be predicted given the UV levels at their latitude. This is probably caused by two factors. The first is that they are relatively recent inhabitants of these climes, having migrated to North America only roughly 5,000 years ago. The second is that the traditional diet of the Inuit is extremely high in foods containing vitamin D, especially fish and marine mammals. This vitamin D-rich diet offsets the problem that they would otherwise have with vitamin D synthesis in their skin at northern latitudes and permits them to remain more darkly pigmented.

Our analysis of the potential to synthesize vitamin D allowed us to understand another trait related to human skin color: women in all populations are generally lighter-skinned than men. (Our data show that women tend to be between 3 and 4 percent lighter than men.) Scientists have often speculated on the reasons, and most have argued that the phenomenon stems from sexual selection—the preference of men for women of lighter color. We contend that although this is probably part of the story, it is not the original reason for the sexual difference. Females have significantly greater needs for calcium throughout their reproductive lives, especially during pregnancy and lactation, and must be able to make the most of the calcium contained in food. We propose, therefore, that women tend to be lighter-skinned than men to allow slightly more UVB rays to penetrate their skin and thereby increase their ability to produce vitamin D. In areas of the world that receive a large amount of UV radiation, women are indeed at the knife's edge of natural selection, needing to maximize the photoprotective function of their skin on the one hand and the ability to synthesize vitamin D on the other.

Where Culture and Biology Meet

As modern humans moved throughout the Old World about 100,000 years ago, their skin adapted to the environmental conditions that prevailed in different regions. The skin color of the indigenous people of Africa has had the longest time to adapt because anatomically modern humans first evolved there. The skin-color changes that modern humans underwent as they moved from one continent to another—first Asia, then Austro-Melanesia, then Europe and, finally, the Americas—can be reconstructed to some extent. It is important to remember, however, that those humans had clothing and shelter to help protect them from the elements. In some places, they also had the ability to harvest foods that were extraordinarily rich in vitamin D, as in the case of the Inuit. These two factors had profound effects on the tempo and degree of skin-color evolution in human populations.

Africa is an environmentally heterogeneous continent. A number of the earliest movements of contemporary humans outside equatorial Africa were into southern Africa. The descendants of some of these early colonizers, the Khoisan (previously known as Hottentots), are still found in southern Africa and have significantly lighter skin than indigenous equatorial Africans do—a clear adaptation to the lower levels of UV radiation that prevail at the southern extremity of the continent.

Interestingly, however, human skin color in southern Africa is not uniform. Populations of Bantu-language speakers who live in southern Africa today are far darker than the Khoisan. We know from the history of this region that Bantu speakers migrated into this region recently—probably within the past 1,000 years—from parts of West Africa near the equator. The skin-color difference between the Khoisan and Bantu speakers such as the Zulu indicates that the length of time that a group has inhabited a particular region is important in understanding why they have the color they do.

Cultural behaviors have probably also strongly influenced the evolution of skin color in recent human history. This effect can be seen in the indigenous peoples who live on the eastern and western banks of the Red Sea. The tribes on the western side, which speak so-called Nilo-Hamitic languages, are thought to have inhabited this region for as long as 6,000 years. These individuals are distinguished by very darkly pigmented skin and long, thin bodies with long limbs, which are excellent biological adaptations for dissipating heat and intense UV radiation. In contrast, modern agricultural and pastoral groups on the eastern bank of the Red Sea, on the Arabian Peninsula, have lived there for only about 2,000 years. These earliest Arab people, of European origin, have adapted to very similar environmental conditions by almost exclusively cultural means—wearing heavy protective clothing and devising portable shade in the form of tents. (Without such clothing, one would have expected their skin to have begun to darken.) Generally speaking, the more recently a group has migrated into an area, the more extensive its cultural, as opposed to biological, adaptations to the area will be.

Perils of Recent Migrations

Despite great improvements in overall human health in the past century, some diseases have appeared or reemerged in populations that had previously been little affected by them. One of these is skin cancer, especially basal and squamous cell carcinomas, among light-skinned peoples. Another is rickets, brought about by severe vitamin D deficiency, in dark-skinned peoples. Why are we seeing these conditions?

As people move from an area with one pattern of UV radiation to another region, biological and cultural adaptations have not been able to keep pace. The light-skinned people of northern European origin who bask in the sun of Florida or northern Australia increasingly pay the price in the form of premature aging of the skin and skin cancers, not to mention the unknown cost in human life of folate depletion. Conversely, a number of dark-skinned people of southern Asian and African origin now living in the northern U.K., northern Europe or the northeastern U.S. suffer from a lack of UV radiation and vitamin D, an insidious problem that manifests itself in high rates of rickets and other diseases related to vitamin D deficiency.

The ability of skin color to adapt over long periods to the various environments to which humans have moved reflects the importance of skin color to our survival. But its unstable nature also makes it one of the least useful characteristics in determining the evolutionary relations between human groups. Early Western scientists used skin color improperly to delineate human races, but the beauty of science is that it can and does correct itself. Our current knowledge of the evolution of human skin indicates that variations in skin color, like most of our physical attributes, can be explained by adaptation to the environment through natural selection. We look ahead to the day when the vestiges of old scientific mistakes will be erased and replaced by a better understanding of human origins and diversity. Our variation in skin color should be celebrated as one of the most visible manifestations of our evolution as a species.

MORE TO EXPLORE

The Evolution of Human Skin Coloration. Nina G. Jablonski and George Chaplin in *Journal of Human Evolution*, Vol. 39, No. 1, pages 57-106; July 1, 2000. An abstract of the article is available online at **www.idealibrary.com/links/doi/10.1006/jhev.2000.0403**

Why Skin Comes in Colors. Blake Edgar in *California Wild*, Vol. 53, No. 1, pages 6-7; Winter 2000. The article is also available at **www.calacademy.org/calwild/winter2000/html/horizons.html**

The Biology of Skin Color: Black and White. Gina Kirchweger in *Discover*, Vol. 22, No. 2, pages 32-33; February 2001. The article is also available at **www.discover.com/feb_01/featbiology.html**

NINA G. JABLONSKI and **GEORGE CHAPLIN** work at the California Academy of Sciences in San Francisco, where Jablonski is Irvine Chair and curator of anthropology, and Chaplin is a research associate in the department of anthropology. Jablonski's research centers on the evolutionary adaptations of monkeys, apes and humans. She is particularly interested in how primates have responded to changes over time in the global environment. Chaplin is a private geographic information systems consultant who specializes in describing and analyzing geographic trends in biodiversity. In 2001 he was awarded the Student of the Year prize by the Association of Geographic Information in London for his master's thesis on the environmental correlates of skin color.

Black, White, Other

Racial categories are cultural constructs masquerading as biology

JONATHAN MARKS

While reading the Sunday edition of the *New York Times* one morning last February, my attention was drawn by an editorial inconsistency. The article I was reading was written by attorney Lani Guinier. (Guinier, you may remember, had been President Clinton's nominee to head the civil rights division at the Department of Justice in 1993. Her name was hastily withdrawn amid a blast of criticism over her views on political representation of minorities.) What had distracted me from the main point of the story was a photo caption that described Guinier as being "half-black." In the text of the article, Guinier had described herself simply as "black."

How can a person be black and half black at the same time? In algebraic terms, this would seem to describe a situation where $x = 1/2\ x$, to which the only solution is $x = 0$.

The inconsistency in the *Times* was trivial, but revealing. It encapsulated a longstanding problem in our use of racial categories—namely, a confusion between biological and cultural heredity. When Guinier is described as "half-black," that is a statement of biological ancestry, for one of her two parents is black. And when Guinier describes herself as black, she is using a cultural category, according to which one can either be black or white, but not both.

Race—as the term is commonly used—is inherited, although not in a strictly biological fashion. It is passed down according to a system of folk heredity, an all-or-nothing system that is different from the quantifiable heredity of biology. But the incompatibility of the two notions of race is sometimes starkly evident—as when the state decides that racial differences are so important that interracial marriages must be regulated or outlawed entirely. Miscegenation laws in this country (which stayed on the books in many states through the 1960s) obliged the legal system to define who belonged in what category. The resulting formula stated that anyone with one-eighth or more black ancestry was a "negro." (A similar formula, defining Jews, was promulgated by the Germans in the Nuremberg Laws of the 1930s.)

Applying such formulas led to the biological absurdity that having one black great-grandparent was sufficient to define a person as black, but having seven white great grandparents was insufficient to define a person as white. Here, race and biology are demonstrably at odds. And the problem is not semantic but conceptual, for race is presented as a category of nature.

Human beings come in a wide variety of sizes, shapes, colors, and forms—or, because we are visually oriented primates, it certainly seems that way. We also come in larger packages called populations; and we are said to belong to even larger and more confusing units, which have long been known as races. The history of the study of human variation is to a large extent the pursuit of those human races—the attempt to identify the small number of fundamentally distinct kinds of people on earth.

This scientific goal stretches back two centuries, to Linnaeus, the father of biological systematics, who radically established *Homo sapiens* as one species within a group of animals he called Primates. Linnaeus's system of naming groups within groups logically implied further breakdown. He consequently sought to establish a number of subspecies within *Homo sapiens*. He identified five: four geographical species (from Europe, Asia, Africa, and America) and one grab-bag subspecies called *monstrosus*. This category was dropped by subsequent researchers (as was Linnaeus's use of criteria such as personality and dress to define his subspecies).

While Linnaeus was not the first to divide humans on the basis of the continents on which they lived, he had given the division a scientific stamp. But in attempting to determine the proper number of subspecies, the heirs of Linnaeus always seemed to find different answers, depending upon the criteria they applied. By the mid-twentieth century, scores of anthropologists—led by Harvard's Earnest Hooton—had expended enormous energy on the problem. But these scholars could not convince one another about the precise nature of the fundamental divisions of our species.

Part of the problem—as with the *Times's* identification of Lani Guinier—was that we humans have two constantly intersecting ways of thinking about the divisions among us. On the one hand, we like to think of "race"—as Linnaeus did—as an objective, biological category. In this sense, being a member of a race is supposed to be the equivalent of being a member of a species or of a phylum—except that race, on the analogy of subspecies, is an even narrower (and presumably more exclusive and precise) biological category.

The other kind of category into which we humans allocate ourselves—when we say "Serb" or "Hutu" or "Jew" or "Chicano" or "Republican" or "Red Sox fan"—is cultural. The label refers to little or nothing in the natural attributes of its members.

These members may not live in the same region and may not even know many others like themselves. What they share is neither strictly nature nor strictly community. The groupings are constructions of human social history.

Membership in these *un*biological groupings may mean the difference between life and death, for they are the categories that allow us to be identified (and accepted or vilified) socially. While membership in (or allegiance to) these categories may be assigned or adopted from birth, the differentia that mark members from nonmembers are symbolic and abstract; they serve to distinguish people who cannot be readily distinguished by nature. So important are these symbolic distinctions that some of the strongest animosities are often expressed between very similar-looking peoples. Obvious examples are Bosnian Serbs and Muslims, Irish and English, Huron and Iroquois.

Obvious natural variation is rarely so important as cultural difference. One simply does not hear of a slaughter of the short people at the hands of the tall, the glabrous at the hands of the hairy, the red-haired at the hands of the brown-haired. When we do encounter genocidal violence between different looking peoples, the two groups are invariably socially or culturally distinct as well. Indeed, the tragic frequency of hatred and genocidal violence between biologically indistinguishable peoples implies that biological differences such as skin color are not motivations but, rather, excuses. They allow nature to be invoked to reinforce group identities and antagonisms that would exist without these physical distinctions. But are there any truly "racial" biological distinctions to be found in our species?

Obviously, if you compare two people from different parts of the world (or whose ancestors came from different parts of the world), they will differ physically, but one cannot therefore define three or four or five basically different kinds of people, as a biological notion of race would imply. The anatomical properties that distinguish people—such as pigmentation, eye form, body build—are not clumped in discrete groups, but distributed along geographical gradients, as are nearly all the genetically determined variants detectable in the human gene pool.

These gradients are produced by three forces. Natural selection adapts populations to local circumstances (like climate) and thereby differentiates them from other populations. Genetic drift (random fluctuations in a gene pool) also differentiates populations from one another, but in non-adaptive ways. And gene flow (via intermarriage and other child-producing unions) acts to homogenize neighboring populations.

In practice, the operations of these forces are difficult to discern. A few features, such as body build and the graduated distribution of the sickle cell anemia gene in populations from western Africa, southern Asia, and the Mediterranean can be plausibly related to the effects of selection. Others, such as the graduated distribution of a small deletion in the mitochondrial DNA of some East Asian, Oceanic, and Native American peoples, or the degree of flatness of the face, seem unlikely to be the result of selection and are probably the results of random biohistorical factors. The cause of the distribution of most features, from nose breadth to blood group, is simply unclear.

The overall result of these forces is evident, however. As Johann Friedrich Blumenbach noted in 1775, "you see that all do so run into one another, and that one variety of mankind does so sensibly pass into the other, that you cannot mark out the limits between them." (Posturing as an heir to Linnaeus, he nonetheless attempted to do so.) But from humanity's gradations in appearance, no defined groupings resembling races readily emerge. The racial categories with which we have become so familiar are the result of our imposing arbitrary cultural boundaries in order to partition gradual biological variation.

Unlike graduated biological distinctions, culturally constructed categories are ultrasharp. One can be French or German, but not both; Tutsi or Hutu, but not both; Jew or Catholic, but not both; Bosnian Muslim or Serb, but not both; black or white, but not both. Traditionally, people of "mixed race" have been obliged to choose one and thereby identify themselves unambiguously to census takers and administrative bookkeepers—a practice that is now being widely called into question.

A scientific definition of race would require considerable homogeneity within each group, and reasonably discrete differences between groups, but three kinds of data militate against this view: First, the groups traditionally described as races are not at all homogeneous. Africans and Europeans, for instance, are each a collection of biologically diverse populations. Anthropologists of the 1920s widely recognized *three* European races: Nordic, Alpine, and Mediterranean. This implied that races could exist within races. American anthropologist Carleton Coon identified *ten* European races in 1939. With such protean use, the term race came to have little value in describing actual biological entities within *Homo sapiens*. The scholars were not only grappling with a broad north-south gradient in human appearance across Europe, they were trying to bring the data into line with their belief in profound and fundamental constitutional differences between groups of people.

But there simply isn't one European race to contrast with an African race, nor three, nor ten: the question (as scientists long posed it) fails to recognize the actual patterning of diversity in the human species. Fieldwork revealed, and genetics later quantified, the existence of far more biological diversity within any group than between groups. Fatter and thinner people exist everywhere, as do people with type O and type A blood. What generally varies from one population to the next is the *proportion* of people in these groups expressing the trait or gene. Hair color varies strikingly among Europeans and native Australians, but little among other peoples. To focus on discovering differences between presumptive races, when the vast majority of detectable variants do not help differentiate them, was thus to define a very narrow—if not largely illusory—problem in human biology. (The fact that Africans are biologically more diverse than Europeans, but have rarely been split into so many races, attests to the cultural basis of these categorizations.)

Second, differences between human groups are only evident when contrasting geographical extremes. Noting these extremes, biologists of an earlier era sought to identify representatives of "pure," primordial races presumably located in Norway, Senegal, and Thailand. At no time, however, was our species composed of a few populations within which everyone looked pretty much the same. Ever since some of our ancestors left Africa to spread out through the Old World, we humans have always lived in the "in-between" places. And human populations have also always been in genetic contact with one another. Indeed, for tens of thousands of years, humans have had trade networks; and where goods flow, so do genes. Consequently, we have no basis for considering *extreme* human forms the most pure, or most representative, of some ancient primordial populations. Instead, they represent populations adapted to the most disparate environments.

And third, between each presumptive "major" race are unclassifiable populations and people. Some populations of India, for example, are darkly pigmented (or "black"), have European-like ("Caucasoid") facial features, but inhabit the continent of Asia (which should make them "Asian"). Americans might tend to ignore these "exceptions" to the racial categories, since immigrants to the United States from West Africa, Southeast Asia, and northwest Europe far outnumber those from India. The very existence of unclassifiable peoples undermines the idea that there are just three human biological groups in the Old World. Yet acknowledging the biological distinctiveness of such groups leads to a rapid proliferation of categories. What about Australians? Polynesians? The Ainu of Japan?

Categorizing people is important to any society. It is, at some basic psychological level, probably necessary to have group identity about who and what you are, in contrast to who and what you are not. The concept of race, however, specifically involves the recruitment of biology to validate those categories of self-identity.

Mice don't have to worry about that the way humans do. Consequently, classifying them into subspecies entails less of a responsibility for a scientist than classifying humans into subspecies does. And by the 1960s, most anthropologists realized they could not defend any classification of *Homo sapiens* into biological subspecies or races that could be considered reasonably objective. They therefore stopped doing it, and stopped identifying the endeavor as a central goal of the field. It was a biologically intractable problem—the old square-peg-in-a-round-hole enterprise; and people's lives, or welfares, could well depend on the ostensibly scientific pronouncement. Reflecting on the social history of the twentieth century, that was a burden anthropologists would no longer bear.

This conceptual divorce in anthropology—of cultural from biological phenomena was one of the most fundamental scientific revolutions of our time. And since it affected assumptions so rooted in our everyday experience, and resulted in conclusions so counterintuitive—like the idea that the earth goes around the sun, and not vice-versa—it has been widely underappreciated.

Kurt Vonnegut, in *Slaughterhouse Five*, describes what he remembered being taught about human variation: "At that time, they were teaching that there was absolutely no difference between anybody. They may be teaching that still." Of course there are biological differences between people, and between populations. The question is: How are those differences patterned? And the answer seems to be: Not racially. Populations are the only readily identifiable units of humans, and even they are fairly fluid, biologically similar to populations nearby, and biologically different from populations far away.

In other words, the message of contemporary anthropology is: You may group humans into a small number of races if you want to, but you are denied biology as a support for it.

New York-born **JONATHAN MARKS** earned an undergraduate degree in natural science at Johns Hopkins. After getting his Ph.D. in anthropology, Marks did a post-doc in genetics at the University of California at Davis and is now an associate professor of anthropology at Yale University. He is the coauthor, with Edward Staski, of the introductory textbook *Evolutionary Anthropology* (San Diego: Harcourt, Brace Jovanovich, 1992). His new book, *Human Biodiversity: Genes, Race, and History* is published (1995) by Aldine de Gruyter.

Reprinted with permission from *Natural History,* December 1994, pp. 32–35. © 1994 by Natural History Magazine, Inc.

Does Race Exist?

A Proponent's Perspective

GEORGE W. GILL

Slightly over half of all biological/physical anthropologists today believe in the traditional view that human races are biologically valid and real. Furthermore, they tend to see nothing wrong in defining and naming the different populations of *Homo sapiens*. The other half of the biological anthropology community believes either that the traditional racial categories for humankind are arbitrary and meaningless, or that at a minimum there are better ways to look at human variation than through the "racial lens."

Are there differences in the research concentrations of these two groups of experts? Yes, most decidedly there are. As pointed out in a recent 2000 edition of a popular physical anthropology textbook, forensic anthropologists (those who do skeletal identification for law-enforcement agencies) are overwhelmingly in support of the idea of the basic biological reality of human races, and yet those who work with blood-group data, for instance, tend to reject the biological reality of racial categories.

I happen to be one of those very few forensic physical anthropologists who actually does research on the particular traits used today in forensic racial identification (i.e., "assessing ancestry," as it is generally termed today). Partly this is because for more than a decade now U.S. national and regional forensic anthropology organizations have deemed it necessary to quantitatively test both traditional and new methods for accuracy in legal cases. I volunteered for this task of testing methods and developing new methods in the late 1980s. What have I found? Where do I now stand in the "great race debate?" Can I see truth on one side or the other—or on both sides—in this argument?

Findings

First, I have found that forensic anthropologists attain a high degree of accuracy in determining geographic racial affinities (white, black, American Indian, etc.) by utilizing both new and traditional methods of bone analysis. Many well-conducted studies were reported in the late 1980s and 1990s that test methods objectively for percentage of correct placement. Numerous individual methods involving midfacial measurements, femur traits, and so on are over 80 percent accurate alone, and in combination produce very high levels of accuracy. No forensic anthropologist would make a racial assessment based upon just *one* of these methods, but in combina-

tion they can make very reliable assessments, just as in determining sex or age. In other words, multiple criteria are the key to success in all of these determinations.

I have a respected colleague, the skeletal biologist C. Loring Brace, who is as skilled as any of the leading forensic anthropologists at assessing ancestry from bones, yet he does not subscribe to the concept of race. Neither does Norman Sauer, a board-certified forensic anthropologist. My students ask, "How can this be? They can identify skeletons as to racial origins but do not believe in race!" My answer is that we can often *function* within systems that we do not believe in.

As a middle-aged male, for example, I am not so sure that I believe any longer in the chronological "age" categories that many of my colleagues in skeletal biology use. Certainly parts of the skeletons of some 45-year-old people look older than corresponding portions of the skeletons of some 55-year-olds. If, however, law enforcement calls upon me to provide "age" on a skeleton, I can provide an answer that will be proven sufficiently accurate should the decedent eventually be identified. I may not believe in society's "age" categories, but I can be very effective at "aging" skeletons. The next question, of course, is how "real" is age biologically? My answer is that if one can use biological criteria to assess age with reasonable accuracy, then age has some basis in biological reality even if the particular "social construct" that defines its limits might be imperfect. I find this true not only for age and stature estimations but for sex and race identification.

The "reality of race" therefore depends more on the definition of reality than on the definition of race. If we choose to accept the system of racial taxonomy that physical anthropologists have traditionally established—major races: black, white, etc.—then one can classify human skeletons within it just as well as one can living humans. The bony traits of the nose, mouth, femur, and cranium are just as revealing to a good osteologist as skin color, hair form, nose form, and lips to the perceptive observer of living humanity. I have been able to prove to myself over the years, in actual legal cases, that I am *more* accurate at assessing race from skeletal remains than from looking at living people standing before me. So those of us in forensic anthropology know that the skeleton reflects race, whether "real" or not, just as well if not better than superficial soft tissue does. The idea that race is "only skin deep" is simply not true, as any experienced forensic anthropologist will affirm.

Position on Race

Where I stand today in the "great race debate" after a decade and a half of pertinent skeletal research is clearly more on the side of the reality of race than on the "race denial" side. Yet I do see why many other physical anthropologists are able to ignore or deny the race concept. Blood-factor analysis, for instance, shows many traits that cut across racial boundaries in a purely *clinal* fashion with very few if any "breaks" along racial boundaries. (A cline is a gradient of change, such as from people with a high frequency of blue eyes, as in Scandinavia, to people with a high frequency of brown eyes, as in Africa.)

Morphological characteristics, however, like skin color, hair form, bone traits, eyes, and lips tend to follow geographic boundaries coinciding often with climatic zones. This is not surprising since the selective forces of climate are probably the primary forces of nature that have shaped human races with regard not only to skin color and hair form but also the underlying bony structures of the nose, cheekbones, etc. (For example, more prominent noses humidify air better.) As far as we know, blood-factor frequencies are *not* shaped by these same climatic factors.

So, serologists who work largely with blood factors will tend to see human variation as clinal and races as not a valid construct, while skeletal biologists, particularly forensic anthropologists, will see races as biologically real. The common person on the street who sees only a person's skin color, hair form, and face shape will also tend to see races as biologically real. They are not incorrect. Their perspective is just different from that of the serologist.

So, yes, I see truth on both sides of the race argument.

Those who believe that the concept of race is valid do not discredit the notion of clines, however. Yet those with the clinal perspective who believe that races are not real do try to discredit the evidence of skeletal biology. Why this bias from the "race denial" faction? This bias seems to stem largely from socio-political motivation and not science at all. For the time being at least, the people in "race denial" are in "reality denial" as well. Their motivation (a positive one) is that they have come to believe that the race concept is socially dangerous. In other words, they have convinced themselves that race promotes racism. Therefore, they have pushed the politically correct agenda that human races are not biologically real, no matter what the evidence.

Consequently, at the beginning of the 21st century, even as a majority of biological anthropologists favor the reality of the race perspective, not one introductory textbook of physical anthropology even presents that perspective as a possibility. In a case as flagrant as this, we are not dealing with science but rather with blatant, politically motivated censorship. But, you may ask, are the politically correct actually correct? Is there a relationship between thinking about race and racism?

Race and Racism

Does discussing human variation in a framework of racial biology promote or reduce racism? This is an important question, but one that does not have a simple answer. Most social scientists over the past decade have convinced themselves that it runs the risk of promoting racism in certain quarters. Anthropologists of the 1950s, 1960s, and early 1970s, on the other hand, believed that they were combating racism by openly discussing race and by teaching courses on human races and racism. Which approach has worked best? What do the intellectuals among racial minorities believe? How do students react and respond?

Three years ago, I served on a NOVA-sponsored panel in New York, in which panelists debated the topic "Is There Such a Thing as Race?" Six of us sat on the panel, three proponents of the race concept and three antagonists. All had authored books or papers on race. Loring Brace and I were the two anthropologists "facing off" in the debate. The ethnic composition of the panel was three white and three black scholars. As our conversations developed, I was struck by how similar many of my concerns regarding racism were to those of my two black teammates. Although recognizing that embracing the race concept can have risks attached, we were (and are) more fearful of the form of racism likely to emerge if race is denied and dialogue about it lessened. We fear that the social taboo about the subject of race has served to suppress open discussion about a very important subject in need of dispassionate debate. One of my teammates, an affirmative-action lawyer, is afraid that a denial that races exist also serves to encourage a denial that racism exists. He asks, "How can we combat racism if no one is willing to talk about race?"

Who Will Benefit?

In my experience, minority students almost invariably have been the strongest supporters of a "racial perspective" on human variation in the classroom. The first-ever black student in my human variation class several years ago came to me at the end of the course and said, "Dr. Gill, I really want to thank you for changing my life with this course." He went on to explain that, "My whole life I have wondered about why I am black, and if that is good or bad. Now I know the reasons why I am the way I am and that these traits are useful and good."

A human-variation course with another perspective would probably have accomplished the same for this student if he had ever noticed it. The truth is, innocuous contemporary human-variation classes with their politically correct titles and course descriptions do not attract the attention of minorities or those other students who could most benefit. Furthermore, the politically correct "race denial" perspective in society as a whole suppresses dialogue, allowing ignorance to replace knowledge and suspicion to replace familiarity. This encourages ethnocentrism and racism more than it discourages it.

DR. GEORGE W. GILL is a professor of anthropology at the University of Wyoming. He also serves as the forensic anthropologist for Wyoming law-enforcement agencies and the Wyoming State Crime Laboratory.

Does Race Exist?

An Antagonist's Perspective

C. LORING BRACE

I am going to start this essay with what may seem to many as an outrageous assertion: There is no such thing as a biological entity that warrants the term "race."

The immediate reaction of most literate people is that this is obviously nonsense. The physician will retort, "What do you mean 'there is no such thing as race'? I see it in my practice everyday!" Jane Doe and John Roe will be equally incredulous. Note carefully, however, that my opening declaration did not claim that "there is no such thing as race." What I said is that there is no "biological entity that warrants the term 'race'." "You're splitting hairs," the reader may retort. "Stop playing verbal games and tell us what you really mean!"

And so I shall, but there is another charge that has been thrown my way, which I need to dispel before explaining the basis for my statement. Given the tenor of our times at the dawn of the new millennium, some have suggested that my position is based mainly on the perception of the social inequities that have accompanied the classification of people into "races." My stance, then, has been interpreted as a manifestation of what is being called "political correctness." My answer is that it is really the defenders of the concept of "race" who are unwittingly shaped by the political reality of American history. [Read a *proponent's perspective,* that of anthropologist George Gill.]

But all of this needs explaining. First, it is perfectly true that the long-term residents of the various parts of the world have patterns of features that we can easily identify as characteristic of the areas from which they come. It should be added that they have to have resided in those places for a couple of hundred thousand years before their regional patterns became established. Well, you may ask, why can't we call those regional patterns "races"? In fact, we can and do, but it does not make them coherent biological entities. "Races" defined in such a way are products of our perceptions. "Seeing is believing" will be the retort, and, after all, aren't we seeing reality in those regional differences?

I should point out that this is the same argument that was made against Copernicus and Galileo almost half a millennium ago. To this day, few have actually made the observations and done the calculations that led those Renaissance scholars to challenge the universal perception that the sun sets in the evening to rise again at the dawn. It was just a matter of common sense to believe that the sun revolves around the Earth, just as it was common sense to "know" that the Earth was flat. Our beliefs concerning "race" are based on the same sort of common sense, and they are just as basically wrong.

The Nature of Human Variation

I would suggest that there are very few who, of their own experience, have actually perceived at first hand the nature of human variation. What we know of the characteristics of the various regions of the world we have largely gained vicariously and in misleadingly spotty fashion. Pictures and the television camera tell us that the people of Oslo in Norway, Cairo in Egypt, and Nairobi in Kenya look very different. And when we actually meet natives of those separate places, which can indeed happen, we can see representations of those differences at first hand. But if one were to walk up beside the Nile from Cairo, across the Tropic of Cancer to Khartoum in the Sudan and on to Nairobi, there would be no visible boundary between one people and another. The same thing would be true if one were to walk north from Cairo, through the Caucasus, and on up into Russia, eventually swinging west across the northern end of the Baltic Sea to Scandinavia. The people at any adjacent stops along the way look like one another more than they look like anyone else since, after all, they are related to one another. As a rule, the boy marries the girl next door throughout the whole world, but next door goes on without stop from one region to another.

We realize that in the extremes of our transit—Moscow to Nairobi, perhaps—there is a major but gradual change in skin color from what we euphemistically call white to black, and that this is related to the latitudinal difference in the intensity of the ultraviolet component of sunlight. What we do not see, however, is the myriad other traits that are distributed in a fashion quite unrelated to the intensity of ultraviolet radiation. Where skin color is concerned, all the northern populations of the Old World are lighter than the long-term inhabitants near the equator. Although Europeans and Chinese are obviously different, in skin color they are closer to each other than either is to equatorial Africans. But if we test the distribution of the widely known ABO blood-group system, then Europeans and Africans are closer to each other than either is to Chinese.

Then if we take that scourge sickle-cell anemia, so often thought of as an African disease, we discover that, while it does reach high frequencies in some parts of sub-Saharan Africa, it did not originate there. Its distribution includes southern Italy, the eastern Mediterranean, parts of the Middle East, and over into India. In fact, it represents a kind of adaptation that aids survival in the face of a particular kind of malaria, and wherever that malaria is a prominent threat, sickle-cell anemia tends to occur in higher frequencies. It would appear that the gene that controls that trait was introduced to sub-Saharan Africa by traders from those parts of the Middle East where it had arisen in conjunction with the conditions created by the early development of agriculture.

Every time we plot the distribution of a trait possessing a survival value that is greater under some circumstances than under others, it will have a different pattern of geographical variation, and no two such patterns will coincide. Nose form, tooth size, relative arm and leg length, and a whole series of other traits are distributed each in accordance with its particular controlling selective force. The gradient of the distribution of each is called a "cline" and those clines are completely independent of one another. This is what lies behind the aphorism, "There are no races, there are only clines." Yes, we can recognize people from a given area. What we are seeing, however, is a pattern of features derived from common ancestry in the area in question, and these are largely without different survival value. To the extent that the people in a given region look more like one another than they look like people from other regions, this can be regarded as "family resemblance writ large." And as we have seen, each region grades without break into the one next door.

There is nothing wrong with using geographic labels to designate people. Major continental terms are just fine, and sub-regional refinements such as Western European, Eastern African, Southeast Asian, and so forth carry no unintentional baggage. In contrast, terms such as "Negroid," "Caucasoid," and "Mongoloid" create more problems than they solve. Those very terms reflect a mix of narrow regional, specific ethnic, and descriptive physical components with an assumption that such separate dimensions have some kind of common tie. Biologically, such terms are worse than useless. Their continued use, then, is in social situations where people think they have some meaning.

America and the Race Concept

The role played by America is particularly important in generating and perpetuating the concept of "race." The human inhabitants of the Western Hemisphere largely derive from three very separate regions of the world—Northeast Asia, Northwest Europe, and Western Africa—and none of them has been in the New World long enough to have been shaped by their experiences in the manner of those long-term residents in the various separate regions of the Old World.

It was the American experience of those three separate population components facing one another on a daily basis under conditions of manifest and enforced inequality that created the concept in the first place and endowed it with the assumption that those perceived "races" had very different sets of capabilities. Those thoughts are very influential and have become enshrined in laws and regulations. This is why I can conclude that, while the word "race" has no coherent biological meaning, its continued grip on the public mind is in fact a manifestation of the power of the historical continuity of the American social structure, which is assumed by all to be essentially "correct."

Finally, because of America's enormous influence on the international scene, ideas generated by the idiosyncrasies of American history have gained currency in ways that transcend American intent or control. One of those ideas is the concept of "race," which we have exported to the rest of the world without any realization that this is what we were doing. The adoption of the biologically indefensible American concept of "race" by an admiring world has to be the ultimate manifestation of political correctness.

DR. C. LORING BRACE is professor anthropology and curator of biological anthropology at the Museum of Anthropology, University of Michigan, Ann Arbor.

The Tall and the Short of It

BARRY BOGIN

Baffled by your future prospects? As a biological anthropologist, I have just one word of advice for you: plasticity. *Plasticity* refers to the ability of many organisms, including humans, to alter themselves—their behavior or even their biology—in response to changes in the environment. We tend to think that our bodies get locked into their final form by our genes, but in fact we alter our bodies as the conditions surrounding us shift, particularly as we grow during childhood. Plasticity is as much a product of evolution's fine-tuning as any particular gene, and it makes just as much evolutionary good sense. Rather than being able to adapt to a single environment, we can, thanks to plasticity, change our bodies to cope with a wide range of environments. Combined with the genes we inherit from our parents, plasticity accounts for what we are and what we can become.

Anthropologists began to think about human plasticity around the turn of the century, but the concept was first clearly defined in 1969 by Gabriel Lasker, a biological anthropologist at Wayne State University in Detroit. At that time scientists tended to consider only those adaptations that were built into the genetic makeup of a person and passed on automatically to the next generation. A classic example of this is the ability of adults in some human societies to drink milk. As children, we all produce an enzyme called lactase, which we need to break down the sugar lactose in our mother's milk. In many of us, however, the lactase gene slows down dramatically as we approach adolescence—probably as the result of another gene that regulates its activity. When that regulating gene turns down the production of lactase, we can no longer digest milk.

Lactose intolerance—which causes intestinal gas and diarrhea—affects between 70 and 90 percent of African Americans, Native Americans, Asians, and people who come from around the Mediterranean. But others, such as people of central and western European descent and the Fulani of West Africa, typically have no problem drinking milk as adults. That's because they are descended from societies with long histories of raising goats and cattle. Among these people there was a clear benefit to being able to drink milk, so natural selection gradually changed the regulation of their lactase gene, keeping it functioning throughout life.

That kind of adaptation takes many centuries to become established, but Lasker pointed out that there are two other kinds of adaptation in humans that need far less time to kick in. If peo-

ple have to face a cold winter with little or no heat, for example, their metabolic rates rise over the course of a few weeks and they produce more body heat. When summer returns, the rates sink again.

Lasker's other mode of adaptation concerned the irreversible, lifelong modification of people as they develop—that is, their plasticity. Because we humans take so many years to grow to adulthood, and because we live in so many different environments, from forests to cities and from deserts to the Arctic, we are among the world's most variable species in our physical form and behavior. Indeed, we are one of the most plastic of all species.

In an age when DNA is king, it's worth considering why Americans are no longer the world's tallest people, and some Guatemalans no longer pygmies.

One of the most obvious manifestations of human malleability is our great range of height, and it is a subject I've made a special study of for the last 25 years. Consider these statistics: in 1850 Americans were the tallest people in the world, with American men averaging 5'6". Almost 150 years later, American men now average 5'8", but we have fallen in the standings and are now only the third tallest people in the world. In first place are the Dutch. Back in 1850 they averaged only 5'4"—the shortest men in Europe—but today they are a towering 5'10". (In these two groups, and just about everywhere else, women average about five inches less than men at all times.)

So what happened? Did all the short Dutch sail over to the United States? Did the Dutch back in Europe get an infusion of "tall genes"? Neither. In both America and the Netherlands life got better, but more so for the Dutch, and height increased as a result. We know this is true thanks in part to studies on how height is determined. It's the product of plasticity in our childhood and in our mothers' childhood as well. If a girl is undernourished and suffers poor health, the growth of her body, including her reproductive system, is usually reduced. With a shortage of raw materials, she can't build more cells to construct a bigger body; at the same time, she has to invest what materials she can get into repairing already existing cells and tissues from the damage

175

caused by disease. Her shorter stature as an adult is the result of a compromise her body makes while growing up.

Such a woman can pass on her short stature to her child, but genes have nothing to do with it for either of them. If she becomes pregnant, her small reproductive system probably won't be able to supply a normal level of nutrients and oxygen to her fetus. This harsh environment reprograms the fetus to grow more slowly than it would if the woman was healthier, so she is more likely to give birth to a smaller baby. Low-birth-weight babies (weighing less than 5.5 pounds) tend to continue their prenatal program of slow growth through childhood. By the time they are teenagers, they are usually significantly shorter than people of normal birth weight. Some particularly striking evidence of this reprogramming comes from studies on monozygotic twins, which develop from a single fertilized egg cell and are therefore identical genetically. But in certain cases, monozygotic twins end up being nourished by unequal portions of the placenta. The twin with the smaller fraction of the placenta is often born with low birth weight, while the other one is normal. Follow-up studies show that this difference between the twins can last throughout their lives.

As such research suggests, we can use the average height of any group of people as a barometer of the health of their society. After the turn of the century both the United States and the Netherlands began to protect the health of their citizens by purifying drinking water, installing sewer systems, regulating the safety of food, and, most important, providing better health care and diets to children. The children responded to their changed environment by growing taller. But the differences in Dutch and American societies determined their differing heights today. The Dutch decided to provide public health benefits to all the public, including the poor. In the United States, meanwhile, improved health is enjoyed most by those who can afford it. The poor often lack adequate housing, sanitation, and health care. The difference in our two societies can be seen at birth: in 1990 only 4 percent of Dutch babies were born at low birth weight, compared with 7 percent in the United States. For white Americans the rate was 5.7 percent, and for black Americans the rate was a whopping 13.3 percent. The disparity between rich and poor in the United States carries through to adulthood: poor Americans are shorter than the better-off by about one inch. Thus, despite great affluence in the United States, our average height has fallen to third place.

People are often surprised when I tell them the Dutch are the tallest people in the world. Aren't they shrimps compared with the famously tall Tutsi (or "Watusi," as you probably first encountered them) of Central Africa? Actually, the supposed great height of the Tutsi is one of the most durable myths from the age of European exploration. Careful investigation reveals that today's Tutsi men average 5'7" and that they have maintained that average for more than 100 years. That means that back in the 1800s, when puny European men first met the Tutsi, the Europeans suffered strained necks from looking up all the time. The two-to-three-inch difference in average height back then could easily have turned into fantastic stories of African giants by European adventures and writers.

The Tutsi could be as tall or taller than the Dutch if equally good health care and diets were available in Rwanda and Burundi, where the Tutsi live. But poverty rules the lives of most African people, punctuated by warfare, which makes the conditions for growth during childhood even worse. And indeed, it turns out that the Tutsi and other Africans who migrate to Western Europe or North America at young ages end up taller than Africans remaining in Africa.

At the other end of the height spectrum, Pygmies tell a similar story. The shortest people in the world today are the Mbuti, the Efe, and other Pygmy peoples of Central Africa. Their average stature is almost 4'9" for adult men and 4'6" for women. Part of the reason Pygmies are short is indeed genetic: some evidently lack the genes for producing the growth-promoting hormones that course through other people's bodies, while others are genetically incapable of using these hormones to trigger the cascade of reactions that lead to growth. But another important reason for their small size is environmental. Pygmies living as hunter-gatherers in the forests of Central African countries appear to be undernourished, which further limits their growth. Pygmies who live on farms and ranches outside the forest are better fed than their hunter-gatherer relatives and are taller as well. Both genes and nutrition thus account for the size of Pygmies.

Peoples in other parts of the world have also been labeled pygmies, such as some groups in Southeast Asia and the Maya of Guatemala. Well-meaning explorers and scientists have often claimed that they are genetically short, but here we encounter another myth of height. A group of extremely short people in New Guinea, for example, turned out to eat a diet deficient in iodine and other essential nutrients. When they were supplied with cheap mineral and vitamin supplements, their supposedly genetic short stature vanished in their children, who grew to a more normal height.

Another way for these so-called pygmies to stop being pygmies is to immigrate to the United States. In my own research, I study the growth of two groups of Mayan children. One group lives in their homeland of Guatemala, and the other is a group of refugees living in the United States. The Maya in Guatemala live in the village of San Pedro, which has no safe source of drinking water. Most of the water is contaminated with fertilizers and pesticides used on nearby agricultural fields. Until recently, when a deep well was dug, the townspeople depended on an unreliable supply of water from rain-swollen streams. Most homes still lack running water and have only pit toilets. The parents of the Mayan children work mostly at clothing factories and are paid only a few dollars a day.

I began working with the schoolchildren in this village in 1979, and my research shows that most of them eat only 80 percent of the food they need. Other research shows that almost 30 percent of the girls and 20 percent of the boys are deficient in iodine, that most of the children suffer from intestinal parasites, and that many have persistent ear and eye infections. As a consequence, their health is poor and their height reflects it: they average about three inches shorter than better-fed Guatemalan children.

The Mayan refugees I work with in the United States live in Los Angeles and in the rural agricultural community of Indiantown in central Florida. Although the adults work mostly in minimum-wage jobs, the children in these communities are generally better off than their counterparts in Guatemala. Most Maya arrived in the 1980s as refugees escaping a civil war as well as a political system that threatened them and their children. In the United States they found security and started new lives, and before long their children began growing faster and bigger. My data show that the average increase in height among the first generation of these immigrants was 2.2 inches, which means that these so-called pygmies have undergone one of the largest single-generation increases in height ever recorded. When people such as my own grandparents migrated from the poverty of rural life in Eastern Europe to the cities of the United States just after World War I, the increase in height of the next generation was only about one inch.

One reason for the rapid increase in stature is that in the United States the Maya have access to treated drinking water and to a reliable supply of food. Especially critical are school breakfast and lunch programs for children from low-income families, as well as public assistance programs such as the federal Woman, Infants, and Children (WIC) program and food stamps. That these programs improve health and growth is no secret. What is surprising is how fast they work. Mayan mothers in the United States tell me that even their babies are bigger and healthier than the babies they raised in Guatemala, and hospital statistics bear them out. These women must be enjoying a level of health so improved from that of their lives in Guatemala that their babies are growing faster in the womb. Of course, plasticity means that such changes are dependent on external conditions, and unfortunately the rising height—and health—of the Maya is in danger from political forces that are attempting to cut funding for food stamps and the WIC program. If that funding is cut, the negative impact on the lives of poor Americans, including the Mayan refugees, will be as dramatic as were the former positive effects.

One way for the so-called pygmies of Guatemala to stop being pygmies is to immigrate to the United States.

Height is only the most obvious example of plasticity's power; there are others to be found everywhere you look. The Andes-dwelling Quechua people of Peru are well-adapted to their high-altitude homes. Their large, barrel-shaped chests house big lungs that inspire huge amounts of air with each breath, and they manage to survive on the lower pressure of oxygen they breathe with an unusually high level of red blood cells. Yet these secrets of mountain living are not hereditary. Instead the bodies of young Quechua adapt as they grow in their particular environment, just as those of European children do when they live at high altitudes.

Plasticity may also have a hand in determining our risks for developing a number of diseases. For example, scientists have long been searching for a cause for Parkinson's disease. Because Parkinson's tends to run in families, it is natural to think there is a genetic cause. But while a genetic mutation linked to some types of Parkinson's disease was reported in mid-1997, the gene accounts for only a fraction of people with the disease. Many more people with Parkinson's do not have the gene, and not all people with the mutated gene develop the disease.

Ralph Garruto, a medical researcher and biological anthropologist at the National Institutes of Health, is investigating the role of the environment and human plasticity not only in Parkinson's but in Lou Gehrig's disease as well. Garruto and his team traveled to the islands of Guam and New Guinea, where rates of both diseases are 50 to 100 times higher than in the United States. Among the native Chamorro people of Guam these diseases kill one person out of every five over the age of 25. The scientists found that both diseases are linked to a shortage of calcium in the diet. This shortage sets off a cascade of events that result in the digestive system's absorbing too much of the aluminum present in the diet. The aluminum wreaks havoc on various parts of the body, including the brain, where it destroys neurons and eventually causes paralysis and death.

The most amazing discovery made by Garruto's team is that up to 70 percent of the people they studied in Guam had some brain damage, but only 20 percent progressed all the way to Parkinson's or Lou Gehrig's disease. Genes and plasticity seem to be working hand in hand to produce these lower-than-expected rates of disease. There is a certain amount of genetic variation in the ability that all people have in coping with calcium shortages—some can function better than others. But thanks to plasticity, it's also possible for people's bodies to gradually develop ways to protect themselves against aluminum poisoning. Some people develop biochemical barriers to the aluminum they eat, while others develop ways to prevent the aluminum from reaching the brain.

An appreciation of plasticity may temper some of our fears about these diseases and even offer some hope. For if Parkinson's and Lou Gehrig's diseases can be prevented among the Chamorro by plasticity, then maybe medical researchers can figure out a way to produce the same sort of plastic changes in you and me. Maybe Lou Gehrig's disease and Parkinson's disease—as well as many other, including some cancers—aren't our genetic doom but a product of our development, just like variations in human height. And maybe their danger will in time prove as illusory as the notion that the Tutsi are giants, or the Maya pygmies—or Americans still the tallest of the tall.

BARRY BOGIN is a professor of anthropology at the University of Michigan in Dearborn and the author of *Patterns of Human Growth*.

UNIT 7
Living With the Past

Unit Selections

Key Points to Consider

- What are the ways to prevent epidemics in the human species?

- What social policy issues are involved in the nature versus nurture debate?

- Does the concept of natural selection have relevance to the treatment of disease? Defend your answer.

- What healthful habits can we learn from studying hunter-gatherers and why?

- Why is Tay-Sachs disease so common among Eastern European Jews?

- What is the "saltshaker's curse" and why are some people more affected by it than others?

Student Website
www.mhcls.com/online

Internet References
Further information regarding these websites may be found in this book's preface or online.

Forensic Science Reference Page
   http://www.lab.fws.gov
Zeno's Forensic Page
   http://forensic.to/forensic.html

Anthropology continues to evolve as a discipline, not only in the tools and techniques of the trade, but also in the application of whatever knowledge we stand to gain about ourselves. Sometimes an awareness of our biological and behavioral past may help us to better understand the present. For instance, in showing how our evolutionary past may make a difference in bodily health, Patricia Gadsby (in "The Inuit Paradox") deals with the question of how traditional hunters, such as the Inuit, could gorge themselves on fat, rarely see a vegetable and be healthier than we are. Lori Oliwenstein (in "Dr. Darwin") then talks about how the symptoms of disease must first be interpreted as to whether they represent part of the aggressive strategy of microbes or the defensive mechanisms of the patient before treatment can be applied. Jared Diamond, in "Curse and Blessing of the Ghetto" and "The Saltshaker's Curse," shows that, while many deleterious genes do get weeded out of the population by means of natural selection, there are other harmful ones that may actually have a good side to them and will therefore be perpetuated.

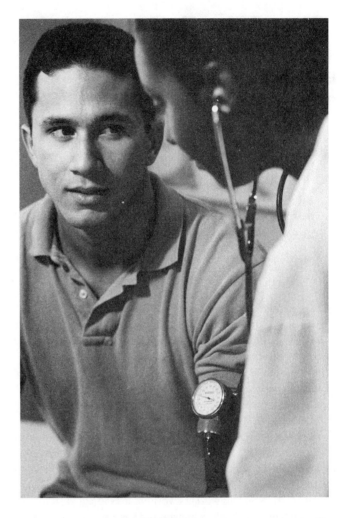

As we reflect upon where we have been and how we came to be as we are in the evolutionary sense, the inevitable question arises as to what will happen next. This is the most difficult issue of all, since our biological future depends so much on long-range environmental trends that no one seems to be able to predict. Take, for example, the sweeping effects of ecological change upon the viruses of the world, which in turn seem to be paving the way for new waves of human epidemics. There is no better example of this problem than the recent explosion of new diseases, as described in "The Viral Superhighway" by George Armelagos.

As we gain a better understanding of the processes of mutation and natural selection and their relevance to human beings, we might even gain some control over the evolutionary direction of our species. However, the issue of just what is a beneficial application of scientific knowledge becomes a matter for debate. Who will have the final word as to how these technological breakthroughs will be employed in the future? Even with the best of intentions, how can we be certain of the long-range consequences of our actions in such a complicated field?

Knowledge itself, of course, is neutral—its potential for good or ill being determined by those who happen to be in a position to use it. Consider, for example, that some men may be dying from a genetically caused overabundance of iron in their blood systems in a trade-off that allows some women to absorb sufficient amounts of the element to guarantee their own survival. The question of whether we should eliminate such a gene that brings about this situation would seem to depend on which sex we decide should reap the benefit.

As we read the essays in this unit and contemplate the significance of "Darwinian medicine" for human evolution, we can hope that a better understanding of the diseases that afflict our species will lead to a reduction of human suffering. At the same time, we must remain aware that someone, at some time, may actually use the same knowledge to increase rather than reduce the misery that exists in the world.

The Viral Superhighway

Environmental disruptions and international travel have brought on a new era in human illness, one marked by diabolical new diseases

GEORGE J. ARMELAGOS

So the Lord sent a pestilence upon Israel from the morning until the appointed time; and there died of the people from Dan to Beer-sheba seventy thousand men.

—2 Sam. 24:15

Swarms of crop-destroying locusts, rivers fouled with blood, lion-headed horses breathing fire and sulfur: the Bible presents a lurid assortment of plagues, described as acts of retribution by a vengeful God. Indeed, real-life epidemics—such as the influenza outbreak of 1918, which killed 21 million people in a matter of months—can be so sudden and deadly that it is easy, even for nonbelievers, to view them as angry messages from the beyond.

How reassuring it was, then, when the march of technology began to give people some control over the scourges of the past. In the 1950s the Salk vaccine, and later, the Sabin vaccine, dramatically reduced the incidence of polio. And by 1980 a determined effort by health workers worldwide eradicated smallpox, a disease that had afflicted humankind since earliest times with blindness, disfigurement and death, killing nearly 300 million people in the twentieth century alone.

But those optimistic years in the second half of our century now seem, with hindsight, to have been an era of inflated expectations, even arrogance. In 1967 the surgeon general of the United States, William H. Stewart, announced that victory over infectious diseases was imminent—a victory that would close the book on modern plagues. Sadly, we now know differently. Not only have deadly and previously unimagined new illnesses such as AIDS and Legionnaires' disease emerged in recent years, but historical diseases that just a few decades ago seemed to have been tamed are returning in virulent, drug-resistant varieties. Tuberculosis, the ancient lung disease that haunted nineteenth-century Europe, afflicting, among others, Chopin, Dostoyevski and Keats, is aggressively mutating into strains that defy the standard medicines; as a result, modern TB victims must undergo a daily drug regimen so elaborate that health-department workers often have to personally monitor patients to make sure they comply [see "A Plague Returns," by Mark Ear-

nest and John A. Sbarbaro, September/October 1993]. Meanwhile, bacteria and viruses in foods from chicken to strawberries to alfalfa sprouts are sickening as many as 80 million Americans each year.

And those are only symptoms of a much more general threat. Deaths from infectious diseases in the United States rose 58 percent between 1980 and 1992. Twenty-nine new diseases have been reported in the past twenty-five years, a few of them so bloodcurdling and bizarre that descriptions of them bring to mind tacky horror movies. Ebola virus, for instance, can in just a few days reduce a healthy person to a bag of teeming flesh spilling blood and organ parts from every orifice. Creutzfeldt-Jakob disease, which killed the choreographer George Balanchine in 1983, eats away at its victims' brains until they resemble wet sponges. Never slow to fan mass hysteria, Hollywood has capitalized on the phenomenon with films such as *Outbreak*, in which a monkey carrying a deadly new virus from central Africa infects unwitting Californians and starts an epidemic that threatens to annihilate the human race.

The reality about infectious disease is less sensational but alarming nonetheless. Gruesome new pathogens such as Ebola are unlikely to cause a widespread epidemic because they sicken and kill so quickly that victims can be easily identified and isolated; on the other hand, the seemingly innocuous practice of overprescribing antibiotics for bad colds could ultimately lead to untold deaths, as familiar germs evolve to become untreatable. We are living in the twilight of the antibiotic era: within our lifetimes, scraped knees and cut fingers may return to the realm of fatal conditions.

Through international travel, global commerce and the accelerating destruction of ecosystems worldwide, people are inadvertently exposing themselves to a Pandora's box of emerging microbial threats. And the recent rumblings of biological terrorism from Iraq highlight the appalling potential of disease organisms for being manipulated to vile ends. But although it may appear that the apocalypse has arrived, the truth is that people today are not facing a unique predicament. Emerging diseases have long loomed like a shadow over the human race.

People and pathogens have a long history together. Infections have been detected in the bones of human ancestors more than a million years old, and evidence from the mummy of the Egyptian pharaoh Ramses V suggests that he may have died from smallpox more than 3,000 years ago. Widespread outbreaks of disease are also well documented. Between 1347 and 1351 roughly a third of the population of medieval Europe was wiped out by bubonic plague, which is carried by fleas that live on rodents. In 1793, 10 percent of the population of Philadelphia succumbed to yellow fever, which is spread by mosquitoes. And in 1875 the son of a Fiji chief came down with measles after a ceremonial trip to Australia. Within four months more than 20,000 Fijians were dead from the imported disease, which spreads through the air when its victims cough or sneeze.

According to conventional wisdom in biology, people and invading microorganisms evolve together: people gradually become more resistant, and the microorganisms become less virulent. The result is either mutualism, in which the relation benefits both species, or commensalism, in which one species benefits without harming the other. Chicken pox and measles, once fatal afflictions, now exist in more benign forms. Logic would suggest, after all, that the best interests of an organism are not served if it kills its host; doing so would be like picking a fight with the person who signs your paycheck.

But recently it has become clear to epidemiologists that the reverse of that cooperative paradigm of illness can also be true: microorganisms and their hosts sometimes exhaust their energies devising increasingly powerful weaponry and defenses. For example, several variants of human immunodeficiency virus (HIV) may compete for dominance within a person's body, placing the immune system under ever-greater siege. As long as a virus has an effective mechanism for jumping from one person to another, it can afford to kill its victims [see "The Deadliest Virus," by Cynthia Mills, January/February 1997].

If the competition were merely a question of size, humans would surely win: the average person is 10^{17} times the size of the average bacterium. But human beings, after all, constitute only one species, which must compete with 5,000 kinds of viruses and more than 300,000 species of bacteria. Moreover, in the twenty years it takes humans to produce a new generation, bacteria can reproduce a half-million times. That disparity enables pathogens to evolve ever more virulent adaptations that quickly outstrip human responses to them. The scenario is governed by what the English zoologist Richard Dawkins of the University of Oxford and a colleague have called the "Red Queen Principle." In Lewis Carroll's *Through the Looking Glass* the Red Queen tells Alice she will need to run faster and faster just to stay in the same place. Staving off illness can be equally elusive.

The Centers For Disease Control and Prevention (CDC) in Atlanta, Georgia, has compiled a list of the most recent emerging pathogens. They include:

- *Campylobacter*, a bacterium widely found in chickens because of the commercial practice of raising them in cramped, unhealthy conditions. It causes between two

million and eight million cases of food poisoning a year in the United States and between 200 and 800 deaths.

- *Escherichia coli* 0157:H7, a dangerously mutated version of an often harmless bacterium. Hamburger meat from Jack in the Box fast-food restaurants that was contaminated with this bug led to the deaths of at least four people in 1993.

- Hantaviruses, a genus of fast-acting, lethal viruses, often carried by rodents, that kill by causing the capillaries to leak blood. A new hantavirus known as *sin nombre* (Spanish for "nameless") surfaced in 1993 in the southwestern United States, causing the sudden and mysterious deaths of thirty-two people.

- HIV, the deadly virus that causes AIDS (acquired immunodeficiency syndrome). Although it was first observed in people as recently as 1981, it has spread like wildfire and is now a global scourge, affecting more than 30 million people worldwide.

- The strange new infectious agent that causes bovine spongiform encephalopathy, or mad cow disease, which recently threw the British meat industry and consumers into a panic. This bizarre agent, known as a prion, or "proteinaceous infectious particle," is also responsible for Creutzfeldt-Jakob disease, the brain-eater I mentioned earlier. A Nobel Prize was awarded last year to the biochemist Stanley B. Prusiner of the University of California, San Francisco, for his discovery of the prion.

- *Legionella pneumophila*, the bacterium that causes Legionnaires' disease. The microorganism thrives in wet environments; when it lodges in air-conditioning systems or the mist machines in supermarket produce sections, it can be expelled into the air, reaching people's lungs. In 1976 thirty-four participants at an American Legion convention in Philadelphia died—the incident that led to the discovery and naming of the disease.

- *Borrelia burgdorferi*, the bacterium that causes Lyme disease. It is carried by ticks that live on deer and white-footed mice. Left untreated, it can cause crippling, chronic problems in the nerves, joints and internal organs.

How ironic, given such a rogues' gallery of nasty characters, that just a quarter-century ago the Egyptian demographer Abdel R. Omran could observe that in many modern industrial nations the major killers were no longer infectious diseases. Death, he noted, now came not from outside but rather from within the body, the result of gradual deterioration. Omran traced the change to the middle of the nineteenth century, when the industrial revolution took hold in the United States and parts of Europe. Thanks to better nutrition, improved public-health measures and medical advances such as mass immunization and the introduction of antibiotics, microorganisms were brought under control. As people began living longer, their aging bodies succumbed to "diseases of civilization": cancer, clogged arteries, diabetes, obesity and osteoporosis. Omran was the first to formally recognize that shift in the disease environment. He called it an "epidemiological transition."

Like other anthropologists of my generation, I learned of Omran's theory early in my career, and it soon became a basic tenet—a comforting one, too, implying as it did an end to the supremacy of microorganisms. Then, three years ago, I began working with the anthropologist Kathleen C. Barnes of Johns Hopkins University in Baltimore, Maryland, to formulate an expansion of Omran's ideas. It occurred to us that his epidemiological transition had not been a unique event. Throughout history human populations have undergone shifts in their relations with disease—shifts, we noted, that are always linked to major changes in the way people interact with the environment. Barnes and I, along with James Lin, a master's student at Johns Hopkins University School of Hygiene and Public Health, have since developed a new theory: that there have been not one but three major epidemiological transitions; that each one has been sparked by human activities; and that we are living through the third one right now.

The first epidemiological transition took place some 10,000 years ago, when people abandoned their nomadic existence and began farming. That profoundly new way of life disrupted ecosystems and created denser living conditions that led, as I will soon detail, to new diseases. The second epidemiological transition was the salutary one Omran singled out in 1971, when the war against infectious diseases seemed to have been won. And in the past two decades the emergence of illnesses such as hepatitis C, cat scratch disease (caused by the bacterium *Bartonella henselae*), Ebola and others on CDC's list has created a third epidemiological transition, a disheartening set of changes that in many ways have reversed the effects of the second transition and coincide with the shift to globalism. Burgeoning population growth and urbanization, widespread environmental degradation, including global warming and tropical deforestation, and radically improved methods of transportation have given rise to new ways of contracting and spreading disease.

We are, quite literally, making ourselves sick.

When early human ancestors moved from African forests onto the savanna millions of years ago, a few diseases came along for the ride. Those "heirloom" species—thus designated by the Australian parasitologist J. F. A. Sprent because they had afflicted earlier primates—included head and body lice; parasitic worms such as pinworms, tapeworms and liver flukes; and possibly herpes virus and malaria.

Global Warming could allow the mosquitoes that carry dengue fever to survive as far north as New York City.

For 99.8 percent of the five million years of human existence, hunting and gathering was the primary mode of subsistence. Our ancestors lived in small groups and relied on wild animals and plants for their survival. In their foraging rounds, early humans would occasionally have contracted new kinds of illnesses through insect bites or by butchering and eating disease-ridden animals. Such events would not have led to widespread epidemics, however, because groups of people were so sparse and widely dispersed.

About 10,000 years ago, at the end of the last ice age, many groups began to abandon their nomadic lifestyles for a more efficient and secure way of life. The agricultural revolution first appeared in the Middle East; later, farming centers developed independently in China and Central America. Permanent villages grew up, and people turned their attention to crafts such as toolmaking and pottery. Thus when people took to cultivating wheat and barley, they planted the seeds of civilization as well.

With the new ways, however, came certain costs. As wild habitats were transformed into urban settings, the farmers who brought in the harvest with their flint-bladed sickles were assailed by grim new ailments. Among the most common was scrub typhus, which is carried by mites that live in tall grasses, and causes a potentially lethal fever. Clearing vegetation to create arable fields brought farmers frequently into mite-infested terrain.

Irrigation brought further hazards. Standing thigh-deep in watery canals, farm workers were prey to the worms that cause schistosomiasis. After living within aquatic snails during their larval stage, those worms emerge in a free-swimming form that can penetrate human skin, lodge in the intestine or urinary tract, and cause bloody urine and other serious maladies. Schistosomiasis was well known in ancient Egypt, where outlying fields were irrigated with water from the Nile River; descriptions of its symptoms and remedies are preserved in contemporary medical papyruses.

The domestication of sheep, goats and other animals cleared another pathway for microorganisms. With pigs in their yards and chickens roaming the streets, people in agricultural societies were constantly vulnerable to pathogens that could cross interspecies barriers. Many such organisms had long since reached commensalism with their animal hosts, but they were highly dangerous to humans. Milk from infected cattle could transmit tuberculosis, a slow killer that eats away at the lungs and causes its victims to cough blood and pus. Wool and skins were loaded with anthrax, which can be fatal when inhaled and, in modern times, has been developed by several nations as a potential agent of biological warfare. Blood from infected cattle, injected into people by biting insects such as the tsetse fly, spread sleeping sickness, an often-fatal disease marked by tremors and protracted lethargy.

A second major effect of agriculture was to spur population growth and, perhaps more important, density. Cities with populations as high as 50,000 had developed in the Near East by 3000 B.C. Scavenger species such as rats, mice and sparrows, which congregate wherever large groups of people live, exposed city dwellers to bubonic plague, typhus and rabies. And now that people were crowded together, a new pathogen could quickly start an epidemic. Larger populations also enabled diseases such as measles, mumps, chicken pox and smallpox to persist in an endemic form—always present, afflicting part of the population while sparing those with acquired immunity.

Thus the birth of agriculture launched humanity on a trajectory that has again and again brought people into contact with new pathogens. Tilling soil and raising livestock led to more energy-intensive ways of extracting resources from the earth—to lumbering, coal mining, oil drilling. New resources led to increasingly complex social organization, and to new and more frequent contacts between various societies. Loggers today who venture into the rain forest disturb previously untouched creatures and give them, for the first time, the chance to attack humans. But there is nothing new about this drama; only the players have changed. Some 2,000 years ago the introduction of iron tools to sub-Saharan Africa led to a slash-and-burn style of agriculture that brought people into contact with *Anopheles gambiae*, a mosquito that transmits malaria.

Improved transportation methods also help diseases extend their reach: microorganisms cannot travel far on their own, but they are expert hitchhikers. When the Spanish invaded Mexico in the early 1500s, for instance, they brought with them diseases that quickly raged through Tenochtitlán, the stately, temple-filled capital of the Aztec Empire. Smallpox, measles and influenza wiped out millions of Central America's original inhabitants, becoming the invisible weapon in the European conquest.

In the past three decades people and their inventions have drilled, polluted, engineered, paved, planted and deforested at soaring rates, changing the biosphere faster than ever before. The combined effects can, without hyperbole, be called a global revolution. After all, many of them have worldwide repercussions: the widespread chemical contamination of waterways, the thinning of the ozone layer, the loss of species diversity. And such global human actions have put people at risk for infectious diseases in newly complex and devastating ways. Global warming, for instance, could expose millions of people for the first time to malaria, sleeping sickness and other insect-borne illnesses; in the United States, a slight overall temperature increase would allow the mosquitoes that carry dengue fever to survive as far north as New York City.

Major changes to the landscape that have become possible in the past quarter-century have also triggered new diseases. After the construction of the Aswan Dam in 1970, for instance, Rift Valley fever infected 200,000 people in Egypt, killing 600. The disease had been known to affect livestock, but it was not a major problem in people until the vast quantities of dammed water became a breeding ground for mosquitoes. The insects bit both cattle and humans, helping the virus jump the interspecies barrier.

In the eastern United States, suburbanization, another relatively recent phenomenon, is a dominant factor in the emergence of Lyme disease—10,000 cases of which are reported annually. Thanks to modern earth-moving equipment, a soaring economy and population pressures, many Americans have built homes in formerly remote, wooded areas. Nourished by lawns and gardens and unchecked by wolves, which were exterminated by settlers long ago, the deer population has exploded, exposing people to the ticks that carry Lyme disease.

Meanwhile, widespread pollution has made the oceans a breeding ground for microorganisms. Epidemiologists have suggested that toxic algal blooms—fed by the sewage, fertilizers and other contaminants that wash into the oceans—harbor countless viruses and bacteria. Thrown together into what amounts to a dirty genetic soup, those pathogens can undergo gene-swapping and mutations, engendering newly antibiotic-resistant strains. Nautical traffic can carry ocean pathogens far and wide: a devastating outbreak of cholera hit Latin America in 1991 after a ship from Asia unloaded its contaminated ballast water into the harbor of Callao, Peru. Cholera causes diarrhea so severe its victims can die in a few days from dehydration; in that outbreak more than 300,000 people became ill, and more than 3,000 died.

The modern world is becoming—to paraphrase the words of the microbiologist Stephen S. Morse of Columbia University—a viral superhighway. Everyone is at risk.

Our newly global society is characterized by huge increases in population, international travel and international trade—factors that enable diseases to spread much more readily than ever before from person to person and from continent to continent. By 2020 the world population will have surpassed seven billion, and half those people will be living in urban centers. Beleaguered third-world nations are already hard-pressed to provide sewers, plumbing and other infrastructure; in the future, clean water and adequate sanitation could become increasingly rare. Meanwhile, political upheavals regularly cause millions of people to flee their homelands and gather in refugee camps, which become petri dishes for germs.

More than 500 million people cross international borders each year on commercial flights. Not only does that traffic volume dramatically increase the chance a sick person will infect the inhabitants of a distant area when she reaches her destination; it also exposes the sick person's fellow passengers to the disease, because of poor air circulation on planes. Many of those passengers can, in turn, pass the disease on to others when they disembark.

The global economy that has arisen in the past two decades has established a myriad of connections between far-flung places. Not too long ago bananas and oranges were rare treats in northern climes. Now you can walk into your neighborhood market and find food that has been flown and trucked in from all over the world: oranges from Israel, apples from New Zealand, avocados from California. Consumers in affluent nations expect to be able to buy whatever they want whenever they want it. What people do not generally realize, however, is that this global network of food production and delivery provides countless pathways for pathogens. Raspberries from Guatemala, carrots from Peru and coconut milk from Thailand have been responsible for recent outbreaks of food poisoning in the United States. And the problem cuts both ways: contaminated radish seeds and frozen beef from the United States have ended up in Japan and South Korea.

183

Finally, the widespread and often indiscriminate use of antibiotics has played a key role in spurring disease. Forty million pounds of antibiotics are manufactured annually in the United States, an eightyfold increase since 1954. Dangerous microorganisms have evolved accordingly, often developing antibiotic-resistant strains. Physicians are now faced with penicillin-resistant gonorrhea, multiple-drug-resistant tuberculosis and E. coli variants such as 0157:H7. And frighteningly, some enterococcus bacteria have become resistant to all known antibiotics. Enterococcus infections are rare, but staphylococcus infections are not, and many strains of staph bacteria now respond to just one antibiotic, vancomycin. How long will it be before run-of-the-mill staph infections—in a boil, for instance, or in a surgical incision—become untreatable?

Although civilization can expose people to new pathogens, cultural progress also has an obvious countervailing effect: it can provide tools—medicines, sensible city planning, educational campaigns about sexually transmitted diseases—to fight the encroachments of disease. Moreover, since biology seems to side with microorganisms anyway, people have little choice but to depend on protective cultural practices to keep pace: vaccinations, for instance, to confer immunity, combined with practices such as hand-washing by physicians between patient visits, to limit contact between people and pathogens.

All too often, though, obvious protective measures such as using only clean hypodermic needles or treating urban drinking water with chlorine are neglected, whether out of ignorance or a wrongheaded emphasis on the short-term financial costs. The worldwide disparity in wealth is also to blame: not surprisingly, the advances made during the second epidemiological transition were limited largely to the affluent of the industrial world.

Such lapses are now beginning to teach the bitter lesson that the delicate balance between humans and invasive microorganisms can tip the other way again. Overconfidence—the legacy of the second epidemiological transition—has made us especially vulnerable to emerging and reemerging diseases. Evolutionary principles can provide this useful corrective: in spite of all our medical and technological hubris, there is no quick fix. If human beings are to overcome the current crisis, it will be through sensible changes in behavior, such as increased condom use and improved sanitation, combined with a commitment to stop disturbing the ecological balance of the planet.

The Bible, in short, was not far from wrong: We do bring plagues upon ourselves—not by sinning, but by refusing to heed our own alarms, our own best judgment. The price of peace—or at least peaceful coexistence—with the microorganisms on this planet is eternal vigilance.

GEORGE J. ARMELAGOS is a professor of anthropology at Emory University in Atlanta, Georgia. He has coedited two books on the evolution of human disease: *Paleopathology at the Origins of Agriculture,* which deals with prehistoric populations, and *Disease in Populations in Transition,* which focuses on contemporary societies.

The Inuit Paradox

How can people who gorge on fat and rarely see a vegetable be healthier than we are?

PATRICIA GADSBY

Patricia Cochran, an Inupiat from Northwestern Alaska, is talking about the native foods of her childhood: "We pretty much had a subsistence way of life. Our food supply was right outside our front door. We did our hunting and foraging on the Seward Peninsula and along the Bering Sea."

"Our meat was seal and walrus, marine mammals that live in cold water and have lots of fat. We used seal oil for our cooking and as a dipping sauce for food. We had moose, caribou, and reindeer. We hunted ducks, geese, and little land birds like quail, called ptarmigan. We caught crab and lots of fish—salmon, whitefish, tomcod, pike, and char. Our fish were cooked, dried, smoked, or frozen. We ate frozen raw whitefish, sliced thin. The elders liked stinkfish, fish buried in seal bags or cans in the tundra and left to ferment. And fermented seal flipper, they liked that too."

Cochran's family also received shipments of whale meat from kin living farther north, near Barrow. Beluga was one she liked; raw muktuk, which is whale skin with its underlying blubber, she definitely did not. "To me it has a chew-on-a-tire consistency," she says, "but to many people it's a mainstay." In the short subarctic summers, the family searched for roots and greens and, best of all from a child's point of view, wild blueberries, crowberries, or salmonberries, which her aunts would mix with whipped fat to make a special treat called *akutuq*—in colloquial English, Eskimo ice cream.

Now Cochran directs the Alaska Native Science Commission, which promotes research on native cultures and the health and environmental issues that affect them. She sits at her keyboard in Anchorage, a bustling city offering fare from Taco Bell to French cuisine. But at home Cochran keeps a freezer filled with fish, seal, walrus, reindeer, and whale meat, sent by her family up north, and she and her husband fish and go berry picking—"sometimes a challenge in Anchorage," she adds, laughing. "I eat fifty-fifty," she explains, half traditional, half regular American.

No one, not even residents of the northernmost villages on Earth, eats an entirely traditional northern diet anymore. Even the groups we came to know as Eskimo—which include the Inupiat and the Yupiks of Alaska, the Canadian Inuit and Inuvialuit, Inuit Greenlanders, and the Siberian Yupiks—have probably seen more changes in their diet in a lifetime than their ancestors did over thousands of years. The closer people live to towns and the more access they have to stores and cash-paying jobs, the more likely they are to have westernized their eating. And with westernization, at least on the North American continent, comes processed foods and cheap carbohydrates—Crisco, Tang, soda, cookies, chips, pizza, fries. "The young and urbanized," says Harriet Kuhnlein, director of the Centre for Indigenous Peoples' Nutrition and Environment at McGill University in Montreal, "are increasingly into fast food." So much so that type 2 diabetes, obesity, and other diseases of Western civilization are becoming causes for concern there too.

Today, when diet books top the best-seller list and nobody seems sure of what to eat to stay healthy, it's surprising to learn how well the Eskimo did on a high-protein, high-fat diet. Shaped by glacial temperatures, stark landscapes, and protracted winters, the traditional Eskimo diet had little in the way of plant food, no agricultural or dairy products, and was unusually low in carbohydrates. Mostly people subsisted on what they hunted and fished. Inland dwellers took advantage of caribou feeding on tundra mosses, lichens, and plants too tough for humans to stomach (though predigested vegetation in the animals' paunches became dinner as well). Coastal people exploited the sea. The main nutritional challenge was avoiding starvation in late winter if primary meat sources became too scarce or lean.

These foods hardly make up the "balanced" diet most of us grew up with, and they look nothing like the mix of grains, fruits, vegetables, meat, eggs, and dairy we're accustomed to seeing in conventional food pyramid diagrams. How could such a diet possibly be adequate? How did people get along on little else but fat and animal protein?

'The diet of the Far North shows that there are no essential foods—only essential nutrients'

What the diet of the Far North illustrates, says Harold Draper, a biochemist and expert in Eskimo nutrition, is that there are no essential foods—only essential nutrients. And humans can get those nutrients from diverse and eye-opening sources.

One might, for instance, imagine gross vitamin deficiencies arising from a diet with scarcely any fruits and vegetables. What furnishes vitamin A, vital for eyes and bones? We derive much of ours from colorful plant foods, constructing it from pigmented plant precursors called carotenoids (as in carrots). But vitamin A, which is oil soluble, is also plentiful in the oils of cold-water fishes and sea mammals, as well as in the animals' livers, where fat is processed. These dietary staples also provide vitamin D, another oil-soluble vitamin needed for bones. Those of us living in temperate and tropical climates, on the other hand, usually make vitamin D indirectly by exposing skin to strong sun—hardly an option in the Arctic winter—and by consuming fortified cow's milk, to which the indigenous northern groups had little access until recent decades and often don't tolerate all that well.

As for vitamin C, the source in the Eskimo diet was long a mystery. Most animals can synthesize their own vitamin C, or ascorbic acid, in their livers, but humans are among the exceptions, along with other primates and oddballs like guinea pigs and bats. If we don't ingest enough of it, we fall apart from scurvy, a gruesome connective-tissue disease. In the United States today we can get ample supplies from orange juice, citrus fruits, and fresh vegetables. But vitamin C oxidizes with time; getting enough from a ship's provisions was tricky for early 18th- and 19th-century voyagers to the polar regions. Scurvy—joint pain, rotting gums, leaky blood vessels, physical and mental degeneration—plagued European and U.S. expeditions even in the 20th century. However, Arctic peoples living on fresh fish and meat were free of the disease.

Impressed, the explorer Vilhjalmur Stefansson adopted an Eskimo-style diet for five years during the two Arctic expeditions he led between 1908 and 1918. "The thing to do is to find your antiscorbutics where you are," he wrote. "Pick them up as you go." In 1928, to convince skeptics, he and a young colleague spent a year on an Americanized version of the diet under medical supervision at Bellevue Hospital in New York City. The pair ate steaks, chops, organ meats like brain and liver, poultry, fish, and fat with gusto. "If you have some fresh meat in your diet every day and don't overcook it," Stefansson declared triumphantly, "there will be enough C from that source alone to prevent scurvy."

In fact, all it takes to ward off scurvy is a daily dose of 10 milligrams, says Karen Fediuk, a consulting dietitian and former graduate student of Harriet Kuhnlein's who did her master's thesis on vitamin C. (That's far less than the U.S. recommended daily allowance of 75 to 90 milligrams—75 for women, 90 for men.) Native foods easily supply those 10 milligrams of scurvy prevention, especially when organ meats—preferably raw—are on the menu. For a study published with Kuhnlein in 2002, Fediuk compared the vitamin C content of 100-gram (3.55-ounce) samples of foods eaten by Inuit women living in the Canadian Arctic: Raw caribou liver supplied almost 24 milligrams, seal brain close to 15 milligrams, and raw kelp more than 28 milligrams. Still higher levels were found in whale skin and muktuk.

As you might guess from its antiscorbutic role, vitamin C is crucial for the synthesis of connective tissue, including the matrix of skin. "Wherever collagen's made, you can expect vitamin C," says Kuhnlein. Thick skinned, chewy, and collagen rich, raw muktuk can serve up an impressive 36 milligrams in a 100-gram piece, according to Fediuk's analyses. "Weight for weight, it's as good as orange juice," she says. Traditional Inuit practices like freezing meat and fish and frequently eating them raw, she notes, conserve vitamin C, which is easily cooked off and lost in food processing.

Hunter-gatherer diets like those eaten by these northern groups and other traditional diets based on nomadic herding or subsistence farming are among the older approaches to human eating. Some of these eating plans might seem strange to us—diets centered around milk, meat, and blood among the East African pastoralists, enthusiastic tuber eating by the Quechua living in the High Andes, the staple use of the mongongo nut in the southern African !Kung—but all proved resourceful adaptations to particular eco-niches. No people, though, may have been forced to push the nutritional envelope further than those living at Earth's frozen extremes. The unusual makeup of the far-northern diet led Loren Cordain, a professor of evolutionary nutrition at Colorado State University at Fort Collins, to make an intriguing observation.

Four years ago, Cordain reviewed the macronutrient content (protein, carbohydrates, fat) in the diets of 229 hunter-gatherer groups listed in a series of journal articles collectively known as the Ethnographic Atlas. These are some of the oldest surviving human diets. In general, hunter-gatherers tend to eat more animal protein than we do in our standard Western diet, with its reliance on agriculture and carbohydrates derived from grains and starchy plants. Lowest of all in carbohydrate, and highest in combined fat and protein, are the diets of peoples living in the Far North, where they make up for fewer plant foods with extra fish. What's equally striking, though, says Cordain, is that these meat-and-fish diets also exhibit a natural "protein ceiling." Protein accounts for no more than 35 to 40 percent of their total calories, which suggests to him that's all the protein humans can comfortably handle.

'Wild-animal fats are different from other fats. Farm animals typically have lots of highly saturated fat'

This ceiling, Cordain thinks, could be imposed by the way we process protein for energy. The simplest, fastest way to make energy is to convert carbohydrates into glucose, our body's primary fuel. But if the body is out of carbs, it can burn fat, or if necessary, break down protein. The name given to the convoluted business of making glucose from protein is gluconeogenesis. It takes place in the liver, uses a dizzying slew of enzymes, and creates nitrogen waste that has to be converted into urea and disposed of through the kidneys. On a truly traditional diet, says Draper, recalling his studies in the 1970s, Arctic people had plenty of protein but little carbohydrate, so they often relied on gluconeogenesis. Not only did they have bigger livers to handle the additional work but their urine volumes were also typically larger to get rid of the extra urea. Nonetheless, there appears to be a limit on how much protein the human liver can safely cope with: Too much overwhelms the liver's waste-disposal system, leading to protein poisoning—nausea, diarrhea, wasting, and death.

Whatever the metabolic reason for this syndrome, says John Speth, an archaeologist at the University of Michigan's Museum of Anthropology, plenty of evidence shows that hunters through the ages avoided protein excesses, discarding fat-depleted animals even when food was scarce. Early pioneers and trappers in North America encountered what looks like a similar affliction, sometimes referred to as rabbit starvation because rabbit meat is notoriously lean. Forced to subsist on fat-deficient meat, the men would gorge themselves, yet wither away. Protein can't be the sole source of energy for humans, concludes Cordain. Anyone eating a meaty diet that is low in carbohydrates must have fat as well.

Stefansson had arrived at this conclusion, too, while living among the Copper Eskimo. He recalled how he and his Eskimo companions had become quite ill after weeks of eating "caribou so skinny that there was no appreciable fat behind the eyes or in the marrow." Later he agreed to repeat the miserable experience at Bellevue Hospital, for science's sake, and for a while ate nothing but defatted meat. "The symptoms brought on at Bellevue by an incomplete meat diet [lean without fat] were exactly the same as in the Arctic … diarrhea and a feeling of general baffling discomfort," he wrote. He was restored with a fat fix but "had lost considerable weight." For the remainder of his year on meat, Stefansson tucked into his rations of chops and steaks with fat intact. "A normal meat diet is not a high-protein diet," he pronounced. "We were really getting three-quarters of our calories from fat." (Fat is more than twice as calorie dense as protein or carbohydrate, but even so, that's a lot of lard. A typical U.S diet provides about 35 percent of its calories from fat.)

Stefansson dropped 10 pounds on his meat-and-fat regimen and remarked on its "slenderizing" aspect, so perhaps it's no surprise he's been co-opted as a posthumous poster boy for Atkins-type diets. No discussion about diet these days can avoid Atkins. Even some researchers interviewed for this article couldn't resist referring to the Inuit way of eating as the "original Atkins." "Superficially, at a macronutrient level, the two diets certainly look similar," allows Samuel Klein, a nutrition researcher at Washington University in St. Louis, who's attempting to study how Atkins stacks up against conventional weight-loss diets. Like the Inuit diet, Atkins is low in carbohydrates and very high in fat. But numerous researchers, including Klein, point out that there are profound differences between the two diets, beginning with the type of meat and fat eaten.

Fats have been demonized in the United States, says Eric Dewailly, a professor of preventive medicine at Laval University in Quebec. But all fats are not created equal. This lies at the heart of a paradox—the Inuit paradox, if you

will. In the Nunavik villages in northern Quebec, adults over 40 get almost half their calories from native foods, says Dewailly, and they don't die of heart attacks at nearly the same rates as other Canadians or Americans. Their cardiac death rate is about half of ours, he says. As someone who looks for links between diet and cardiovascular health, he's intrigued by that reduced risk. Because the traditional Inuit diet is "so restricted," he says, it's easier to study than the famously heart-healthy Mediterranean diet, with its cornucopia of vegetables, fruits, grains, herbs, spices, olive oil, and red wine.

A key difference in the typical Nunavik Inuit's diet is that more than 50 percent of the calories in Inuit native foods come from fats. Much more important, the fats come from wild animals.

Wild-animal fats are different from both farm-animal fats and processed fats, says Dewailly. Farm animals, cooped up and stuffed with agricultural grains (carbohydrates) typically have lots of solid, highly saturated fat. Much of our processed food is also riddled with solid fats, or so-called trans fats, such as the re-engineered vegetable oils and shortenings cached in baked goods and snacks. "A lot of the packaged food on supermarket shelves contains them. So do commercial french fries," Dewailly adds.

Trans fats are polyunsaturated vegetable oils tricked up to make them more solid at room temperature. Manufacturers do this by hydrogenating the oils—adding extra hydrogen atoms to their molecular structures—which "twists" their shapes. Dewailly makes twisting sound less like a chemical transformation than a perversion, an act of public-health sabotage: "These man-made fats are dangerous, even worse for the heart than saturated fats." They not only lower high-density lipoprotein cholesterol (HDL, the "good" cholesterol) but they also raise low-density lipoprotein cholesterol (LDL, the "bad" cholesterol) and triglycerides, he says. In the process, trans fats set the stage for heart attacks because they lead to the increase of fatty buildup in artery walls.

Wild animals that range freely and eat what nature intended, says Dewailly, have fat that is far more healthful. Less of their fat is saturated, and more of it is in the monounsaturated form (like olive oil). What's more, cold-water fishes and sea mammals are particularly rich in polyunsaturated fats called n-3 fatty acids or omega-3 fatty acids. These fats appear to benefit the heart and vascular system. But the polyunsaturated fats in most Americans' diets are the omega-6 fatty acids supplied by vegetable oils. By contrast, whale blubber consists of 70 percent monounsaturated fat and close to 30 percent omega-3s, says Dewailly.

'Dieting is the price we pay for too little exercise and too much mass-produced food'

Omega-3s evidently help raise HDL cholesterol, lower triglycerides, and are known for anticlotting effects. (Ethnographers have remarked on an Eskimo propensity for nosebleeds.) These fatty acids are believed to protect the heart from life-threatening arrhythmias that can lead to sudden cardiac death. And like a "natural aspirin," adds Dewailly, omega-3 polyunsaturated fats help put a damper on runaway inflammatory processes, which play a part in atherosclerosis, arthritis, diabetes, and other so-called diseases of civilization.

You can be sure, however, that Atkins devotees aren't routinely eating seal and whale blubber. Besides the acquired taste problem, their commerce is extremely restricted in the United States by the Marine Mammal Protection Act, says Bruce Holub, a nutritional biochemist in the department of human biology and nutritional sciences at the University of Guelph in Ontario.

"In heartland America it's probable they're not eating in an Eskimo-like way," says Gary Foster, clinical director of the Weight and Eating Disorders Program at the Pennsylvania School of Medicine. Foster, who describes himself as open-minded about Atkins, says he'd nonetheless worry if people saw the diet as a green light to eat all the butter and bacon—saturated fats—they want. Just before rumors surfaced that Robert Atkins had heart and weight problems when he died, Atkins officials themselves were stressing saturated fat should account for no more than 20 percent of dieters' calories. This seems to be a clear retreat from the diet's original don't-count-the-calories approach to bacon and butter and its happy exhortations to "plow into those prime ribs." Furthermore, 20 percent of calories from saturated fats is *double* what most nutritionists advise. Before plowing into those prime ribs, readers of a recent edition of the *Dr. Atkins' New Diet Revolution* are urged to take omega-3 pills to help protect their hearts. "If you watch carefully," says Holub wryly, "you'll see many popular U.S. diets have quietly added omega-3 pills, in the form of fish oil or flaxseed capsules, as supplements."

Needless to say, the subsistence diets of the Far North are not "dieting." Dieting is the price we pay for too little exercise and too much mass-produced food. Northern diets were a way of life in places too cold for agriculture, where food, whether hunted, fished, or foraged, could not be taken for granted. They were about keeping weight on.

This is not to say that people in the Far North were fat: Subsistence living requires exercise—hard physical work. Indeed, among the good reasons for native people to maintain their old way of eating, as far as it's possible today, is that it provides a hedge against obesity, type 2 diabetes, and heart disease. Unfortunately, no place on Earth is immune to the spreading taint of growth and development. The very well-being of the northern food chain is coming under threat from global warming, land development, and industrial pollutants in the marine environment. "I'm a pragmatist," says Cochran, whose organization is involved in pollution monitoring and disseminating food-safety information to native villages. "Global warming we don't have control over. But we can, for example, do cleanups of military sites in Alaska or of communication cables leaching lead into fish-spawning areas. We can help communities make informed food choices. A young woman of childbearing age may choose not to eat certain organ meats that concentrate contaminants. As individuals, we do have options. And eating our salmon and our seal is still a heck of a better option than pulling something processed that's full of additives off a store shelf."

Not often in our industrial society do we hear someone speak so familiarly about "our" food animals. We don't talk of "our pig" and "our beef." We've lost that creature feeling, that sense of kinship with food sources. "You're taught to think in boxes," says Cochran. "In our culture the connectivity between humans, animals, plants, the land they live on, and the air they share is ingrained in us from birth.

"You truthfully can't separate the way we get our food from the way we live," she says. "How we get our food is intrinsic to our culture. It's how we pass on our values and knowledge to the young. When you go out with your aunts and uncles to hunt or to gather, you learn to smell the air, watch the wind, understand the way the ice moves, know the land. You get to know where to pick which plant and what animal to take."

"It's part, too, of your development as a person. You share food with your community. You show respect to your elders by offering them the first catch. You give thanks to the animal that gave up its life for your sustenance. So you get all the physical activity of harvesting your own food, all the social activity of sharing and preparing it, and all the spiritual aspects as well," says Cochran. "You certainly don't get all that, do you, when you buy prepackaged food from a store."

"That's why some of us here in Anchorage are working to protect what's ours, so that others can continue to live back home in the villages," she adds. "Because if we don't take care of our food, it won't be there for us in the future. And if we lose our foods, we lose who we are." The word Inupiat means "the real people." "That's who we are," says Cochran.

Dr. Darwin

With a nod to evolution's god, physicians are looking at illness through the lens of natural selection to find out why we get sick and what we can do about it.

Lori Oliwenstein

Paul Ewald knew from the beginning that the Ebola virus outbreak in Zaire would fizzle out. On May 26, after eight days in which only six new cases were reported, that fizzle became official. The World Health Organization announced it would no longer need to update the Ebola figures daily (though sporadic cases continued to be reported until June 20).

The virus had held Zaire's Bandundu Province in its deadly grip for weeks, infecting some 300 people and killing 80 percent of them. Most of those infected hailed from the town of Kikwit. It was all just as Ewald predicted. "When the Ebola outbreak occurred," he recalls, "I said, as I have before, these things are going to pop up, they're going to smolder, you'll have a bad outbreak of maybe 100 or 200 people in a hospital, maybe you'll have the outbreak slip into another isolated community, but then it will peter out on its own."

> **"If you look at it from an evolutionary point of view, you can sort out the 95 percent of disease organisms that aren't a major threat from the 5 percent that are."**

Ewald is no soothsayer. He's an evolutionary biologist at Amherst College in Massachusetts and perhaps the world's leading expert on how infectious diseases—and the organisms that cause them—evolve. He's also a force behind what some are touting as the next great medical revolution: the application of Darwin's theory of natural selection to the understanding of human diseases.

A Darwinian view can shed some light on how Ebola moves from human to human once it has entered the population. (Between human outbreaks, the virus resides in some as yet unknown living reservoir.) A pathogen can survive in a population, explains Ewald, only if it can easily transmit its progeny from one host to another. One way to do this is to take a long time to disable a host, giving him plenty of time to come into contact with other potential victims. Ebola, however, kills quickly, usually in less than a week. Another way is to survive

for a long time outside the human body, so that the pathogen can wait for new hosts to find it. But the Ebola strains encountered thus far are destroyed almost at once by sunlight, and even if no rays reach them, they tend to lose their infectiousness outside the human body within a day. "If you look at it from an evolutionary point of view, you can sort out the 95 percent of disease organisms that aren't a major threat from the 5 percent that are," says Ewald. "Ebola really isn't one of those 5 percent."

The earliest suggestion of a Darwinian approach to medicine came in 1980, when George Williams, an evolutionary biologist at the State University of New York at Stony Brook, read an article in which Ewald discussed using Darwinian theory to illuminate the origins of certain symptoms of infectious disease—things like fever, low iron counts, diarrhea. Ewald's approach struck a chord in Williams. Twenty-three years earlier he had written a paper proposing an evolutionary framework for senescence, or aging. "Way back in the 1950s I didn't worry about the practical aspects of senescence, the medical aspects," Williams notes. "I was pretty young then." Now, however, he sat up and took notice.

While Williams was discovering Ewald's work, Randolph Nesse was discovering Williams's. Nesse, a psychiatrist and a founder of the University of Michigan Evolution and Human Behavior Program, was exploring his own interest in the aging process, and he and Williams soon got together. "He had wanted to find a physician to work with on medical problems," says Nesse, "and I had long wanted to find an evolutionary biologist, so it was a very natural match for us." Their collaboration led to a 1991 article that most researchers say signaled the real birth of the field.

Nesse and Williams define Darwinian medicine as the hunt for evolutionary explanations of vulnerabilities to disease. It can, as Ewald noted, be a way to interpret the body's defenses, to try to figure out, say, the reasons we feel pain or get runny noses when we have a cold, and to determine what we should—or shouldn't—be doing about those defenses. For instance, Darwinian researchers like physiologist Matthew Kluger of the Lovelace Institute in Albuquerque now say that a moderate rise in body temperature is more than just a symptom of disease; it's an evolutionary adaptation the body uses to fight infection by making itself inhospitable to invading microbes. It would seem, then, that if you lower the fever, you may prolong the infection. Yet no one is ready to say whether we should toss out our aspirin bottles. "I would love to see a dozen proper studies of whether it's wise to bring fever down when someone has influenza," says Nesse. "It's never been done, and it's just astounding that it's never been done."

Diarrhea is another common symptom of disease, one that's sometimes the result of a pathogen's manipulating your body for its own good purposes, but it may also be a defense mechanism mounted by your body. Cholera bacteria, for example, once they invade the human body, induce diarrhea by producing toxins that make the intestine's cells leaky. The resultant diarrhea then both flushes competing beneficial bacteria from the gut and gives the cholera bacteria a ride into the world, so that they can find another hapless victim. In the case of cholera, then, it seems clear that stopping the diarrhea can only do good.

But the diarrhea that results from an invasion of shigella bacteria—which cause various forms of dysentery—seems to be more an intestinal defense than a bacterial offense. The infection causes the muscles surrounding the gut to contract more frequently, apparently in an attempt to flush out the bacteria as quickly as possible. Studies done more than a decade ago showed that using drugs like Lomotil to decrease the gut's contractions and cut down the diarrheal output actually prolong infection. On the other hand, the ingredients in over-the-counter preparations like Pepto Bismol, which don't affect how frequently the gut contracts, can be used to stem the diarrheal flow without prolonging infection.

Seattle biologist Margie Profet points to menstruation as another "symptom" that may be more properly viewed as an evolutionary defense. As Profet points out, there must be a good reason for the body to engage in such costly activities as shedding the uterine lining and letting blood flow away. That reason, she claims, is to rid the uterus of any organisms that might arrive with sperm in the seminal fluid. If an egg is fertilized, infection may be worth risking. But if there is no fertilized egg, says Profet, the body defends itself by ejecting the uterine cells, which might have been infected. Similarly, Profet has theorized that morning sickness during pregnancy causes the mother to avoid foods that might contain chemicals harmful to a developing fetus. If she's right, blocking that nausea with drugs could result in higher miscarriage rates or more birth defects.

D arwinian medicine isn't simply about which symptoms to treat and which to ignore. It's a way to understand microbes—which, because they evolve so much more quickly than we do, will probably always beat us unless we figure out how to harness their evolutionary power for our own benefit. It's also a way to realize how disease-causing genes that persist in the population are often selected for, not against, in the long run.

Sickle-cell anemia is a classic case of how evolution tallies costs and benefits. Some years ago, researchers discovered that people with one copy of the sickle-cell gene are better able to resist the protozoans that cause malaria than are people with no copies of the gene. People with two copies of the gene may die, but in malaria-plagued regions such as tropical Africa, their numbers will be more than made up for by the offspring left by the disease-resistant kin.

Cystic fibrosis may also persist through such genetic logic. Animal studies indicate that individuals with just one copy of the cystic fibrosis gene may be more resistant to the effects of the cholera bacterium. As is the case with malaria and sickle-cell, cholera is much more prevalent than cystic fibrosis; since there are many more people with a single, resistance-conferring copy of the gene than with a disease-causing double dose, the gene is stably passed from generation to generation.

"I used to hunt saber-toothed tigers all the time, thousands of years ago. I got lots of exercise and all that sort of stuff. Now I sit in front of a computer and don't get exercise, so I've changed my body chemistry."

"With our power to do gene manipulations, there will be temptations to find genes that do things like cause aging, and get rid of them," says Nesse. "If we're sure about everything a gene does, that's fine. But an evolutionary approach cautions us not to go too fast, and to expect that every gene might well have some benefit as well as costs, and maybe some quite unrelated benefit."

Darwinian medicine can also help us understand the problems encountered in the New Age by a body designed for the Stone Age. As evolutionary psychologist Charles Crawford of Simon Fraser University in Burnaby, British Columbia, put it: "I used to hunt saber-toothed tigers all the time, thousands of years ago. I got lots of exercise and all that sort of stuff. Now I sit in front of a computer, and all I do is play with a mouse, and I don't get exercise. So I've changed my body biochemistry in all sorts of unknown ways, and it could affect me in all sorts of ways, and we have no idea what they are."

Radiologist Boyd Eaton of Emory University and his colleagues believe such biochemical changes are behind today's breast cancer epidemic. While it's impossible to study a Stone Ager's biochemistry, there are still groups of hunter-gatherers around—such as the San of Africa—who make admirable stand-ins. A foraging life-style, notes Eaton, also means a lifestyle in which menstruation begins later, the first child is born earlier, there are more children altogether, they are breast-fed for years rather than months, and menopause comes somewhat earlier. Overall, he says, American women today probably experience 3.5 times more menstrual cycles than our ancestors did 10,000 years ago. During each cycle a woman's body is flooded with the hormone estrogen, and breast cancer, as research has found, is very much estrogen related. The more frequently the breasts are exposed to the hormone, the greater the chance that a tumor will take seed.

Depending on which data you choose, women today are somewhere between 10 and 100 times more likely to be stricken with breast cancer than our ancestors were. Eaton's proposed solutions are pretty radical, but he hopes people will at least entertain them; they include delaying puberty with hormones and using hormones to create pseudopregnancies, which offer a woman the biochemical advantages of pregnancy at an early age without requiring her to bear a child.

In general, Darwinian medicine tells us that the organs and systems that make up our bodies result not from the pursuit of perfection but from millions of years of evolutionary compromises designed to get the greatest reproductive benefit at the lowest cost. We walk upright with a spine that evolved while we scampered on four limbs; balancing on two legs leaves our hands free, but we'll probably always suffer some back pain as well.

"What's really different is that up to now people have used evolutionary theory to try to explain why things work, why they're normal," explains Nesse. "The twist—and I don't know if it's simple or profound—is to say we're trying to understand the abnormal, the vulnerability to disease. We're trying to understand why natural selection has not made the body better, why natural selection has left the body with vulnerabilities. For every single disease, there is an answer to that question. And for very few of them is the answer very clear yet."

One reason those answers aren't yet clear is that few physicians or medical researchers have done much serious surveying from Darwin's viewpoint. In many cases, that's because evolutionary theories are hard to test. There's no way to watch human evolution in progress—at best it works on a time scale involving hundreds of thousands of years. "Darwinian medicine is mostly a guessing game about how we think evolution worked in the past on humans, what it designed for us," say evolutionary biologist James Bull of the University of Texas at Austin. "It's almost impossible to test ideas that we evolved to respond to this or that kind of environment. You can make educated guesses, but no one's going to go out and do an experiment to show that yes, in fact humans will evolve this way under these environmental conditions."

Yet some say that these experiments can, should, and will be done. Howard Howland, a sensory physiologist at Cornell, is setting up just such an evolutionary experiment, hoping to interfere with the myopia, or nearsightedness, that afflicts a full quarter of all Americans. Myopia is thought to be the result of a delicate feedback loop that tries to keep images focused on the eye's retina. There's not much room for error: if the length of your eyeball is off by just a tenth of a millimeter, your vision will be blurry. Research has shown that when the eye perceives an image as fuzzy, it compensates by altering its length.

This loop obviously has a genetic component, notes Howland, but what drives it is the environment. During the Stone Age, when we were chasing buffalo in the field, the images we saw were usually sharp and clear. But with modern civilization came a lot of close work. When your eye focuses on something nearby, the lens has to bend, and since bending that lens is hard

work, you do as little bending as you can get away with. That's why, whether you're conscious of it or not, near objects tend to be a bit blurry. "Blurry image?" says the eye. "Time to grow." And the more it grows, the fuzzier those buffalo get. Myopia seems to be a disease of industrial society.

To prevent that disease, Howland suggests going back to the Stone Age—or at least convincing people's eyes that that's where they are. If you give folks with normal vision glasses that make their eyes think they're looking at an object in the distance when they're really looking at one nearby, he says, you'll avoid the whole feedback loop in the first place. "The military academies induct young men and women with twenty-twenty vision who then go through four years of college and are trained to fly an airplane or do some difficult visual task. But because they do so much reading, they come out the other end nearsighted, no longer eligible to do what they were hired to do," Howland notes. "I think these folks would very much like not to become nearsighted in the course of their studies." He hopes to be putting glasses on them within a year.

The numbing pace of evolution is a much smaller problem for researchers interested in how the bugs that plague us do their dirty work. Bacteria are present in such large numbers (one person can carry around more pathogens than there are people on the planet) and evolve so quickly (a single bacterium can reproduce a million times in one human lifetime) that experiments we couldn't imagine in humans can be carried out in microbes in mere weeks. We might even, says Ewald, be able to use evolutionary theory to tame the human immunodeficiency virus.

"HIV is mutating so quickly that surely we're going to have plenty of sources of mutants that are mild as well as severe," he notes. "So now the question is, which of the variants will win?" As in the case of Ebola, he says, it will all come down to how well the virus manages to get from one person to another.

"If there's a great potential for sexual transmission to new partners, then the viruses that reproduce quickly will spread," Ewald says. "And since they're reproducing in a cell type that's critical for the well-being of the host—the helper T cell—then that cell type will be decimated, and the host is likely to suffer from it." On the other hand, if you lower the rate of transmission—through abstinence, monogamy, condom use—then the more severe strains might well die out before they have a chance to be passed very far. "The real question," says Ewald, "is, exactly how mild can you make this virus as a result of reducing the rate at which it could be transmitted to new partners, and how long will it take for this change to occur?" There are already strains of HIV in Senegal with such low virulence, he points out, that most people infected will die of old age. "We don't have all the answers. But I think we're going to be living with this virus for a long time, and if we have to live with it, let's live with a really mild virus instead of a severe virus."

Though condoms and monogamy are not a particularly radical treatment, that they might be used not only to stave off the virus but to tame it is a radical notion—and one that some researchers find suspect. "If it becomes too virulent, it will end up cutting off its own transmission by killing its host too quickly," notes James Bull. "But the speculation is that people transmit HIV primarily within one to five months of infection, when they spike a high level of virus in the blood. So with HIV, the main period of transmission occurs a few months into the infection, and yet the virulence—the death from it—occurs years later. The major stage of transmission is decoupled from the virulence." So unless the protective measures are carried out by everyone, all the time, we won't stop most instances of transmission; after all, most people don't even know they're infected when they pass the virus on.

But Ewald thinks these protective measures are worth a shot. After all, he says, pathogen taming has occurred in the past. The forms of dysentery we encounter in the United States are quite mild because our purified water supplies have cut off the main route of transmission for virulent strains of the bacteria. Not only did hygienic changes reduce the number of cases, they selected for the milder shigella organisms, those that leave their victim well enough to get out and about. Diphtheria is another case in point. When the diphtheria vaccine was invented, it targeted only the most severe form of diphtheria toxin, though for economic rather than evolutionary reasons. Over the years, however, that choice has weeded out the most virulent strains of diphtheria, selecting for the ones that cause few or no symptoms. Today those weaker strains act like another level of vaccine to protect us against new, virulent strains.

"We did with diphtheria what we did with wolves. We took an organism that caused harm, and unknowingly, we domesticated it into an organism that protects us."

"You're doing to these organisms what we did to wolves," says Ewald. "Wolves were dangerous to us, we domesticated them into dogs, and then they helped us, they warned us against the wolves that were out there ready to take our babies. And by doing that, we've essentially turned what was a harmful organism into a helpful organism. That's the same thing we did with diphtheria; we took an organism that was causing harm, and without knowing it, we domesticated it into an organism that is protecting us against harmful ones."

Putting together a new scientific discipline—and getting it recognized—is in itself an evolutionary process. Though Williams and Neese say there are hundreds of researchers working (whether they know it or not) within this newly built framework, they realize the field is still in its infancy. It may take some time before *Darwinian medicine* is a household term. Nesse tells how the editor of a prominent medical journal, when asked about the field, replied, "Darwinian medicine? I haven't heard of it, so it can't be very important."

But Darwinian medicine's critics don't deny the field's legitimacy; they point mostly to its lack of hard-and-fast answers, its lack of clear clinical guidelines. "I think this idea will eventually establish itself as a basic science for medicine," answers Nesse. "What did people say, for instance, to the biochemists back in 1900 as they were playing out the Krebs cycle? People would say, 'So what does biochemistry really have to do with medicine? What can you cure now that you couldn't before you knew about the Krebs cycle?' And the biochemists could only say, 'Well, gee, we're not sure, but we know what we're doing is answering important scientific questions, and eventually this will be useful.' And I think exactly the same applies here."

LORI OLIWENSTEIN, a former *Discover* senior editor, is now a freelance journalist based in Los Angeles.

Curse and Blessing of the Ghetto

Tay-Sachs disease is a choosy killer, one that for centuries targeted Eastern European Jews above all others. By decoding its lethal logic, we can learn a lot about how genetic diseases evolve— and how they can be conquered.

JARED DIAMOND

Marie and I hated her at first sight, even though she was trying hard to be helpful. As our obstetrician's genetics counselor, she was just doing her job, explaining to us the unpleasant results that might come out of the genetic tests we were about to have performed. As a scientist, though, I already knew all I wanted to know about Tay-Sachs disease, and I didn't need to be reminded that the baby sentenced to death by it could be my own.

Fortunately, the tests would reveal that my wife and I were not carriers of the Tay-Sachs gene, and our preparenthood fears on that matter at least could be put to rest. But at the time I didn't yet know that. As I glared angrily at that poor genetics counselor, so strong was my anxiety that now, four years later, I can still clearly remember what was going through my mind: If I were an evil deity, I thought, trying to devise exquisite tortures for babies and their parents, I would be proud to have designed Tay-Sachs disease.

Tay-Sachs is completely incurable, unpreventable, and preprogrammed in the genes. A Tay-Sachs infant usually appears normal for the first few months after birth, just long enough for the parents to grow to love him. An exaggerated "startle reaction" to sounds is the first ominous sign. At about six months the baby starts to lose control of his head and can't roll over or sit without support. Later he begins to drool, breaks out into unmotivated bouts of laughter, and suffers convulsions. Then his head grows abnormally large, and he becomes blind. Perhaps what's most frightening for the parents is that their baby loses all contact with his environment and becomes virtually a vegetable. By the child's third birthday, if he's still alive, his skin will turn yellow and his hands pudgy. Most likely he will die before he's four years old.

My wife and I were tested for the Tay-Sachs gene because at the time we rated as high-risk candidates, for two reasons. First, Marie was carrying twins, so we had double the usual chance to bear a Tay-Sachs baby. Second, both she and I are of Eastern European Jewish ancestry, the population with by far the world's highest Tay-Sachs frequency.

In peoples around the world Tay-Sachs appears once in every 400,000 births. But it appears a hundred times more frequently—about once in 3,600 births—among descendants of Eastern European Jews, people known as Ashkenazim. For descendants of most other groups of Jews—Oriental Jews, chiefly from the Middle East, or Sephardic Jews, from Spain and other Mediterranean countries—the frequency of Tay-Sachs disease is no higher than in non-Jews. Faced with such a clear correlation, one cannot help but wonder: What is it about this one group of people that produces such an extraordinarily high risk of this disease?

Finding the answer to this question concerns all of us, regardless of our ancestry. Every human population is especially susceptible to certain diseases, not only because of its life-style but also because of its genetic inheritance. For example, genes put European whites at high risk for cystic fibrosis, African blacks for sickle-cell disease, Pacific Islanders for diabetes—and Eastern European Jews for ten different diseases, including Tay-Sachs. It's not that Jews are notably susceptible to genetic diseases in general; but a combination of historical factors has led to Jews' being intensively studied, and so their susceptibilities are far better known than those of, say, Pacific Islanders.

Tay-Sachs exemplifies how we can deal with such diseases; it has been the object of the most successful screening program to date. Moreover, Tay-Sachs is helping us understand how ethnic diseases evolve. Within the past couple of years discoveries by molecular biologists have provided tantalizing clues to precisely how a deadly gene can persist and spread over the centuries. Tay-Sachs may be primarily a disease of Eastern European Jews, but through this affliction of one group of people, we gain a window on how our genes simultaneously curse and bless us all.

The disease's hyphenated name comes from the two physicians—British ophthalmologist W. Tay and New York neurologist B. Sachs—who independently first recognized the disease, in 1881 and 1887, respectively. By 1896 Sachs had seen enough cases to realize that the disease was most common among Jewish children.

Not until 1962, however, were researchers able to trace the cause of the affliction to a single biochemical abnormality: the excessive accumulation in nerve cells of a fatty substance called G_{M2} ganglioside. Normally G_{M2} ganglioside is present at only modest levels in cell membranes, because it is constantly being broken down as well as synthesized. The breakdown depends on the enzyme hexosaminidase A, which is found in the tiny structures within our cells known as lysosomes. In the unfortunate Tay-Sachs victims this enzyme is lacking, and without it the ganglioside piles up and produces all the symptoms of the disease.

We have two copies of the gene that programs our supply of hexosaminidase A, one inherited from our father, the other from our mother; each of our parents, in turn, has two copies derived from their own parents. As long as we have one good copy of the gene, we can produce enough hexosaminidase A to prevent a buildup of G_{M2} ganglioside and we won't get Tay-Sachs. This genetic disease is of the sort termed recessive rather than dominant—meaning that to get it, a child must inherit a defective gene not just from one parent but from both of them. Clearly, each parent must have had one good copy of the gene along with the defective copy—if either had had two defective genes, he or she would have died of the disease long before reaching the age of reproduction. In genetic terms the diseased child is homozygous for the defective gene and both parents are heterozygous for it.

None of this yet gives any hint as to why the Tay-Sachs gene should be most common among Eastern European Jews. To come to grips with that question, we must take a short detour into history.

From their biblical home of ancient Israel, Jews spread peacefully to other Mediterranean lands, Yemen, and India. They were also dispersed violently through conquest by Assyrians, Babylonians, and Romans. Under the Carolingian kings of the eighth and ninth centuries Jews were invited to settle in France and Germany as traders and financiers. In subsequent centuries, however, persecutions triggered by the Crusades gradually drove Jews out of Western Europe; the process culminated in their total expulsion from Spain in 1492. Those Spanish Jews—called Sephardim—fled to other lands around the Mediterranean. Jews of France and Germany—the Ashkenazim—fled east to Poland and from there to Lithuania and western Russia, where they settled mostly in towns, as businessmen engaged in whatever pursuit they were allowed.

There the Jews stayed for centuries, through periods of both tolerance and oppression. But toward the end of the nineteenth century and the beginning of the twentieth, waves of murderous anti-Semitic attacks drove millions of Jews out of Eastern Europe, with most of them heading for the United States. My mother's parents, for example, fled to New York from Lithuanian pogroms of the 1880s, while my father's parents fled from the Ukrainian pogroms of 1903–6. The more modern history of Jewish migration is probably well known to you all: most Jews who remained in Eastern Europe were exterminated during World War II, while most the survivors immigrated to the United States and Israel. Of the 13 million Jews alive today, more than three-quarters are Ashkenazim, the descendants of the Eastern European Jews and the people most at risk for Tay-Sachs.

Have these Jews maintained their genetic distinctness through the thousands of years of wandering? Some scholars claim that there has been so much intermarriage and conversion that Ashkenazic Jews are now just Eastern Europeans who adopted Jewish culture. However, modern genetic studies refute that speculation.

First of all, there are those ten genetic diseases that the Ashkenazim have somehow acquired, by which they differ both from other Jews and from Eastern European non-Jews. In addition, many Ashkenazic genes turn out to be ones typical of Palestinian Arabs and other peoples of the Eastern Mediterranean areas where Jews originated. (In fact, by genetic standards the current Arab-Israeli conflict is an internecine civil war.) Other Ashkenazic genes have indeed diverged from Mediterranean ones (including genes of Sephardic and Oriental Jews) and have evolved to converge on genes of Eastern European non-Jews subject to the same local forces of natural selection. But the degree to which Ashkenazim prove to differ genetically from Eastern European non-Jews implies an intermarriage rate of only about 15 percent.

Can history help explain why the Tay-Sachs gene in particular is so much more common in Ashkenazim than in their non-Jewish neighbors or in other Jews? At the risk of spoiling a mystery, I'll tell you now that the answer is yes, but to appreciate it, you'll have to understand the four possible explanations for the persistence of the Tay-Sachs gene.

First, new copies of the gene might be arising by mutation as fast as existing copies disappear with the death of Tay-Sachs children. That's the most likely explanation for the gene's persistence in most of the world, where the disease frequency is only one in 400,000 births—that frequency reflects a typical human mutation rate. But for this explanation to apply to the Ashkenazim would require a mutation rate of at least one per 3,600 births—far above the frequency observed for any human gene. Furthermore, there would be no precedent for one particular gene mutating so much more often in one human population than in others.

As a second possibility, the Ashkenazim might have acquired the Tay-Sachs gene from some other people who already had the gene at high frequency. Arthur Koestler's controversial book *The Thirteenth Tribe*, for example, popularized the view that the Ashkenazim are really not a Semitic people but are instead descended from the Khazar, a Turkic tribe whose rulers converted to Judaism in the eighth century. Could the Khazar have brought the Tay-Sachs gene to Eastern Europe? This speculation makes good romantic reading, but there is no good evidence to support it. Moreover, it fails to explain why deaths of Tay-Sachs children didn't eliminate the gene by natural selection in the past 1,200 years, nor how the Khazar acquired high frequencies of the gene in the first place.

The third hypothesis was the one preferred by a good many geneticists until recently. It invokes two genetic processes, termed the founder effect and genetic drift, that may operate in small populations. To understand these concepts, imagine that 100 couples settle in a new land and found a population that then increases. Imagine further that one parent among those original 100 couples happens to have some rare gene, one, say, that normally occurs at a frequency of one in a million. The gene's frequency in the new population will now be one in 200 as a result of the accidental presence of that rare founder.

Or suppose again that 100 couples found a population, but that one of the 100 men happens to have lots of kids by his wife or that he is exceptionally popular with other women, while the other 99 men are childless or have few kids or are simply less popular. That one man may thereby father 10 percent rather than a more representative one percent of the next generation's babies, and their genes will disproportionately reflect that man's genes. In other words, gene frequencies will have drifted between the first and second generation.

Through these two types of genetic accidents a rare gene may occur with an unusually high frequency in a small expanding population. Eventually, if the gene is harmful, natural selection will bring its frequency back to normal by killing off gene bearers. But if the resultant disease is recessive—if heterozygous individuals don't get the disease and only the rare, homozygous individuals die of it—the gene's high frequency may persist for many generations.

These accidents do in fact account for the astonishingly high Tay-Sachs gene frequency found in one group of Pennsylvania Dutch: out of the 333 people in this group, 98 proved to carry the Tay-Sachs gene. Those 333 are all descended from one couple who settled in the United States in the eighteenth century and had 13 children. Clearly, one of that founding couple must have carried the gene. A similar accident may explain why Tay-Sachs is also relatively common among French Canadians, who number 5 million today but are descended from fewer than 6,000 French immigrants who arrived in the New World between 1638 and 1759. In the two or three centuries since both these founding events, the high Tay-Sachs gene frequency among Pennsylvania Dutch and French Canadians has not yet had enough time to decline to normal levels.

The same mechanisms were one proposed to explain the high rate of Tay-Sachs disease among the Ashkenazim. Perhaps, the reasoning went, the gene just happened to be overrepresented in the founding Jewish population that settled in Germany or Eastern Europe. Perhaps the gene just happened to drift up in frequency in the Jewish populations scattered among the isolated towns of Eastern Europe.

It seems unlikely that genetic accidents would have pumped up the frequency of the same gene not once but twice in the same population.

But geneticists have long questioned whether the Ashkenazim population's history was really suitable for these genetic accidents to have been significant. Remember, the founder effect and genetic drift become significant only in small populations, and the founding populations of Ashkenazim may have been quite large. Moreover, Ashkenazic communities were considerably widespread; drift would have sent gene frequencies up in some towns but down in others. And, finally, natural selection has by now had a thousand years to restore gene frequencies to normal.

Granted, those doubts are based on historical data, which are not always as precise or reliable as one might want. But within the past several years the case against those accidental explanations for Tay-Sachs disease in the Ashkenazim has been bolstered by discoveries by molecular biologists.

Like all proteins, the enzyme absent in Tay-Sachs children is coded for by a piece of our DNA. Along that particular stretch of DNA there are thousands of different sites where a mutation could occur that would result in no enzyme and hence in the same set of symptoms. If molecular biologists had discovered that all cases of Tay-Sachs in Ashkenazim involved damage to DNA at the same site, that would have been strong evidence that in Ashkenazim the disease stems from a single mutation that has been multiplied by the founder effect or genetic drift—in other words, the high incidence of Tay-Sachs among Eastern European Jews is accidental.

In reality, though, several different mutations along this stretch of DNA have been identified in Ashkenazim, and two of them occur much more frequently than in non-Ashkenazim populations. It seems unlikely that genetic accidents would have pumped up the frequency of the same gene not once but twice in the same population.

And that's not the sole unlikely coincidence arguing against accidental explanations. Recall that Tay-Sachs is caused by the excessive accumulation of one fatty substance, G_{M2} ganglioside, from a defect in one enzyme, hexosaminidase A. But Tay-Sachs is one of ten genetic diseases characteristic of Ashkenazim. Among those other nine, two—Gaucher's disease and Niemann-Pick disease—result from the accumulation of two other fatty substances similar to G_{M2} ganglioside, as a result of defects in two other enzymes similar to hexosaminidase A. Yet our bodies contain thousands of different enzymes. It would have been an incredible roll of the genetic dice if, by nothing more than chance, Ashkenazim had independently acquired mutations in three closely related enzymes—and had acquired mutations in one of those enzymes twice.

All these facts bring us to the fourth possible explanation of why the Tay-Sachs gene is so prevalent among Ashkenazim: namely, that something about them favored accumulation of G_{M2} ganglioside and related fats.

For comparison, suppose that a friend doubles her money on one stock while you are getting wiped out with your investments. Taken alone, that could just mean she was lucky on that one occasion. But suppose that she doubles her money on each of two different stocks and at the same time rings up big profits in real estate while also making a killing in bonds. That implies more than lady luck; it suggests that something about your friend—like shrewd judgment—favors financial success.

What could be the blessings of fat accumulation in Eastern European Jews? At first this question sounds weird. After all, that fat accumulation was noticed only because of the curses it bestows: Tay-Sachs, Gaucher's, or Niemann-Pick disease. But many of our common genetic diseases may persist because they bring both blessings and curses (see "The Cruel Logic of Our

Genes," *Discover*, November 1989). They kill or impair individuals who inherit two copies of the faulty gene, but they help those who receive only one defective gene by protecting them against other diseases. The best understood example is the sickle-cell gene of African blacks, which often kills homozygotes but protects heterozygotes against malaria. Natural selection sustains such genes because more heterozygotes than normal individuals survive to pass on their genes, and those extra gene copies offset the copies lost through the deaths of homozygotes.

So let us refine our question and ask, What blessing could the Tay-Sachs gene bring to those individuals who are heterozygous for it? A clue first emerged back in 1972, with the publication of the results of a questionnaire that had asked U.S. Ashkenzaic parents of Tay-Sachs children what their own Eastern European-born parents had died of. Keep in mind that since these unfortunate children had to be homozygotes, with two copies of the Tay-Sachs gene, all their parents had to be heterozygotes, with one copy, and half of the parents' parents also had to be heterozygotes.

As it turned out, most of those Tay-Sachs grandparents had died of the usual causes: heart disease, stroke, cancer, and diabetes. But strikingly, only one of the 306 grandparents had died of tuberculosis, even though TB was generally one of the big killers in these grandparents' time. Indeed, among the general population of large Eastern European cities in the early twentieth century, TB caused up to 20 percent of all deaths.

This big discrepancy suggested that Tay-Sachs heterozygotes might somehow have been protected against TB. Interestingly, it was already well known that Ashkenazim in general had some such protection: even when Jews and non-Jews were compared within the same European city, class, and occupational group (for example, Warsaw garment workers), Jews had only half the TB death rate of non-Jews, despite their being equally susceptible to infection. Perhaps, one could reason, the Tay-Sachs gene furnished part of that well-established Jewish resistance.

We're not a melting pot, and we won't be for a long time. Each ethnic group has some characteristic genes of its own, a legacy of its distinct history.

A second clue to a heterozygote advantage conveyed by the Tay-Sachs gene emerged in 1983, with a fresh look at the data concerning the distributions of TB and the Tay-Sachs gene within Europe. The statistics showed that the Tay-Sachs gene was nearly three times more frequent among Jews originating from Austria, Hungary, and Czechoslovakia—areas where an amazing 9 to 10 percent of the population were heterozygotes—than among Jews from Poland, Russia, and Germany. At the same time records from an old Jewish TB sanatorium in Denver in 1904 showed that among patients born in Europe

between 1860 and 1910, Jews from Austria and Hungary were overrepresented.

Initially, in putting together these two pieces of information, you might be tempted to conclude that because the highest frequency of the Tay-Sachs gene appeared in the same geographic region that produced the most cases of TB, the gene in fact offers no protection whatsoever. Indeed, this was precisely the mistaken conclusion of many researchers who had looked at these data before. But you have to pay careful attention to the numbers here: even at its highest frequency the Tay-Sachs gene was carried by far fewer people than would be infected by TB. What the statistics really indicate is that where TB is the biggest threat, natural selection produces the biggest response.

Think of it this way: You arrive at an island where you find that all the inhabitants of the north end wear suits of armor, while all the inhabitants of the south end wear only cloth shirts. You'd be pretty safe in assuming that warfare is more prevalent in the north—and that war-related injuries account for far more deaths there than in the south. Thus, if the Tay-Sachs gene does indeed lend heterozygotes some protection against TB, you would expect to find the gene most often precisely where you find TB most often. Similarly, the sickle-cell gene reaches its highest frequencies in those parts of Africa where malaria is the biggest risk.

But you may believe there's still a hole in the argument: If Tay-Sachs heterozygotes are protected against TB, you may be asking, why is the gene common just in the Ashkenazim? Why did it not become common in the non-Jewish populations also exposed to TB in Austria, Hungary, and Czechoslovakia?

At this point we must recall the peculiar circumstances in which the Jews of Eastern Europe were forced to live. They were unique among the world's ethnic groups in having been virtually confined to towns for most of the past 2,000 years. Being forbidden to own land, Eastern European Jews were not peasant farmers living in the countryside, but businesspeople forced to live in crowded ghettos, in an environment where tuberculosis thrived.

Of course, until recent improvements in sanitation, these towns were not very healthy places for non-Jews either. Indeed, their populations couldn't sustain themselves: deaths exceeded births, and the number of dead had to be balanced by continued emigration from the countryside. For non-Jews, therefore, there was no genetically distinct urban population. For ghetto-bound Jews, however, there could be no emigration from the countryside; thus the Jewish population was under the strongest selection to evolve genetic resistance to TB.

Those are the conditions that probably led to Jewish TB resistance, whatever particular genetic factors prove to underlie it. I'd speculate that G_{M2} and related fats accumulate at slightly higher-than-normal levels in heterozygotes, although not at the lethal levels seen in homozygotes. (The fat accumulation in heterozygotes probably takes place in the cell membrane, the cell's "armor.") I'd also speculate that the accumulation provides heterozygotes with some protection against TB, and that that's why the genes for Tay-Sachs, Gaucher's, and Niemann-Pick disease reached high frequencies in the Ashkenazim.

Having thus stated the case, let me make clear that I don't want to overstate it. The evidence is still speculative. Depending on how you do the calculation, the low frequency of TB deaths in Tay-Sachs grandparents either barely reaches or doesn't quite reach the level of proof that statisticians require to accept an effect as real rather than as one that's arisen by chance. Moreover, we have no idea of the biochemical mechanism by which fat accumulation might confer resistance against TB. For the moment, I'd say that the evidence points to some selective advantage of Tay-Sachs heterozygotes among the Ashkenazim, and that TB resistance is the only plausible hypothesis yet proposed.

For now Tay-Sachs remains a speculative model for the evolution of ethnic diseases. But it's already a proven model of what to do about them. Twenty years ago a test was developed to identify Tay-Sachs heterozygotes, based on their lower-than-normal levels of hexosaminidase A. The test is simple, cheap, and accurate: all I did was to donate a small sample of my blood, pay $35, and wait a few days to receive the results.

If that test shows that at least one member of a couple is not a Tay-Sachs heterozygote, then any child of theirs can't be a Tay-Sachs homozygote. If both parents prove to be heterozygotes, there's a one-in-four chance of their child being a homozygote; that can then be determined by other tests performed on the mother early in pregnancy. If the results are positive, it's early enough for her to abort, should she choose to. That critical bit of knowledge has enabled parents who had gone through the agony of bearing a Tay-Sachs baby and watching him die to find the courage to try again.

The Tay-Sachs screening program launched in the United States in 1971 was targeted at the high-risk population: Ashkenazic Jewish couples of childbearing age. So successful has this approach been that the number of Tay-Sachs babies born each year in this country has declined tenfold. Today, in fact, more Tay-Sachs cases appear here in non-Jews than in Jews, because only the latter couples are routinely tested. Thus, what used to be the classic genetic disease of Jews is so no longer.

There's also a broader message to the Tay-Sachs story. We commonly refer to the United States as a melting pot, and in many ways that metaphor is apt. But in other ways we're not a melting pot, and we won't be for a long time. Each ethnic group has some characteristic genes of its own, a legacy of its distinct history. Tuberculosis and malaria are not major causes of death in the United States, but the genes that some of us evolved to protect ourselves against them are still frequent. Those genes are frequent only in certain ethnic groups, though, and they'll be slow to melt through the population.

With modern advances in molecular genetics, we can expect to see more, not less, ethnically targeted practice of medicine. Genetic screening for cystic fibrosis in European whites, for example, is one program that has been much discussed recently; when it comes, it will surely be based on the Tay-Sachs experience. Of course, what that may mean someday is more anxiety-ridden parents-to-be glowering at more dedicated genetics counselors. It will also mean fewer babies doomed to the agonies of diseases we may understand but that we'll never be able to accept.

Contributing editor **JARED DIAMOND** is a professor of physiology at the UCLA School of Medicine.

The Saltshaker's Curse

Physiological adaptations that helped American blacks survive slavery may now be predisposing their descendants to hypertension

JARED DIAMOND

On the walls of the main corridor at UCLA Medical School hang thirty-seven photographs that tell a moving story. They are the portraits of each graduating class, from the year that the school opened (Class of 1955) to the latest crop (Class of 1991). Throughout the 1950s and early 1960s the portraits are overwhelmingly of young white men, diluted by only a few white women and Asian men. The first black student graduated in 1961, an event not repeated for several more years. When I came to UCLA in 1966, I found myself lecturing to seventy-six students, of whom seventy-four were white. Thereafter the numbers of blacks, Hispanics, and Asians exploded, until the most recent photos show the number of white medical students declining toward a minority.

In these changes of racial composition, there is of course nothing unique about UCLA Medical School. While the shifts in its student body mirror those taking place, at varying rates, in other professional groups throughout American society, we still have a long way to go before professional groups truly mirror society itself. But ethnic diversity among physicians is especially important because of the dangers inherent in a profession composed of white practitioners for whom white biology is the norm.

Different ethnic groups face different health problems, for reasons of genes as well as of life style. Familiar examples include the prevalence of skin cancer and cystic fibrosis in whites, stomach cancer and stroke in Japanese, and diabetes in Hispanics and Pacific islanders. Each year, when I teach a seminar course in ethnically varying disease patterns, these by-now-familiar textbook facts assume a gripping reality, as my various students choose to discuss some disease that affects themselves or their relatives. To read about the molecular biology of sickle-cell anemia is one thing. It's quite another thing when one of my students, a black man homozygous for the sickle-cell gene, describes the pain of his own sickling attacks and how they have affected his life.

Sickle-cell anemia is a case in which the evolutionary origins of medically important genetic differences among peoples are well understood. (It evolved only in malarial regions because it

confers resistance against malaria.) But in many other cases the evolutionary origins are not nearly so transparent. Why is it, for example, that only some human populations have a high frequency of the Tay-Sachs gene or of diabetes?...

Compared with American whites of the same age and sex, American blacks have, on the average, higher blood pressure, double the risk of developing hypertension, and nearly ten times the risk of dying of it. By age fifty, nearly half of U.S. black men are hypertensive. For a given age and blood pressure, hypertension more often causes heart disease and especially kidney failure and strokes in U.S. blacks than whites. Because the frequency of kidney disease in U.S. blacks is eighteen times that in whites, blacks account for about two-thirds of U.S. patients with hypertensive kidney failure, even though they make up only about one-tenth of the population. Around the world, only Japanese exceed U.S. blacks in their risk of dying from stroke. Yet it was not until 1932 that the average difference in blood pressure between U.S. blacks and whites was clearly demonstrated, thereby exposing a major health problem outside the norms of white medicine.

What is it about American blacks that makes them disproportionately likely to develop hypertension and then to die of its consequences? While this question is of course especially "interesting" to black readers, it also concerns all Americans, because other ethnic groups in the United States are not so far behind blacks in their risk of hypertension. If *Natural History* readers are a cross section of the United States, then about one-quarter of you now have high blood pressure, and more than half of you will die of a heart attack or stroke to which high blood pressure predisposes. Thus, we all have valid reasons for being interested in hypertension.

First, some background on what those numbers mean when your doctor inflates a rubber cuff about your arm, listens, deflates the cuff, and finally pronounces, "Your blood pressure is 120 over 80." The cuff device is called a sphygmomanometer, and it measures the pressure in your artery in units of millimeters of mercury (that's the height to which your blood pressure would force up a column of mercury in case, God forbid, your artery were suddenly connected to a vertical mercury column). Naturally, your blood pressure varies with each stroke of your heart, so the first and second numbers refer, respectively, to the peak pressure at each heartbeat (systolic pressure) and to the minimum pressure between beats (diastolic pressure). Blood pressure varies somewhat with position, activity, and anxiety level, so the measurement is usually made while you are resting flat on your back. Under those conditions, 120 over 80 is an average reading for Americans.

There is no magic cutoff between normal blood pressure and high blood pressure. Instead, the higher your blood pressure, the more likely you are to die of a heart attack, stroke, kidney failure, or ruptured aorta. Usually, a pressure reading higher than 140 over 90 is arbitrarily defined as constituting hypertension, but some people with lower readings will die of a stroke at age fifty, while others with higher readings will die in a car accident in good health at age ninety.

Why do some of us have much higher blood pressure than others? In about 5 percent of hypertensive patients there is an identifiable single cause, such as hormonal imbalance or use of oral contraceptives. In 95 percent of such cases, though, there is no such obvious cause. The clinical euphemism for our ignorance in such cases is "essential hypertension."

Nowadays, we know that there is a big genetic component in essential hypertension, although the particular genes involved have not yet been identified. Among people living in the same household, the correlation coefficient for blood pressure is 0.63 between identical twins, who share all of their genes. (A correlation coefficient of 1.00 would mean that the twins share identical blood pressures as well and would suggest that pressure is determined entirely by genes and not at all by environment.) Fraternal twins or ordinary siblings or a parent and child, who share half their genes and whose blood pressure would therefore show a correlation coefficient of 0.5 if purely determined genetically, actually have a coefficient of about 0.25. Finally, adopted siblings or a parent and adopted child, who have no direct genetic connection, have a correlation coefficient of only 0.05. Despite the shared household environment, their blood pressures are barely more similar than those of two people pulled randomly off the street. In agreement with this evidence for genetic factors underlying blood pressure itself, your risk of actually developing hypertensive disease increases from 4 percent to 20 percent to 35 percent if, respectively, none or one or both of your parents were hypertensive.

But these same facts suggest that environmental factors also contribute to high blood pressure, since identical twins have similar but not identical blood pressures. Many environmental or life style factors contributing to the risk of hypertension have been identified by epidemiological studies that compare hypertension's frequency in groups of people living under different conditions. Such contributing factors include obesity, high intake of salt or alcohol or saturated fats, and low calcium intake. The proof of this approach is that hypertensive patients who modify their life styles so as to minimize these putative factors often succeed in reducing their blood pressure. Patients are especially advised to reduce salt intake and stress, reduce intake of cholesterol and saturated fats and alcohol, lose weight, cut out smoking, and exercise regularly.

Here are some examples of the epidemiological studies pointing to these risk factors. Around the world, comparisons within and between populations show that both blood pressure and the frequency of hypertension increase hand in hand with salt intake. At the one extreme, Brazil's Yanomamö Indians have the world's lowest-known salt consumption (somewhat above 10 milligrams per day!), lowest average blood pressure (95 over 61!), and lowest incidence of hypertension (no cases!). At the opposite extreme, doctors regard Japan as the "land of apoplexy" because of the high frequency of fatal strokes (Japan's leading cause of death, five times more frequent than in the United States), linked with high blood pressure and notoriously salty food. Within Japan itself these factors reach their extremes in Akita Prefecture, famous for its tasty rice, which Akita farmers flavor with salt, wash down with salty miso soup, and alternate with salt pickles between meals. Of 300 Akita adults studied, not one consumed less than five grams of salt daily, the average consumption was twenty-seven grams, and the most salt-loving individual consumed an incredible sixty-one grams—enough to devour the contents of the usual twenty-six-ounce supermarket salt container in a mere twelve days. The average blood pressure in Akita by age fifty is 151 over 93, making hypertension (pressure higher than 140 over 90) the norm. Not surprisingly, Akitas' frequency of death by stroke is more than double even the Japanese average, and in some Akita villages 99 percent of the population dies before age seventy.

Why salt intake often (in about 60 percent of hypertensive patients) leads to high blood pressure is not fully understood. One possible interpretation is that salt intake triggers thirst, leading to an increase in blood volume. In response, the heart increases its output and blood pressure rises, causing the kidneys to filter more salt and water under that increased pressure. The result is a new steady state, in which salt and water excretion again equals intake, but more salt and water are stored in the body and blood pressure is raised.

At this point, let's contrast hypertension with a simple genetic disease like Tay-Sachs disease. Tay-Sachs is due to a defect in a single gene; every Tay Sachs patient has a defect in that same gene. Everybody in whom that gene is defective is certain to die of Tay-Sachs, regardless of their life style or environment. In contrast, hypertension involves several different genes whose molecular products remain to be identified. Because there are many causes of raised blood pressure, different hypertensive patients may owe their condition to different gene combinations. Furthermore, whether someone genetically predisposed to hypertension actually develops symptoms depends a lot on life style. Thus, hypertension is not one of those uncommon, homogeneous, and intellectually elegant diseases that geneticists prefer to study. Instead, like diabetes and ulcers, hypertension is a shared set of symptoms produced by heterogeneous causes, all involving an interaction between environmental agents and a susceptible genetic background.

Since U.S. blacks and whites differ on the average in the conditions under which they live, could those differences account for excess hypertension in U.S. blacks? Salt intake, the dietary factor that one thinks of first, turns out on the average not to differ between U.S. blacks and whites. Blacks do consume less potassium and calcium, do experience more stress associated with more difficult socioeconomic conditions, have much less access to medical care, and are therefore much less likely to be diagnosed or treated until it is too late. Those factors surely contribute to the frequency and severity of hypertension in blacks.

However, those factors don't seem to be the whole explanation: hypertensive blacks aren't merely like severely hypertensive whites. Instead, physiological differences seem to contribute as well. On consuming salt, blacks retain it on average far longer before excreting it into the urine, and they experience a greater rise in blood pressure on a high-salt diet. Hypertension is more likely to be "salt-sensitive" in blacks than in whites, meaning that blood pressure is more likely to rise and fall with rises and falls in dietary salt intake. By the same token, black hypertension is more likely to be treated successfully by drugs that cause the kidneys to excrete salt (the so-called thiazide diuretics) and less likely to respond to those drugs that reduce heart rate and cardiac output (so-called beta blockers, such as propanolol). These facts suggest that there are some qualitative differences between the causes of black and white hypertension, with black hypertension more likely to involve how the kidneys handle salt.

Physicians often refer to this postulated feature as a "defect": for example, "kidneys of blacks have a genetic defect in excreting sodium." As an evolutionary biologist, though, I hear warning bells going off inside me whenever a seemingly harmful trait that occurs frequently in an old and large human population is dismissed as a "defect." Given enough generations, genes that greatly impede survival are extremely unlikely to spread, unless their net effect is to increase survival and reproductive success. Human medicine has furnished the best examples of seemingly defective genes being propelled to high frequency by counterbalancing benefits. For example, sickle-cell hemoglobin protects far more people against malaria than it kills of anemia, while the Tay-Sachs gene may have protected far more Jews against tuberculosis than it killed of neurological disease. Thus, to understand why U.S. blacks now are prone to die as a result of their kidneys' retaining salt, we need to ask under what conditions people might have benefited from kidneys good at retaining salt.

That question is hard to understand from the perspective of modern Western society, where saltshakers are on every dining table, salt (sodium chloride) is cheap, and our bodies' main problem is getting rid of it. But imagine what the world used to be like before saltshakers became ubiquitous. Most plants contain very little sodium, yet animals require sodium at high concentrations in all their extracellular fluids. As a result, carnivores readily obtain their needed sodium by eating herbivores, but herbivores themselves face big problems in acquiring that sodium. That's why the animals that one sees coming to salt licks are deer and antelope, not lions and tigers. Similarly, some human hunter-gatherers obtained enough salt

from the meat that they ate. But when we began to take up farming ten thousand years ago, we either had to evolve kidneys superefficient at conserving salt or learn to extract salt at great effort or trade for it at great expense.

Examples of these various solutions abound. I already mentioned Brazil's Yanomamö Indians, whose staple food is low-sodium bananas and who excrete on the average only 10 milligrams of salt daily—barely one-thousandth the salt excretion of the typical American. A single Big Mac hamburger analyzed by *Consumer Reports* contained 1.5 grams (1,500 milligrams) of salt, representing many weeks of intake for a Yanomamö. The New Guinea highlanders with whom I work, and whose diet consists up to 90 percent of low-sodium sweet potatoes, told me of the efforts to which they went to make salt a few decades ago, before Europeans brought it as trade goods. They gathered leaves of certain plant species, burned them, scraped up the ash, percolated water through it to dissolve the solids, and finally evaporated the water to obtain small amounts of bitter salt.

Thus, salt has been in very short supply for much of recent human evolutionary history. Those of us with efficient kidneys able to retain salt even on a low-sodium diet were better able to survive our inevitable episodes of sodium loss (of which more in a moment). Those kidneys proved to be a detriment only when salt became routinely available, leading to excessive salt retention and hypertension with its fatal consequences. That's why blood pressure and the frequency of hypertension have shot up recently in so many populations around the world as they have made the transition from being self-sufficient subsistence farmers to members of the cash economy and patrons of supermarkets.

This evolutionary argument has been advanced by historian-epidemiologist Thomas Wilson and others to explain the current prevalence of hypertension in American blacks in particular. Many West African blacks, from whom most American blacks originated via the slave trade, must have faced the chronic problem of losing salt through sweating in their hot environment. Yet in West Africa, except on the coast and certain inland areas, salt was traditionally as scarce for African farmers as it has been for Yanomamö and New Guinea farmers. (Ironically, those Africans who sold other Africans as slaves often took payment in salt traded from the Sahara.) By this argument, the genetic basis for hypertension in U.S. blacks was already widespread in many of their West African ancestors. It required only the ubiquity of saltshakers in twentieth-century America for that genetic basis to express itself as hypertension. This argument also predicts that as Africa's life style becomes increasingly Westernized, hypertension could become as prevalent in West Africa as it now is among U.S. blacks. In this view, American blacks would be no different from the many Polynesian, Melanesian, Kenyan, Zulu, and other populations that have recently developed high blood pressure under a Westernized life style.

But there's an intriguing extension to this hypothesis, proposed by Wilson and physician Clarence Grim, collaborators at the Hypertension Research Center of Drew University in Los Angeles. They suggest a scenario in which New World blacks may now be at more risk for hypertension than their African ancestors. That scenario involves very recent selection for superefficient kidneys, driven by massive mortality of black slaves from salt loss.

Grim and Wilson's argument goes as follows. Black slavery in the Americas began about 1517, with the first imports of slaves from West Africa, and did not end until Brazil freed its slaves barely a century ago in 1888. In the course of the slave trade an estimated 12 million Africans were brought to the Americas. But those imports were winnowed by deaths at many stages, from an even larger number of captives and exports.

First, slaves captured by raids in the interior of West Africa were chained together, loaded with heavy burdens, and marched for one or two months, with little food and water, to the coast. About 25 percent of the captives died en route. While awaiting purchase by slave traders, the survivors were held on the coast in hot, crowded buildings called barracoons, where about 12 percent of them died. The traders went up and down the coast buying and loading slaves for a few weeks or months until a ship's cargo was full (5 percent more died). The dreaded Middle Passage across the Atlantic killed 10 percent of the slaves, chained together in a hot, crowded, unventilated hold without sanitation. (Picture to yourself the result of those toilet "arrangements.") Of those who lived to land in the New World, 5 percent died while awaiting sale, and 12 percent died while being marched or shipped from the sale yard to the plantation. Finally, of those who survived, between 10 and 40 percent died during the first three years of plantation life, in a process euphemistically called seasoning. At that stage, about 70 percent of the slaves initially captured were dead, leaving 30 percent as seasoned survivors.

Even the end of seasoning, however, was not the end of excessive mortality. About half of slave infants died within a year of birth because of the poor nutrition and heavy workload of their mothers. In plantation terminology, slave women were viewed as either "breeding units" or "work units," with a built-in conflict between those uses: "These Negroes breed the best, whose labour is least," as an eighteenth-century observer put it. As a result, many New World slave populations depended on continuing slave imports and couldn't maintain their own numbers because death rates exceeded birth rates. Since buying new slaves cost less than rearing slave children for twenty years until they were adults, slave owners lacked economic incentive to change this state of affairs.

Recall that Darwin discussed natural selection and survival of the fittest with respect to animals. Since many more animals die than survive to produce offspring, each generation becomes enriched in the genes of those of the preceding generation that were among the survivors. It should now be clear that slavery represented a tragedy of unnatural selection in humans on a gigantic scale. From examining accounts of slave mortality, Grim and Wilson argue that death was indeed selective: much of it was related to unbalanced salt loss, which quickly brings on collapse. We think immediately of salt loss by sweating under hot conditions: while slaves were working, marching, or confined in unventilated barracoons or ships' holds. More body salt may have been spilled with vomiting

from seasickness. But the biggest salt loss at every stage was from diarrhea due to crowding and lack of sanitation—ideal conditions for the spread of gastrointestinal infections. Cholera and other bacterial diarrheas kill us by causing sudden massive loss of salt and water. (Picture your most recent bout of *turista*, multiplied to a diarrheal fluid output of twenty quarts in one day, and you'll understand why.) All contemporary accounts of slave ships and plantation life emphasized diarrhea, or "fluxes" in eighteenth-century terminology, as one of the leading killers of slaves.

Grim and Wilson reason, then, that slavery suddenly selected for superefficient kidneys surpassing the efficient kidneys already selected by thousands of years of West African history. Only those slaves who were best able to retain salt could survive the periodic risk of high salt loss to which they were exposed. Salt supersavers would have had the further advantage of building up, under normal conditions, more of a salt reserve in their body fluids and bones, thereby enabling them to survive longer or more frequent bouts of diarrhea. Those superkidneys became a disadvantage only when modern medicine began to reduce diarrhea's lethal impact, thereby transforming a blessing into a curse.

Thus, we have two possible evolutionary explanations for salt retention by New World blacks. One involves slow selection by conditions operating in Africa for millennia; the other, rapid recent selection by slave conditions within the past few centuries. The result in either case would make New World blacks more susceptible than whites to hypertension, but the second explanation would, in addition, make them more susceptible than African blacks. At present, we don't know the relative importance of these two explanations. Grim and Wilson's provocative hypothesis is likely to stimulate medical and physiological comparisons of American blacks with African blacks and thereby to help resolve the question.

While this piece has focused on one medical problem in one human population, it has several larger morals. One, of course, is that our differing genetic heritages predispose us to different diseases, depending on the part of the world where our ancestors lived. Another is that our genetic differences reflect not only ancient conditions in different parts of the world but also recent episodes of migration and mortality. A well-established example is the decrease in the frequency of the sickle-cell hemoglobin gene in U.S. blacks compared with African blacks, because selection for resistance to malaria is now unimportant in the United States. The example of black hypertension that Grim and Wilson discuss opens the door to considering other possible selective effects of the slave experience. They note that occasional periods of starvation might have selected slaves for superefficient sugar metabolism, leading under modern conditions to a propensity for diabetes.

Finally, consider a still more universal moral. Almost all people alive today exist under very different conditions from those under which every human lived 10,000 years ago. It's remarkable that our old genetic heritage now permits us to survive at all under such different circumstances. But our heritage still catches up with most of us, who will die of life style related diseases such as cancer, heart attack, stroke, and diabetes. The risk factors for these diseases are the strange new conditions prevailing in modern Western society. One of the hardest challenges for modern medicine will be to identify for us which among all those strange new features of diet, life style, and environment are the ones getting us into trouble. For each of us, the answers will depend on our particular genes, hence on our ancestry.

Only with such individually tailored advice can we hope to reap the benefits of modern living while still housed in bodies designed for life before saltshakers.

JARED DIAMOND is a professor of physiology at UCLA Medical School.

Reprinted with permission from *Natural History,* October 1991, pp. 20, 22–26. © 1991 by Natural History Magazine, Inc.

Index

Index

Test Your Knowledge Form

We encourage you to photocopy and use this page as a tool to assess how the articles in *Annual Editions* expand on the information in your textbook. By reflecting on the articles you will gain enhanced text information. You can also access this useful form on a product's book support Web site at *http://www.mhcls.com/online/*.

NAME:

DATE:

TITLE AND NUMBER OF ARTICLE:

BRIEFLY STATE THE MAIN IDEA OF THIS ARTICLE:

LIST THREE IMPORTANT FACTS THAT THE AUTHOR USES TO SUPPORT THE MAIN IDEA:

WHAT INFORMATION OR IDEAS DISCUSSED IN THIS ARTICLE ARE ALSO DISCUSSED IN YOUR TEXTBOOK OR OTHER READINGS THAT YOU HAVE DONE? LIST THE TEXTBOOK CHAPTERS AND PAGE NUMBERS:

LIST ANY EXAMPLES OF BIAS OR FAULTY REASONING THAT YOU FOUND IN THE ARTICLE:

LIST ANY NEW TERMS/CONCEPTS THAT WERE DISCUSSED IN THE ARTICLE, AND WRITE A SHORT DEFINITION:

We Want Your Advice

ANNUAL EDITIONS revisions depend on two major opinion sources: one is our Advisory Board, listed in the front of this volume, which works with us in scanning the thousands of articles published in the public press each year; the other is you—the person actually using the book. Please help us and the users of the next edition by completing the prepaid article rating form on this page and returning it to us. Thank you for your help!

ANNUAL EDITIONS: Physical Anthropology 07/08

ARTICLE RATING FORM

Here is an opportunity for you to have direct input into the next revision of this volume.
We would like you to rate each of the articles listed below, using the following scale:

1. **Excellent: should definitely be retained**
2. **Above average: should probably be retained**
3. **Below average: should probably be deleted**
4. **Poor: should definitely be deleted**

Your ratings will play a vital part in the next revision.
Please mail this prepaid form to us as soon as possible.
Thanks for your help!

RATING	ARTICLE	RATING	ARTICLE
	1. The Growth of Evolutionary Science		23. African Trailblazers
	2. Darwin's Influence on Modern Thought		24. Hunting the First Hominid
	3. Evolution in Action		25. Digital Ancestors Walk Again
	4. 15 Answers to Creationist Nonsense		26. Scavenger Hunt
	5. Why Should Students Learn Evolution?		27. The Scavenging of "Peking Man"
	6. The Illusion of Design		28. *Erectus* Rising
	7. Designer Thinking		29. Hard Times Among the Neanderthals
	8. The Perimeter of Ignorance		30. Rethinking Neanderthals
	9. The 2% Difference		31. A Caveful of Clues About Early Humans
	10. The Mind of the Chimpanzee		32. The Gift *of* Gab
	11. Got Culture?		33. We Are All Africans
	12. Dim Forest, Bright Chimps		34. The Littlest Human
	13. Why Are Some Animals So Smart?		35. Skin Deep
	14. How Animals Do Business		36. Black, White, Other
	15. Are We in Anthropodenial?		37. Does Race Exist? A Proponent's Perspective
	16. A Telling Difference		38. Does Race Exist? An Antagonist's Perspective
	17. What Are Friends For?		39. The Tall and the Short of It
	18. What's Love Got to Do With It?		40. The Viral Superhighway
	19. Apes of Wrath		41. The Inuit Paradox
	20. Mothers and Others		42. Dr. Darwin
	21. Had King Henry VIII's Wives Only Known		43. Curse and Blessing of the Ghetto
	22. The Salamander's Tale		44. The Saltshaker's Curse

BUSINESS REPLY MAIL
FIRST CLASS MAIL PERMIT NO. 551 DUBUQUE IA

POSTAGE WILL BE PAID BY ADDRESEE

McGraw-Hill Contemporary Learning Series
2460 KERPER BLVD
DUBUQUE, IA 52001-9902

NO POSTAGE
NECESSARY
IF MAILED
IN THE
UNITED STATES

ABOUT YOU

Name   Date

Are you a teacher? ❑   A student? ❑
Your school's name

Department

Address   City   State   Zip

School telephone #

YOUR COMMENTS ARE IMPORTANT TO US!

Please fill in the following information:
For which course did you use this book?

Did you use a text with this ANNUAL EDITION?   ❑   yes ❑   no
What was the title of the text?

What are your general reactions to the *Annual Editions* concept?

Have you read any pertinent articles recently that you think should be included in the next edition? Explain.

Are there any articles that you feel should be replaced in the next edition? Why?

Are there any World Wide Web sites that you feel should be included in the next edition? Please annotate.

May we contact you for editorial input?   ❑   yes ❑   no
May we quote your comments? ❑   yes ❑   no